Numerical Recipes Routines and Examples in BASIC

Numerical Recipes
Routines and Examples
in BASIC

companion manual to
Numerical Recipes: The Art of Scientific Computing

Julien C. Sprott
University of Wisconsin - Madison

in association with **Numerical Recipes Software**

CAMBRIDGE
UNIVERSITY PRESS

Published by the Press Syndicate of the University of Cambridge
The Pitt Building, Trumpington Street, Cambridge CB2 1RP
40 West 20th Street, New York, NY 10011-4211, USA
10 Stamford Road, Oakleigh, Melbourne 3166, Australia

First published 1991
Reprinted 1992, 1994, 1996, 1998

Printed in the United States of America
Typeset in TEX

Library of Congress Cataloging-in-Publication Data

Sprott, Julien C.
Numerical recipes : routines and examples in BASIC / Julien C. Sprott.
p. cm.
"Comparison manual to Numerical recipes : the art of scientific computing."
"In association with Numerical Recipes Software."
Includes index.
ISBN 0-521-40689-7 paperback. – ISBN 0-521-40688-9 diskette.
1. Numerical analysis–Computer programs. 2. Science–Mathematics–Computer
programs. 3. BASIC (Computer program language) I. Numerical Recipes Software
(Firm) II. Numerical recipes. III. Title.
QA297.S68 1991   90-26323
519.4'02085'5133–dc20   CIP

ISBN 0-521-40689-7 paperback
ISBN 0-521-40688-9 diskette

CONTENTS

Foreword by Numerical Recipes Software

We were pleased when Professor J.C. Sprott suggested that he might translate into BASIC the routines in our book *Numerical Recipes: The Art of Scientific Computing*, along with the demonstration programs from the coordinated *Numerical Recipes Example Book*. Professor Sprott has a long involvement in scientific computing; he is one of a limited, but significant, group of scientists who use BASIC in substantial scientific calculations.

BASIC was the first programming language to have been widely available on personal computers, and it continues to hold a place of affection in the hearts of many computer users. Its modernization in recent years – notably as Microsoft Corporation's QuickBASIC and Turbo BASIC from Borland International, Inc. – have given the language the essential structural features common to other widely used languages, FORTRAN, C, or Pascal.

Professor Sprott's efforts, contained in this book, address the needs of the scientific programmer in BASIC. This book is not intended to stand by itself. It is not a full edition of *Numerical Recipes: The Art of Scientific Computing*. In the full editions (FORTRAN, C, or Pascal) the algorithms and methods of scientific computation are developed in considerable detail, starting with basic mathematical analysis and working through to actual implementation in the form of computer code. This book is intended to supplement *any one* of the full editions with, in effect, a "simultaneous translation" of computer code into the preferred language of the BASIC user. It is a matter of individual taste which full edition the user should choose. BASIC is syntactically closer to FORTRAN than to the other languages, but this is perhaps only because that language, like BASIC, has relatively ancient roots in the computer world. C seems to us to be the language of most rapid growth within the scientific community.

The text of this book is largely drawn from the *Numerical Recipes Example Book*. There is thus no need for a separate Example Book in BASIC. Correspondingly, the single BASIC diskette combines both the routines (translated from the main book) and the demonstration programs (translated from the Example Book). These cost-saving measures are meant to make up, in part, for the financial setback of buying both this book and a separate full edition.

Reader response to *Numerical Recipes: The Art of Scientific Computing* has been overwhelmingly positive, and very helpful to us. Professor Sprott has prepared his BASIC translations from the most current versions of our FORTRAN programs, in which known bugs have been corrected. Informal statistics suggest that the vast majority of bugs in the FORTRAN *Numerical Recipes* have by now been found; nearly all true bugs reported to us are ones that we already knew about, and these have been corrected. Undoubtedly, some errors have crept into the process of translation into BASIC, and we encourage readers to communicate bug reports to us.

William H. Press
Brian P. Flannery
Saul A. Teukolsky
William T. Vetterling

DISCLAIMER OF WARRANTY

Author's Preface

By inspecting this book you have shown a willingness, shared by only a part of the scientific community, to consider BASIC as a working language. The very name "BASIC" connotes a language for beginners, and its use as the primary language for primitive personal computers has reinforced that view. Early versions of the language were interpreted rather than compiled, lacked many useful features, and neither encouraged nor permitted the writing of well-structured code.

All that has changed. New versions of BASIC have most of the features of languages like FORTRAN, Pascal, and C while retaining backward compatibility with BASIC code written in the 1970's and early 80's. However, we continue to face the difficulty that there are many versions of BASIC and no universal standard. Probably the closest thing to a current standard, and arguably the most powerful BASIC at the time of this writing, is Microsoft BASIC for the IBM family of personal computers and compatibles. The particular version under which the routines in this book run without modification is Microsoft QuickBASIC 4.5.

This book contains translations into BASIC of the routines contained in *Numerical Recipes: The Art of Scientific Computing* by William H. Press, Brian P. Flannery, Saul A. Teukolsky, and William T. Vetterling (Cambridge University Press, 1985; revised 1989). The routines in that book, numbering more than 200, are meant to be incorporated into user applications; they are subroutines (or functions), not stand-alone programs.

Accompanying *Numerical Recipes: The Art of Scientific Computing*, is the *Numerical Recipes Example Book*, which contains source programs to exercise and demonstrate all of the Numerical Recipes subroutines and functions. These have also been translated into BASIC and are included here alongside the routines themselves. In the interests of clarity, the *Numerical Recipes* subroutines and functions are demonstrated in simple ways. As a consequence, the demonstration programs in this book do not generally test all possible regimes of input data, or even all lines of subroutine source code.

I am indebted to W.H. Press for his encouragement and guidance throughout this project, to W.T. Vetterling for a careful reading of the manuscript, and to Nancy Lee Snyder, who transformed my disk files and cryptic notes into a camera-ready manuscript.

> *Julien C. Sprott*
> Madison, Wisconsin
> November, 1990

Important Note on Dialects of BASIC

The programs in this book will *not* run under older, or more elementary, implementations of the BASIC language. In particular, they will *not* run under the ROM-BASIC that is built into most IBM personal computers, *or* under the BASIC interpreter that is supplied with IBM's PC DOS, *or* under Microsoft Corp.'s GW-BASIC (often supplied with PC clones).

The programs in this book *will* run under more advanced BASIC dialects. They should run without modification under Microsoft Corp.'s QuickBASIC 4.5 or later versions, and (with minor modifications) under Borland International's Turbo BASIC and its compatible successors.

All the programs in this book were translated from the FORTRAN version using a translator written in Turbo BASIC 1.1, with a considerable amount of hand-work to correct translation difficulties. Indeed, the programs were translated first into Turbo BASIC and then into Quick-BASIC. This double translation results in some quirks in the code, but it should simplify the task for users who want to convert the code to other versions of BASIC. The notation and style follow rather closely the original FORTRAN.

All variable names have been left the same as in the FORTRAN version except where they conflicted with a reserved word in BASIC. In such cases, the last letter of the variable name has been changed to "Q." Even two of the routines (LOCATE and SHELL) had to be renamed (to LOCATQ and SHELQ). One implication of this procedure is that all variables that begin with the letters I through N can be declared as integers to speed the computation, but this convention is not forced on the user. Actually, FORTRAN integers are 4-byte rather than 2-byte as in most versions of BASIC, and thus in some cases, integer overflow will occur unless BASIC long integers (DEFLNG I-N) are used. In cases where it really matters, a type-declaration suffix character (%, &, !, # or $) has been added to the variable name. The use of array subscripts also follows the FORTRAN convention; the first element is 1 not 0. A small amount of memory can thus be reclaimed by adding an OPTION BASE 1 statement.

The program structure also follows the FORTRAN closely. Although BASIC allows line numbers, and old versions even require them, the modern trend is to avoid them altogether. This has been attempted by liberal use of the WHILE...WEND and DO...LOOP structures. If your version of BASIC doesn't support these, however, it should be relatively easy to convert back to the old IF...THEN GOTO forms. In a few places, line numbers remain, where clarity would have been sacrificed. A good exercise for the reader is to eliminate them all. One place where line numbers can be very useful in BASIC is where a data array is read by a subroutine. If the subroutine is called more than once, a RESTORE statement is required to reread the data. However a RESTORE without a line number will begin reading the first DATA statement in the program, which may not be what is desired. EXIT statements are also used liberally. Watch these carefully since their support is not universal or consistent. For example, Turbo BASIC uses EXIT LOOP where QuickBASIC uses EXIT DO.

A number of quirks in the BASIC code are worth mentioning. The input and output statements may look unnecessarily complicated, but that was done so that data files formatted for the FORTRAN version could be read by the BASIC program and so that output written to the screen would be in essentially the same format as the FORTRAN version to facilitate comparison. In most cases, the source code has been written so that all lines are 80 columns or less in width so as to fit on conventional monitors (despite the QuickBASIC editor's annoying habit of inserting blanks in the lines). In the few cases where this could not be done, the lines wrap around when printed, and you should be careful to distinguish between the continuation of a previous line and a new line. Most BASIC versions don't allow passing the name of a function as an argument of a subroutine call. In such cases the argument has been replaced with a dummy variable (usually DUM) so that the argument count is the same as in the FORTRAN version, and you must take care to name explicitly the functions in an appropriate manner. Other incompatibilities may arise from the fact that some versions of BASIC don't allow passing variables or arrays by reference in function calls (and thus the function cannot change the value of the variable) and different conventions are used for whether variables default to local or to shared in functions.

Improving on the FORTRAN, the BASIC version uses dynamic dimensioning of arrays whenever possible and erases the array when done. In some cases it is important to erase arrays in exactly the inverse order in which they were dimensioned in order to reclaim memory properly. If speed is more important than memory use, and if the subroutine is called repeatedly, you should consider dimensioning all arrays as static.

Finally, it is important to understand that this book can be effectively used only in combination with a full edition of *Numerical Recipes: The Art of Scientific Computing*. The comments in this book's code refer to equations and descriptions contained in the full edition. The blind use of routines in this book without an understanding of their weaknesses and limitations is a certain numerical recipe for trouble!

Chapter 1: Preliminaries

The routines in Chapter 1 of *Numerical Recipes : The Art of Scientific Computing* are introductory and less general in purpose than those in the remainder of the book. This chapter's routines serve primarily to expose the book's notational conventions, to illustrate control structures, and perhaps to amuse. You may even find them useful, but we hope that you will use BADLUK for no serious purpose.

★ ★ ★ ★

Subroutine FLMOON calculates the phases of the moon, or more exactly, the Julian day and fraction thereof on which a given phase will occur or has occurred. The program D1R1 asks the present date and compiles a list of upcoming phases. We have compared the predictions to lunar tables, with happy results. Shown are the results of a test run, which you may replicate as a check. In this program, notice that we have set TZONE (the time zone) to 5.0 to signify the five-hour separation of the Eastern Standard time zone from Greenwich, England. Our parochial viewpoint requires you to use negative values of TZONE if you are east of Greenwich. The Julian day results are converted to calendar dates through the use of CALDAT, which appears later in the chapter. The fractional Julian day and time zone combine to form a correction that can possibly change the calendar date by one day.

```
    Date       Time(EST)    Phase
1   9  1982     3 PM    full moon
1  16  1982     7 PM    last quarter
1  24  1982    11 PM    new moon
2   1  1982    10 AM    first quarter
2   8  1982     2 AM    full moon
2  15  1982     3 PM    last quarter
2  23  1982     4 PM    new moon
3   2  1982     6 PM    first quarter
3   9  1982     3 PM    full moon
3  17  1982    12 AM    last quarter
3  25  1982     5 AM    new moon
4   1  1982     0 AM    first quarter
4   8  1982     5 AM    full moon
4  16  1982     7 AM    last quarter
4  23  1982     4 PM    new moon
4  30  1982     7 AM    first quarter
5   7  1982     8 PM    full moon
5  16  1982     0 AM    last quarter
5  23  1982     0 AM    new moon
5  29  1982     3 PM    first quarter
```

Here is the recipe FLMOON:

```
SUB FLMOON (N, NPH, JD&, FRAC)
```
Our programs begin with an introductory comment summarizing their purpose and explaining their calling sequence. This routine calculates the phases of the moon. Given an integer N and a code NPH for the phase desired (NPH= 0 for new moon, 1 for first quarter, 2 for full, 3 for last quarter), the routine returns the Julian Day Number JD&, and the fractional part of a day FRAC to be added to it, of the Nth such phase since January, 1900. Coordinated Universal Time is assumed.

```
RAD = .017453293#
C = N + NPH / 4!                    This is how we comment an individual line.
T = C / 1236.85
T2 = T ^ 2
AQ = 359.2242 + 29.105356# * C      You aren't really intended to understand this algorithm,
AM = 306.0253 + 385.816918# * C + .01073 * T2      but it does work!
JD& = 2415020 + 28 * N + 7 * NPH
XTRA = .75933 + 1.53058868# * C + (.0001178 - 1.55E-07 * T) * T2
IF NPH = 0 OR NPH = 2 THEN
    XTRA = XTRA + (.1734 - .000393 * T) * SIN(RAD * AQ) - .4068 * SIN(RAD * AM)
ELSEIF NPH = 1 OR NPH = 3 THEN
    XTRA = XTRA + (.1721 - .0004 * T) * SIN(RAD * AQ) - .628 * SIN(RAD * AM)
ELSE
    PRINT "NPH is unknown."
    EXIT SUB                        This is how we will indicate error conditions.
END IF
IF XTRA >= 0! THEN
    I = INT(XTRA)
ELSE
    I = INT(XTRA - 1!)
END IF
JD& = JD& + I
FRAC = XTRA - I
END SUB
```

A sample program using FLMOON is the following:

```
DECLARE SUB FLMOON (N!, NPH!, JD&, FRAC!)
DECLARE SUB CALDAT (JULIAN&, MM!, ID!, IYYY!)
DECLARE FUNCTION JULDAY& (IM!, ID!, IY!)


'PROGRAM D1R1
'Driver for routine FLMOON
CLS
TZONE = 5!
DIM PHASE$(4), TIMSTR$(2)
FOR I = 1 TO 4
  READ PHASE$(I)
NEXT I
DATA "new moon","first quarter","full moon","last quarter"
FOR I = 1 TO 2
  READ TIMSTR$(I)
NEXT I
DATA " AM"," PM"
PRINT "Date of the next few phases of the moon"
PRINT "Enter today's date (e.g. 1,31,1982)"
INPUT IM, ID, IY
```

```
PRINT
TIMZON = -TZONE / 24!
'Approximate number of full moons since January 1900
N = INT(12.37 * (IY - 1900 + (IM - .5) / 12!))
NPH = 2
J1& = JULDAY&(IM, ID, IY)
CALL FLMOON(N, NPH, J2&, FRAC)
N = INT(N + (J1& - J2&) / 28!)
PRINT "   Date", "  Time(EST)", " Phase"
FOR I = 1 TO 20
  CALL FLMOON(N, NPH, J2&, FRAC)
  IFRAC = CINT(24! * (FRAC + TIMZON))
  IF IFRAC < 0 THEN
    J2& = J2& - 1
    IFRAC = IFRAC + 24
  END IF
  IF IFRAC >= 12 THEN
    J2& = J2& + 1
    IFRAC = IFRAC - 12
  ELSE
    IFRAC = IFRAC + 12
  END IF
  IF IFRAC > 12 THEN
    IFRAC = IFRAC - 12
    ISTR = 2
  ELSE
    ISTR = 1
  END IF
  CALL CALDAT(J2&, IM, ID, IY)
  PRINT USING "##"; IM;
  PRINT USING "###"; ID;
  PRINT USING "#####"; IY;
  PRINT USING "#########"; IFRAC;
  PRINT TIMSTR$(ISTR); "     "; PHASE$(NPH + 1)
  IF NPH = 3 THEN
    NPH = 0
    N = N + 1
  ELSE
    NPH = NPH + 1
  END IF
NEXT I
END
```

Program **JULDAY**, our exemplar of the **IF** control structure, converts calendar dates to Julian dates. Not many people know the Julian date of their birthday or any other convenient reference point, for that matter. To remedy this, we offer a list of checkpoints, which appears at the end of this chapter as the file **DATES.DAT**. The program **D1R2** lists the Julian date for each historic event for comparison. Then it allows you to make your own choices for entertainment.

Here is the recipe **JULDAY**:

```
FUNCTION JULDAY& (MM, ID, IYYY)
```
   In this routine JULDAY& returns the Julian Day Number which begins at noon of the calendar
   date specified by month MM, day ID, and year IYYY. Positive year signifies A.D.; negative, B.C.
   Remember that the year after 1 B.C. was 1 A.D.

```
IGREG& = 588829                        Gregorian Calendar was adopted on Oct. 15, 1582.
IF IYYY = 0 THEN PRINT "There is no Year Zero.": EXIT FUNCTION
IF IYYY < 0 THEN IYYY = IYYY + 1
IF MM > 2 THEN                         Here is an example of a block IF-structure.
  JY = IYYY
  JM = MM + 1
ELSE
  JY = IYYY - 1
  JM = MM + 13
END IF
JD& = INT(365.25 * JY) + INT(30.6001 * JM) + ID + 1720995
IF ID + 31 * (MM + 12 * IYYY) >= IGREG& THEN          Change to Gregorian
  JA = INT(.01 * JY)                                  Calendar?
  JD& = JD& + 2 - JA + INT(.25 * JA)
END IF
JULDAY& = JD&
END FUNCTION
```

A sample program using JULDAY is the following:

```
DECLARE FUNCTION JULDAY& (IM!, ID!, IY!).

'PROGRAM D1R2
'Driver for JULDAY
CLS
DIM NAMQ$(12)
FOR I = 1 TO 12
  READ NAMQ$(I)
NEXT I
DATA "January","February","March","April","May","June","July","August"
DATA "September","October","November","December"
OPEN "DATES.DAT" FOR INPUT AS #1
LINE INPUT #1, DUM$
LINE INPUT #1, DUM$
N = VAL(DUM$)
PRINT "Month          Day  Year   Julian Day    Event"
PRINT
FOR I = 1 TO N
  LINE INPUT #1, DUM$
  IM = VAL(MID$(DUM$, 1, 2))
  ID = VAL(MID$(DUM$, 3, 3))
  IY = VAL(MID$(DUM$, 6, 5))
  TXT$ = MID$(DUM$, 11)
  PRINT NAMQ$(IM),
  PRINT USING "###"; ID;
  PRINT USING "######"; IY,
  PRINT USING "###########"; JULDAY&(IM, ID, IY);
  PRINT "     "; TXT$
NEXT I
CLOSE #1
```

```
PRINT
PRINT "Month,Day,Year (e.g. 1,13,1905)"
PRINT
FOR I = 1 TO 20
  PRINT "MM,DD,YYYY"
  INPUT IM, ID, IY
  IF IM < 0 THEN EXIT FOR
  PRINT "Julian Day: "; JULDAY&(IM, ID, IY)
  PRINT
NEXT I
END
```

The next program in *Numerical Recipes* is BADLUK, an infamous code that combines the best and worst instincts of man. We include no demonstration program for BADLUK, not just because we fear it, but also because it is self-contained, with sample results appearing in the text. The purpose of BADLUK is to find all dates Friday the Thirteenth on which the moon is full.

Here is the recipe BADLUK:

```
DECLARE SUB FLMOON (N!, NPH!, JD&, FRAC!)
DECLARE FUNCTION JULDAY& (IM!, ID!, IY!)

'PROGRAM BADLUK
CLS
TIMZON = -5! / 24              Time zone −5 is Eastern Standard Time.
READ IYBEG, IYEND              The range of dates to be searched.
DATA 1900,2000
PRINT "Full moons on Friday the 13th from"; IYBEG; "to"; IYEND
PRINT
FOR IYYY = IYBEG TO IYEND      Loop over each year,
  FOR IM = 1 TO 12             and each month.
    JDAY& = JULDAY&(IM, 13, IYYY)  Is the thirteenth a Friday?
    IDWK = (JDAY& + 1) - 7 * INT((JDAY& + 1) / 7)
    IF IDWK = 5 THEN
      N = INT(12.37 * (IYYY - 1900 + (IM - .5) / 12!))
```

This value N is a first approximation to how many full moons have occurred since 1900. We will feed it into the phase routine and adjust it up or down until we determine that our desired 13th was or was not a full moon. The variable ICON signals the direction of adjustment.

```
      ICON = 0
      DO
        CALL FLMOON(N, 2, JD&, FRAC)      Get date of full moon N.
        IFRAC = CINT(24! * (FRAC + TIMZON))   Convert to hours in correct time zone.
        IF IFRAC < 0 THEN                 Convert from Julian Days beginning at noon
          JD& = JD& - 1                   to civil days beginning at midnight.
          IFRAC = IFRAC + 24
        END IF
        IF IFRAC > 12 THEN
          JD& = JD& + 1
          IFRAC = IFRAC - 12
        ELSE
          IFRAC = IFRAC + 12
        END IF
        IF JD& = JDAY& THEN               Did we hit our target day?
          PRINT IM; "/"; 13; "/"; IYYY
```

```
          PRINT "Full moon"; IFRAC; "hrs after midnight (EST)."
          PRINT
          EXIT DO              Part of the break-structure, case of a match.
        ELSE                   Didn't hit it.
          IC = SGN(JDAY& - JD&)
          IF IC = -ICON THEN EXIT DO           Another break, case of no match.
          ICON = IC
          N = N + IC
        END IF
      LOOP
    END IF
  NEXT IM
NEXT IYYY
END
```

Chapter 1 closes with routine CALDAT, which illustrates no new points, but complements JULDAY by doing conversions from Julian day number to the month, day, and year on which the given Julian day began. This offers an opportunity, grasped by the demonstration program D1R4, to push dates through both JULDAY and CALDAT in succession, to see if they survive intact. This, of course, tests only your authors' ability to make mistakes backward as well as forward, but we hope you will share our optimism that correct results here speak well for both routines. (We have checked them a bit more carefully in other ways.)

Here is the recipe CALDAT:

```
SUB CALDAT (JULIAN&, MM, ID, IYYY)
```
Inverse of the function JULDAY given above. Here JULIAN& is input as a Julian Day Number, and the routine outputs MM,ID, and IYYY as the month, day, and year on which the specified Julian Day started at noon.
```
IGREG& = 2299161                  Cross-over to Gregorian Calendar
IF JULIAN& >= IGREG& THEN         produces this correction,
  JALPHA& = INT(((JULIAN& - 1867216) - .25) / 36524.25)
  JA& = JULIAN& + 1 + JALPHA& - INT(.25 * JALPHA&)
ELSE                              or else no correction.
  JA& = JULIAN&
END IF
JB& = JA& + 1524
JC& = INT(6680! + ((JB& - 2439870) - 122.1) / 365.25)
JD& = 365 * JC& + INT(.25 * JC&)
JE& = INT((JB& - JD&) / 30.6001)
ID = JB& - JD& - INT(30.6001 * JE&)
MM = JE& - 1
IF MM > 12 THEN MM = MM - 12
IYYY = JC& - 4715
IF MM > 2 THEN IYYY = IYYY - 1
IF IYYY <= 0 THEN IYYY = IYYY - 1
END SUB
```

A sample program using `CALDAT` is the following:

```
DECLARE SUB CALDAT (JULIAN&, MM!, ID!, IYYY!)
DECLARE FUNCTION JULDAY& (IM!, ID!, IY!)

'PROGRAM D1R4
'Driver for routine CALDAT
CLS
DIM NAMQ$(12)
'Check whether CALDAT properly undoes the operation of JULDAY
FOR I = 1 TO 12
  READ NAMQ$(I)
NEXT I
DATA "January","February","March","April","May","June","July","August"
DATA "September","October","November","December"
OPEN "DATES.DAT" FOR INPUT AS #1
LINE INPUT #1, DUM$
LINE INPUT #1, DUM$
N = VAL(DUM$)
PRINT "Original Date:                      Reconstructed Date:"
PRINT "Month      Day  Year    Julian Day    Month      Day  Year"
PRINT
FOR I = 1 TO N
  LINE INPUT #1, DUM$
  IM = VAL(MID$(DUM$, 1, 2))
  ID = VAL(MID$(DUM$, 3, 3))
  IY = VAL(MID$(DUM$, 6, 10))
  IYCOPY = IY
  J& = JULDAY&(IM, ID, IYCOPY)
  CALL CALDAT(J&, IMM, IDD, IYY)
  PRINT NAMQ$(IM); TAB(12);
  PRINT USING "##"; ID;
  PRINT USING "######"; IY;
  PRINT "     "; J&; "     "; NAMQ$(IMM); TAB(50);
  PRINT USING "##"; IDD;
  PRINT USING "######"; IYY
NEXT I
END
```

Appendix

File `DATES.DAT`:

```
List of dates for testing routines in Chapter 1
16 entries
12 31    -1 End of millennium
01 01     1 One day later
10 14 1582 Day before Gregorian calendar
10 15 1582 Gregorian calendar adopted
01 17 1706 Benjamin Franklin born
04 14 1865 Abraham Lincoln shot
04 18 1906 San Francisco earthquake
05 07 1915 Sinking of the Lusitania
07 20 1923 Pancho Villa assassinated
05 23 1934 Bonnie and Clyde eliminated
07 22 1934 John Dillinger shot
```

```
04 03 1936 Bruno Hauptmann electrocuted
05 06 1937 Hindenburg disaster
07 26 1956 Sinking of the Andrea Doria
06 05 1976 Teton dam collapse
05 23 1968 Julian Day 2440000
```

Chapter 2: Linear Algebraic Equations

Numerical Recipes Chapter 2 begins the "true grit" of numerical analysis by considering the solution of linear algebraic equations. This is done first by Gauss-Jordan elimination (GAUSSJ), and then by LU decomposition with forward and back substitution (LUDCMP and LUBKSB). For singular or nearly singular matrices the best choice is singular value decomposition with back substitution (SVDCMP and SVBKSB). Several linear systems of special form, represented by tridiagonal, Vandermonde, and Toeplitz matrices, may be treated with subroutines TRIDAG, VANDER, and TOEPLZ respectively. Linear systems with relatively few nonzero coefficients, so-called "sparse" matrices, are handled by routine SPARSE.

⋆   ⋆   ⋆   ⋆

GAUSSJ performs Gauss-Jordan elimination with full pivoting to find the solution of a set of linear equations for a collection of right-hand side vectors. The demonstration routine D2R1 checks its operation with reference to a group of test input matrices printed at the end of this chapter as file MATRX1.DAT. Each matrix is subjected to inversion by GAUSSJ, and then multiplication by its own inverse to see that a unit matrix is produced. Then the solution vectors are each checked through multiplication by the original matrix and comparison with the right-hand side vectors that produced them.

Here is the recipe GAUSSJ:

```
SUB GAUSSJ (A(), N, NP, B(), M, MP)
            Linear equation solution by Gauss-Jordan elimination, equation (2.1.1). A is an input matrix
            of N by N elements, stored in an array of physical dimensions NP by NP. B is an input matrix of
            N by M containing the M right-hand side vectors, stored in an array of physical dimensions NP
            by MP. On output, A is replaced by its matrix inverse, and B is replaced by the corresponding
            set of solution vectors.
DIM IPIV(N), INDXR(N), INDXC(N)
            The arrays IPIV, INDXR, and INDXC are used for bookkeeping on the pivoting.
FOR J = 1 TO N
  IPIV(J) = 0
NEXT J
FOR I = 1 TO N                      This is the main loop over the columns to be reduced.
  BIG = 0!
  FOR J = 1 TO N                    This is the outer loop of the search for a pivot element.
    IF IPIV(J) <> 1 THEN
      FOR K = 1 TO N
        IF IPIV(K) = 0 THEN
          IF ABS(A(J, K)) >= BIG THEN
            BIG = ABS(A(J, K))
            IROW = J
            ICOL = K
          END IF
```

9

```
          ELSEIF IPIV(K) > 1 THEN
            PRINT "Singular matrix"
            EXIT SUB
          END IF
        NEXT K
      END IF
    NEXT J
    IPIV(ICOL) = IPIV(ICOL) + 1
```

We now have the pivot element, so we interchange rows, if needed, to put the pivot element on the diagonal. The columns are not physically interchanged, only relabeled: INDXC(I), the column of the Ith pivot element, is the Ith column that is reduced, while INDXR(I) is the row in which that pivot element was originally located. If INDXR(I)$\neq$INDXC(I) there is an implied column interchange. With this form of bookkeeping, the solution B's will end up in the correct order, and the inverse matrix will be scrambled by columns.

```
    IF IROW <> ICOL THEN
      FOR L = 1 TO N
        DUM = A(IROW, L)
        A(IROW, L) = A(ICOL, L)
        A(ICOL, L) = DUM
      NEXT L
      FOR L = 1 TO M
        DUM = B(IROW, L)
        B(IROW, L) = B(ICOL, L)
        B(ICOL, L) = DUM
      NEXT L
    END IF
    INDXR(I) = IROW            We are now ready to divide the pivot row by the pivot element,
    INDXC(I) = ICOL            located at IROW and ICOL.
    IF A(ICOL, ICOL) = 0! THEN PRINT "Singular matrix.": EXIT SUB
    PIVINV = 1! / A(ICOL, ICOL)
    A(ICOL, ICOL) = 1!
    FOR L = 1 TO N
      A(ICOL, L) = A(ICOL, L) * PIVINV
    NEXT L
    FOR L = 1 TO M
      B(ICOL, L) = B(ICOL, L) * PIVINV
    NEXT L
    FOR LL = 1 TO N            Next, we reduce the rows...
      IF LL <> ICOL THEN       ...except for the pivot one, of course.
        DUM = A(LL, ICOL)
        A(LL, ICOL) = 0!
        FOR L = 1 TO N
          A(LL, L) = A(LL, L) - A(ICOL, L) * DUM
        NEXT L
        FOR L = 1 TO M
          B(LL, L) = B(LL, L) - B(ICOL, L) * DUM
        NEXT L
      END IF
    NEXT LL
  NEXT I                       This is the end of the main loop over columns of the reduction.
  FOR L = N TO 1 STEP -1       It only remains to unscramble the solution in view of the
    IF INDXR(L) <> INDXC(L) THEN   column interchanges. We do this by interchanging pairs of
      FOR K = 1 TO N           columns in the reverse order that the permutation was built
        DUM = A(K, INDXR(L))   up.
        A(K, INDXR(L)) = A(K, INDXC(L))
```

```
      A(K, INDXC(L)) = DUM
    NEXT K
  END IF
NEXT L
ERASE INDXC, INDXR, IPIV          And we are done.
END SUB
```

A sample program using **GAUSSJ** is the following:

```
DECLARE SUB GAUSSJ (A!(), N!, NP!, B!(), M!, MP!)

'PROGRAM D2R1
'Driver program for subroutine GAUSSJ
'Reads Matrices from file MATRX1.DAT and feeds them to GAUSSJ
CLS
NP = 20
DIM A(NP, NP), B(NP, NP), AI(NP, NP), X(NP, NP)
DIM U(NP, NP), T(NP, NP)
OPEN "MATRX1.DAT" FOR INPUT AS #1
DO
  LINE INPUT #1, DUM$
  IF DUM$ = "END" THEN CLOSE #1: END
  LINE INPUT #1, DUM$
  LINE INPUT #1, DUM$
  N = VAL(MID$(DUM$, 1, 2))
  M = VAL(MID$(DUM$, 3, 2))
  LINE INPUT #1, DUM$
  FOR K = 1 TO N
    LINE INPUT #1, DUM$
    FOR L = 1 TO N
      A(K, L) = VAL(MID$(DUM$, 4 * L - 3, 4))
    NEXT L
  NEXT K
  LINE INPUT #1, DUM$
  FOR L = 1 TO M
    LINE INPUT #1, DUM$
    FOR K = 1 TO N
      B(K, L) = VAL(MID$(DUM$, 4 * K - 3, 4))
    NEXT K
  NEXT L
  'Save Matrices for later testing of results
  FOR L = 1 TO N
    FOR K = 1 TO N
      AI(K, L) = A(K, L)
    NEXT K
    FOR K = 1 TO M
      X(L, K) = B(L, K)
    NEXT K
  NEXT L
  'Invert Matrix
  PRINT
  CALL GAUSSJ(AI(), N, NP, X(), M, NP)
  PRINT "Inverse of Matrix A : "
  FOR K = 1 TO N
    FOR L = 1 TO N
```

```
      PRINT USING "#####.######"; AI(K, L);
    NEXT L
    PRINT
  NEXT K
  'Test Results
  'Check Inverse
  PRINT "A times A-inverse (compare with unit matrix)"
  FOR K = 1 TO N
    FOR L = 1 TO N
      U(K, L) = 0!
      FOR J = 1 TO N
        U(K, L) = U(K, L) + A(K, J) * AI(J, L)
      NEXT J
    NEXT L
    FOR L = 1 TO N
      PRINT USING "#####.######"; U(K, L);
    NEXT L
    PRINT
  NEXT K
  'Check Vector Solutions
  PRINT "Check the following vectors for equality:"
  PRINT "          Original    Matrix*Sol'n"
  FOR L = 1 TO M
    PRINT "Vector "; STR$(L); ":"
    FOR K = 1 TO N
      T(K, L) = 0!
      FOR J = 1 TO N
        T(K, L) = T(K, L) + A(K, J) * X(J, L)
      NEXT J
      PRINT "        ";
      PRINT USING "#####.######"; B(K, L); T(K, L)
    NEXT K
  NEXT L
  PRINT "**********************************"
  PRINT "Press RETURN for next problem:";
  LINE INPUT DUM$
LOOP
END
```

The demonstration program for routine LUDCMP relies on the same package of test matrices, but just performs an LU decomposition of each. The performance is checked by multiplying the lower and upper matrices of the decomposition and comparing with the original matrix. The array INDX keeps track of the scrambling done by LUDCMP to effect partial pivoting. We had to do the unscrambling here, but you will normally not be called upon to do so, since LUDCMP is used with the descrambler-containing routine LUBKSB.

Here is the recipe LUDCMP:

```
SUB LUDCMP (A(), N, NP, INDX(), D)
```
Given an $N \times N$ matrix A, with physical dimension NP, this routine replaces it by the LU decomposition of a rowwise permutation of itself. A and N are input. A is output, arranged as in equation (2.3.14); INDX is an output vector which records the row permutation effected by the partial pivoting; D is output as ± 1 depending on whether the number of row interchanges was even or odd, respectively. This routine is used in combination with LUBKSB to solve linear equations or invert a matrix.

```
TINY = 1E-20                          A small number.
DIM VV(N)                             VV stores the implicit scaling of each row.
D = 1!                                No row interchanges yet.
FOR I = 1 TO N                        Loop over rows to get the implicit scaling information.
  AAMAX = 0!
  FOR J = 1 TO N
    IF ABS(A(I, J)) > AAMAX THEN AAMAX = ABS(A(I, J))
  NEXT J
  IF AAMAX = 0! THEN PRINT "Singular matrix.": EXIT SUB    No nonzero largest el-
  VV(I) = 1! / AAMAX                  Save the scaling.                ement.
NEXT I
FOR J = 1 TO N                        This is the loop over columns of Crout's method.
  FOR I = 1 TO J - 1                  This is equation 2.3.12 except for i = j.
    SUM = A(I, J)
    FOR K = 1 TO I - 1
      SUM = SUM - A(I, K) * A(K, J)
    NEXT K
    A(I, J) = SUM
  NEXT I
  AAMAX = 0!                          Initialize for the search for largest pivot element.
  FOR I = J TO N                      This is i = j of equation 2.3.12 and i = j + 1...N of
    SUM = A(I, J)                     equation 2.3.13.
    FOR K = 1 TO J - 1
      SUM = SUM - A(I, K) * A(K, J)
    NEXT K
    A(I, J) = SUM
    DUM = VV(I) * ABS(SUM)            Figure of merit for the pivot.
    IF DUM >= AAMAX THEN              Is it better than the best so far?
      IMAX = I
      AAMAX = DUM
    END IF
  NEXT I
  IF J <> IMAX THEN                   Do we need to interchange rows?
    FOR K = 1 TO N                    Yes, do so...
      DUM = A(IMAX, K)
      A(IMAX, K) = A(J, K)
      A(J, K) = DUM
    NEXT K
    D = -D                           ...and change the parity of D.
    VV(IMAX) = VV(J)                 Also interchange the scale factor.
  END IF
  INDX(J) = IMAX                     Now, finally, divide by the pivot element.
  IF A(J, J) = 0! THEN A(J, J) = TINY
```
If the pivot element is zero the matrix is singular (at least to the precision of the algorithm). For some applications on singular matrices, it is desirable to substitute TINY for zero.

```
  IF J <> N THEN
    DUM = 1! / A(J, J)
    FOR I = J + 1 TO N
```

```
      A(I, J) = A(I, J) * DUM
    NEXT I
  END IF
NEXT J                          Go back for the next column in the reduction.
ERASE VV
END SUB
```

A sample program using LUDCMP is the following:

```
DECLARE SUB LUDCMP (A!(), N!, NP!, INDX!(), D!)

'PROGRAM D2R2
'Driver for routine LUDCMP
CLS
NP = 20
DIM A(NP, NP), XL(NP, NP), XU(NP, NP)
DIM INDX(NP), JNDX(NP), X(NP, NP)
OPEN "MATRX1.DAT" FOR INPUT AS #1
LINE INPUT #1, DUM$
DO
  LINE INPUT #1, DUM$
  LINE INPUT #1, DUM$
  N = VAL(MID$(DUM$, 1, 2))
  M = VAL(MID$(DUM$, 3, 2))
  LINE INPUT #1, DUM$
  FOR K = 1 TO N
    LINE INPUT #1, DUM$
    FOR L = 1 TO N
      A(K, L) = VAL(MID$(DUM$, 4 * L - 3, 4))
    NEXT L
  NEXT K
  LINE INPUT #1, DUM$
  FOR L = 1 TO M
    LINE INPUT #1, DUM$
    FOR K = 1 TO N
      X(K, L) = VAL(MID$(DUM$, 4 * K - 3, 4))
    NEXT K
  NEXT L
  'Print out A-matrix for comparison with product of lower
  'and upper decomposition matrices.
  PRINT "Original matrix:"
  FOR K = 1 TO N
    FOR L = 1 TO N
      PRINT USING "#####.######"; A(K, L);
    NEXT L
    PRINT
  NEXT K
  'Perform the decomposition
  CALL LUDCMP(A(), N, NP, INDX(), D)
  'Compose separately the lower and upper matrices
  FOR K = 1 TO N
    FOR L = 1 TO N
      IF L > K THEN
        XU(K, L) = A(K, L)
        XL(K, L) = 0!
```

```
      ELSEIF L < K THEN
        XU(K, L) = 0!
        XL(K, L) = A(K, L)
      ELSE
        XU(K, L) = A(K, L)
        XL(K, L) = 1!
      END IF
    NEXT L
  NEXT K
  'Compute product of lower and upper matrices for
  'comparison with original matrix.
  FOR K = 1 TO N
    JNDX(K) = K
    FOR L = 1 TO N
      X(K, L) = 0!
      FOR J = 1 TO N
        X(K, L) = X(K, L) + XL(K, J) * XU(J, L)
      NEXT J
    NEXT L
  NEXT K
  PRINT "Product of lower and upper matrices (unscrambled):"
  FOR K = 1 TO N
    DUM = JNDX(INDX(K))
    JNDX(INDX(K)) = JNDX(K)
    JNDX(K) = DUM
  NEXT K
  FOR K = 1 TO N
    FOR J = 1 TO N
      IF JNDX(J) = K THEN
        FOR L = 1 TO N
          PRINT USING "#####.######"; X(J, L);
        NEXT L
        PRINT
      END IF
    NEXT J
  NEXT K
  PRINT "Lower matrix of the decomposition:"
  FOR K = 1 TO N
    FOR L = 1 TO N
      PRINT USING "#####.######"; XL(K, L);
    NEXT L
    PRINT
  NEXT K
  PRINT "Upper matrix of the decomposition:"
  FOR K = 1 TO N
    FOR L = 1 TO N
      PRINT USING "#####.######"; XU(K, L);
    NEXT L
    PRINT
  NEXT K
  PRINT "************************************"
  PRINT "Press RETURN for next problem:"
  LINE INPUT DUM$
  LINE INPUT #1, TXT$
LOOP WHILE TXT$ <> "END"
```

```
CLOSE #1
END
```

Our example driver for LUBKSB makes calls to both LUDCMP and LUBKSB in order to solve the linear equation problems posed in file MATRX1.DAT (see discussion of GAUSSJ). The original matrix of coefficients is applied to the solution vectors to check that the result matches the right-hand side vectors posed for each problem. We apologize for using routine LUDCMP in a test of LUBKSB, but LUDCMP has been tested independently, and anyway, LUBKSB is nothing without this partner program, so a test of the combination is more to the point.

Here is the recipe LUBKSB:

```
SUB LUBKSB (A(), N, NP, INDX(), B())
```
>   Solves the set of N linear equations $A \cdot X = B$. Here A is input, not as the matrix A but rather as its LU decomposition, determined by the routine LUDCMP. INDX is input as the permutation vector returned by LUDCMP. B is input as the right-hand side vector B, and returns with the solution vector X. A, N, NP and INDX are not modified by this routine and can be left in place for successive calls with different right-hand sides B. This routine takes into account the possibility that B will begin with many zero elements, so it is efficient for use in matrix inversion.

```
II = 0                              When II is set to a positive value, it will become the index
FOR I = 1 TO N                      of the first nonvanishing element of b. We now do the for-
  LL = INDX(I)                      ward substitution, equation 2.3.6. The only new wrinkle is to
  SUM = B(LL)                       unscramble the permutation as we go.
  B(LL) = B(I)
  IF II <> 0 THEN
    FOR J = II TO I - 1
      SUM = SUM - A(I, J) * B(J)
    NEXT J
  ELSEIF SUM <> 0! THEN
    II = I                          A nonzero element was encountered, so from now on we will
  END IF                            have to do the sums in the loop above.
  B(I) = SUM
NEXT I
FOR I = N TO 1 STEP -1              Now we do the backsubstitution, equation 2.3.7.
  SUM = B(I)
  FOR J = I + 1 TO N
    SUM = SUM - A(I, J) * B(J)
  NEXT J
  B(I) = SUM / A(I, I)              Store a component of the solution vector X.
NEXT I
END SUB                            All done!
```

A sample program using LUBKSB is the following:

```
DECLARE SUB LUDCMP (A!(), N!, NP!, INDX!(), D!)
DECLARE SUB LUBKSB (A!(), N!, NP!, INDX!(), B!())

'PROGRAM D2R3
'Driver for routine LUBKSB
CLS
NP = 20
DIM A(NP, NP), B(NP, NP), INDX(NP)
DIM C(NP, NP), X(NP)
OPEN "MATRX1.DAT" FOR INPUT AS #1
```

```
LINE INPUT #1, DUM$
DO
  LINE INPUT #1, DUM$
  LINE INPUT #1, DUM$
  N = VAL(MID$(DUM$, 1, 2))
  M = VAL(MID$(DUM$, 3, 2))
  LINE INPUT #1, DUM$
  FOR K = 1 TO N
    LINE INPUT #1, DUM$
    FOR L = 1 TO N
      A(K, L) = VAL(MID$(DUM$, 4 * L - 3, 4))
    NEXT L
  NEXT K
  LINE INPUT #1, DUM$
  FOR L = 1 TO M
    LINE INPUT #1, DUM$
    FOR K = 1 TO N
      B(K, L) = VAL(MID$(DUM$, 4 * K - 3, 4))
    NEXT K
  NEXT L
  'Save matrix A for later testing
  FOR L = 1 TO N
    FOR K = 1 TO N
      C(K, L) = A(K, L)
    NEXT K
  NEXT L
  'Do LU decomposition
  CALL LUDCMP(C(), N, NP, INDX(), P)
  'Solve equations for each right-hand vector
  FOR K = 1 TO M
    FOR L = 1 TO N
      X(L) = B(L, K)
    NEXT L
    CALL LUBKSB(C(), N, NP, INDX(), X())
    'Test results with original matrix
    PRINT "Right-hand side vector:"
    FOR L = 1 TO N
      PRINT USING "#####.######"; B(L, K);
    NEXT L
    PRINT
    PRINT "Result of matrix applied to sol'n vector"
    FOR L = 1 TO N
      B(L, K) = 0!
      FOR J = 1 TO N
        B(L, K) = B(L, K) + A(L, J) * X(J)
      NEXT J
    NEXT L
    FOR L = 1 TO N
      PRINT USING "#####.######"; B(L, K);
    NEXT L
    PRINT
    PRINT "************************************"
  NEXT K
  PRINT "Press RETURN for next problem:"
  LINE INPUT DUM$
```

```
LINE INPUT #1, TXT$
LOOP WHILE TXT$ <> "END"
CLOSE #1
END
```

Subroutine TRIDAG solves linear equations with coefficients that form a tridiagonal matrix. We provide at the end of this chapter a second file of matrices MATRIX2.DAT for the demonstration driver. In all other respects, the demonstration program D2R4 operates in the same fashion as D2R3.

Here is the recipe TRIDAG:

```
SUB TRIDAG (A(), B(), C(), R(), U(), N)
```
> Solves for a vector U of length N the tridiagonal linear set given by equation (2.6.1). A, B, C and R are input vectors and are not modified.

```
DIM GAM(N)                              One vector of workspace, GAM is needed.
IF B(1) = 0! THEN PRINT "Abnormal exit": EXIT SUB    If this happens then you should
BET = B(1)                              rewrite your equations as a set of order N − 1, with u₂ trivially
U(1) = R(1) / BET                       eliminated.
FOR J = 2 TO N                          Decomposition and forward substitution.
  GAM(J) = C(J - 1) / BET
  BET = B(J) - A(J) * GAM(J)
  IF BET = 0! THEN PRINT "Abnormal exit": EXIT SUB   Algorithm fails.
  U(J) = (R(J) - A(J) * U(J - 1)) / BET
NEXT J
FOR J = N - 1 TO 1 STEP -1                           Backsubstitution.
  U(J) = U(J) - GAM(J + 1) * U(J + 1)
NEXT J
ERASE GAM
END SUB
```

A sample program using TRIDAG is the following:

```
DECLARE SUB TRIDAG (A!(), B!(), C!(), R!(), U!(), N!)

'PROGRAM D2R4
'Driver for routine TRIDAG
CLS
NP = 20
DIM DIAG(NP), SUPERD(NP), SUBD(NP), RHS(NP), U(NP)
OPEN "MATRX2.DAT" FOR INPUT AS #1
DO
  LINE INPUT #1, TXT$
  IF TXT$ = "END" THEN EXIT DO
  LINE INPUT #1, DUM$
  LINE INPUT #1, DUM$
  N = VAL(DUM$)
  LINE INPUT #1, DUM$
  LINE INPUT #1, DUM$
  FOR K = 1 TO N
    DIAG(K) = VAL(MID$(DUM$, 4 * K - 3, 4))'
  NEXT K
  LINE INPUT #1, DUM$
  LINE INPUT #1, DUM$
```

```
FOR K = 1 TO N - 1
  SUPERD(K) = VAL(MID$(DUM$, 4 * K - 3, 4))
NEXT K
LINE INPUT #1, DUM$
LINE INPUT #1, DUM$
FOR K = 2 TO N
  SUBD(K) = VAL(MID$(DUM$, 4 * K - 7, 4))
NEXT K
LINE INPUT #1, DUM$
LINE INPUT #1, DUM$
FOR K = 1 TO N
  RHS(K) = VAL(MID$(DUM$, 4 * K - 3, 4))
NEXT K
'Carry out solution
CALL TRIDAG(SUBD(), DIAG(), SUPERD(), RHS(), U(), N)
PRINT "The solution vector is:"
FOR K = 1 TO N
  PRINT USING "#####.######"; U(K);
NEXT K
PRINT
'Test solution
PRINT "(matrix)*(sol'n vector) should be:"
FOR K = 1 TO N
  PRINT USING "#####.######"; RHS(K);
NEXT K
PRINT
PRINT "Actual result is:"
FOR K = 1 TO N
  IF K = 1 THEN
    RHS(K) = DIAG(1) * U(1) + SUPERD(1) * U(2)
  ELSEIF K = N THEN
    RHS(K) = SUBD(N) * U(N - 1) + DIAG(N) * U(N)
  ELSE
    RHS(K) = SUBD(K) * U(K - 1) + DIAG(K) * U(K) + SUPERD(K) * U(K + 1)
  END IF
NEXT K
FOR K = 1 TO N
  PRINT USING "#####.######"; RHS(K);
NEXT K
PRINT
PRINT "*************************************"
PRINT "Press RETURN for next problem:"
LINE INPUT DUM$
LOOP
CLOSE #1
END
```

MPROVE is a short routine for improving the solution vector for a set of linear equations, providing that an LU decomposition has been performed on the matrix of coefficients. Our test of this function is to use LUDCMP and LUBKSB to solve a set of equations specified in the DATA statements at the beginning of the program. The solution vector is then corrupted by the addition of random values to each component. MPROVE works on the corrupted vector to recover the original.

Here is the recipe MPROVE:

```
DECLARE SUB LUBKSB (A!(), N!, NP!, INDX!(), B!())

SUB MPROVE (A(), ALUD(), N, NP, INDX(), B(), X())
```
Improves a solution vector X of the linear set of equations $A \cdot X = B$. The matrix A, and the vectors B and X are input, as is the dimension N. Also input is ALUD, the LU decomposition of A as returned by LUDCMP, and the vector INDX also returned by that routine. On output, only X is modified, to an improved set of values.
```
DIM R(N)
FOR I = 1 TO N                        Calculate the right-hand side, accumulating the residual in
  SDP# = -B(I)                        double precision.
  FOR J = 1 TO N
    SDP# = SDP# + CDBL(A(I, J)) * CDBL(X(J))
  NEXT J
  R(I) = SDP#
NEXT I
CALL LUBKSB(ALUD(), N, NP, INDX(), R())    Solve for the error term,
FOR I = 1 TO N                             and subtract it from the old solution.
  X(I) = X(I) - R(I)
NEXT I
ERASE R
END SUB
```

A sample program using MPROVE is the following:

```
DECLARE SUB LUDCMP (A!(), N!, NP!, INDX!(), D!)
DECLARE SUB LUBKSB (A!(), N!, NP!, INDX!(), B!())
DECLARE SUB MPROVE (A!(), ALUD!(), N!, NP!, INDX!(), B!(), X!())
DECLARE FUNCTION RAN3! (IDUM&)

'PROGRAM D2R5
'Driver for routine MPROVE
CLS
N = 5
NP = 5
DIM A(NP, NP), INDX(N), B(N), X(N), AA(NP, NP)
FOR J = 1 TO NP
  FOR I = 1 TO NP
    READ A(I, J)
  NEXT I
NEXT J
DATA 1.0,2.0,1.0,4.0,5.0,2.0,3.0,1.0,5.0,1.0,3.0,4.0,1.0,1.0,2.0,4.0,5.0,1.0
DATA 2.0,3.0,5.0,1.0,1.0,3.0,4.0
FOR I = 1 TO N
  READ B(I)
NEXT I
DATA 1.0,1.0,1.0,1.0,1.0
FOR I = 1 TO N
  X(I) = B(I)
  FOR J = 1 TO N
    AA(I, J) = A(I, J)
  NEXT J
NEXT I
CALL LUDCMP(AA(), N, NP, INDX(), D)
```

```
CALL LUBKSB(AA(), N, NP, INDX(), X())
PRINT "Solution vector for the equations:"
FOR I = 1 TO N
  PRINT USING "#####.######"; X(I);
NEXT I
PRINT
PRINT
'Now phoney up X and let MPROVE fit it
IDUM& = -13
FOR IQ = 1 TO N
  X(IQ) = X(IQ) * (1! + .2 * RAN3(IDUM&))
NEXT IQ
PRINT "Solution vector with noise added:"
FOR I = 1 TO N
  PRINT USING "#####.######"; X(I);
NEXT I
PRINT
PRINT
CALL MPROVE(A(), AA(), N, NP, INDX(), B(), X())
PRINT "Solution vector recovered by MPROVE:"
FOR I = 1 TO N
  PRINT USING "#####.######"; X(I);
NEXT I
PRINT
PRINT
END
```

Vandermonde matrices of dimension $N \times N$ have elements that are entirely integer powers of N arbitrary numbers $x_1 \ldots x_N$. (See *Numerical Recipes* for details.) In the demonstration program **D2R6** we provide five such numbers to specify a 5×5 matrix, and five elements of a right-hand side vector Q. Routine **VANDER** is used to find the solution vector W. This vector is tested by applying the matrix to W and comparing the result to Q.

Here is the recipe **VANDER**:

```
SUB VANDER (X(), W(), Q(), N)
```
Solves the Vandermonde linear system $\sum_{i=1}^{N} x_i^{k-1} w_i = q_k$ $(k = 1, \ldots, N)$. Input consists of the vectors X and Q, each of length N; the vector W is output. Make constants double precision if you convert the program to double precision — which is a good idea.

```
ZERO = 0!
ONE = 1!
DIM C(N)
IF N = 1 THEN
  W(1) = Q(1)
ELSE
  FOR I = 1 TO N              Initialize array.
    C(I) = ZERO
  NEXT I
  C(N) = -X(1)                Coefficients of the master polynomial are found by recursion.
  FOR I = 2 TO N
    XX = -X(I)
    FOR J = N + 1 - I TO N - 1
      C(J) = C(J) + XX * C(J + 1)
    NEXT J
```

```
   C(N) = C(N) + XX
 NEXT I
 FOR I = 1 TO N                    Each subfactor in turn
   XX = X(I)
   T = ONE
   B = ONE
   S = Q(N)
   K = N
   FOR J = 2 TO N                  is synthetically divided,
     K1 = K - 1
     B = C(K) + XX * B
     S = S + Q(K1) * B             matrix-multiplied by the right-hand side,
     T = XX * T + B
     K = K1
   NEXT J
   W(I) = S / T                    and supplied with a denominator.
 NEXT I
END IF
ERASE C
END SUB
```

A sample program using `VANDER` is the following:

```
DECLARE SUB VANDER (X!(), W!(), Q!(), N!)

'PROGRAM D2R6
'Driver for routine VANDER
CLS
N = 5
DIM X(N), Q(N), W(N), TERM(N)
FOR I = 1 TO N
  READ X(I)
NEXT I
DATA 1.0,1.5,2.0,2.5,3.0
FOR I = 1 TO N
  READ Q(I)
NEXT I
DATA 1.0,1.5,2.0,2.5,3.0
CALL VANDER(X(), W(), Q(), N)
PRINT "Solution vector:"
FOR I = 1 TO N
  PRINT "     W(";
  PRINT USING "#"; I;
  PRINT ") = ";
  PRINT USING "#.######^^^^"; W(I)
NEXT I
PRINT
PRINT "Test of solution vector:"
PRINT "    mtrx*sol'n    original"
SUM = 0!
FOR I = 1 TO N
  TERM(I) = W(I)
  SUM = SUM + W(I)
NEXT I
PRINT USING "#######.####"; SUM; Q(1)
```

```
FOR I = 2 TO N
  SUM = 0!
  FOR J = 1 TO N
    TERM(J) = TERM(J) * X(J)
    SUM = SUM + TERM(J)
  NEXT J
  PRINT USING "#######.####"; SUM; Q(I)
NEXT I
END
```

A very similar test is applied to TOEPLZ, which operates on Toeplitz matrices. The $N \times N$ Toeplitz matrix is specified by $2N - 1$ numbers r_i, in this case taken to be simply a linear progression of values. A right-hand side y_i is chosen likewise. TOEPLZ finds the solution vector x_i, and checks it in the usual fashion.

Here is the recipe TOEPLZ:

```
SUB TOEPLZ (R(), X(), Y(), N)
```
> Solves the Toeplitz system $\sum_{j=1}^{N} R_{(N+i-j)}x_j = y_i$ $(i = 1, \dots, N)$. The Toeplitz matrix need not be symmetric. Y and R are input arrays of length N and 2*N−1 respectively. X is the output array, of length N.

```
DIM G(N), H(N)
IF R(N) <> 0! THEN
  X(1) = Y(1) / R(N)                   Initialize for the recursion.
  IF N = 1 THEN ERASE H, G: EXIT SUB
  G(1) = R(N - 1) / R(N)
  H(1) = R(N + 1) / R(N)
  FOR M = 1 TO N                       Main loop over the recursion.
    M1 = M + 1
    SXN = -Y(M1)                       Compute numerator and denominator for x,
    SD = -R(N)
    FOR J = 1 TO M
      SXN = SXN + R(N + M1 - J) * X(J)
      SD = SD + R(N + M1 - J) * G(M - J + 1)
    NEXT J
    IF SD = 0! THEN
      PRINT "Levinson method fails:  singular principal minor"
      EXIT SUB
    END IF
    X(M1) = SXN / SD                   whence x.
    FOR J = 1 TO M
      X(J) = X(J) - X(M1) * G(M - J + 1)
    NEXT J
    IF M1 = N THEN ERASE H, G: EXIT SUB
    SGQ = -R(N - M1)                   Compute numerator and denominator for G and H,
    SHN = -R(N + M1)
    SGD = -R(N)
    FOR J = 1 TO M
      SGQ = SGQ + R(N + J - M1) * G(J)
      SHN = SHN + R(N + M1 - J) * H(J)
      SGD = SGD + R(N + J - M1) * H(M - J + 1)
    NEXT J
    IF SD = 0! OR SGD = 0! THEN
      PRINT "Levinson method fails:  singular principal minor"
```

```
      EXIT SUB
      END IF
      G(M1) = SGQ / SGD              whence G and H.
      H(M1) = SHN / SD
      K = M
      M2 = INT((M + 1) / 2)
      PP = G(M1)
      QQ = H(M1)
      FOR J = 1 TO M2
        PT1 = G(J)
        PT2 = G(K)
        QT1 = H(J)
        QT2 = H(K)
        G(J) = PT1 - PP * QT2
        G(K) = PT2 - PP * QT1
        H(J) = QT1 - QQ * PT2
        H(K) = QT2 - QQ * PT1
        K = K - 1
      NEXT J
    NEXT M                           Back for another recurrence.
    PRINT "never get here"
    EXIT SUB
  END IF
  PRINT "Levinson method fails:  singular principal minor"
END SUB
```

A sample program using TOEPLZ is the following:

```
DECLARE SUB TOEPLZ (R!(), X!(), Y!(), N!)

'PROGRAM D2R7
'Driver for routine TOEPLZ
CLS
N = 5
N2 = 2 * N
DIM X(N), Y(N), R(N2)
FOR I = 1 TO N
  Y(I) = .1 * I
NEXT I
FOR I = 1 TO 2 * N - 1
  R(I) = .1 * I
NEXT I
CALL TOEPLZ(R(), X(), Y(), N)
PRINT "Solution vector:"
FOR I = 1 TO N
  PRINT "    X(";
  PRINT USING "#"; I;
  PRINT ") = ";
  PRINT USING "#.######^^^^"; X(I)
NEXT I
PRINT
PRINT "Test of solution:"
PRINT "    mtrx*soln     original"
FOR I = 1 TO N
  SUM = 0!
```

```
FOR J = 1 TO N
  SUM = SUM + R(N + I - J) * X(J)
NEXT J
PRINT USING "#######.####"; SUM; Y(I)
NEXT I
PRINT
END
```

The pair SVDCMP, SVBKSB are tested in the same manner as LUDCMP, LUBKSB. That is, SVDCMP is checked independently to see that it yields proper decomposition of matrices. Then the pair of programs is tested as a unit to see that they provide correct solutions to some linear sets. (Note: Because of the order of programs in *Numerical Recipes*, the test of the pair in this case comes first.) The matrices and solution vectors are given in the Appendix as file MATRX3.DAT.

Driver D2R8 brings in matrices A and right-hand side vectors B from MATRX3.DAT. Matrix A, itself, is saved for later use. It is copied into matrix U for processing by SVDCMP. The results of the processing are the three arrays U, W, V which form the singular value decomposition of A. The right-hand side vectors are fed one at a time to vector C, and the resulting solution vectors X are checked for accuracy through application of the saved matrix A.

Here is the recipe SVBKSB:

```
SUB SVBKSB (U(), W(), V(), M, N, MP, NP, B(), X())
    Solves A · X = B for a vector X, where A is specified by the arrays U, W, V as returned
    by SVDCMP. M and N are the logical dimensions of A, and will be equal for square matrices.
    MP and NP are the physical dimensions of A. B is the input right-hand side. X is the output
    solution vector. No input quantities are destroyed, so the routine may be called sequentially
    with different B's. M must be greater than or equal to N; see SVDCMP.
DIM TMP(N)
FOR J = 1 TO N                      Calculate U^T B.
  S = 0!
  IF W(J) <> 0! THEN                Nonzero result only if w_j is nonzero.
    FOR I = 1 TO M
      S = S + U(I, J) * B(I)
    NEXT I
    S = S / W(J)                    This is the divide by w_j.
  END IF
  TMP(J) = S
NEXT J
FOR J = 1 TO N                      Matrix multiply by V to get answer.
  S = 0!
  FOR JJ = 1 TO N
    S = S + V(J, JJ) * TMP(JJ)
  NEXT JJ
  X(J) = S
NEXT J
ERASE TMP
END SUB
```

A sample program using SVBKSB is the following:

```
DECLARE SUB SVDCMP (A!(), M!, N!, MP!, NP!, W!(), V!())
DECLARE SUB SVBKSB (U!(), W!(), V!(), M!, N!, MP!, NP!, B!(), X!())

'PROGRAM D2R8
'Driver for routine SVBKSB, which calls routine SVDCMP
CLS
NP = 20
DIM A(NP, NP), B(NP, NP), U(NP, NP), W(NP)
DIM V(NP, NP), C(NP), X(NP)
OPEN "MATRX1.DAT" FOR INPUT AS #1
DO
  LINE INPUT #1, DUM$
  IF (DUM$ = "END") THEN EXIT DO
  LINE INPUT #1, DUM$
  LINE INPUT #1, DUM$
  N = VAL(MID$(DUM$, 1, 2))
  M = VAL(MID$(DUM$, 3, 2))
  LINE INPUT #1, DUM$
  FOR K = 1 TO N
    LINE INPUT #1, DUM$
    FOR L = 1 TO N
      A(K, L) = VAL(MID$(DUM$, 4 * L - 3, 4))
    NEXT L
  NEXT K
  LINE INPUT #1, DUM$
  FOR L = 1 TO M
    LINE INPUT #1, DUM$
    FOR K = 1 TO N
      B(K, L) = VAL(MID$(DUM$, 4 * K - 3, 4))
    NEXT K
  NEXT L
  'Copy A into U
  FOR K = 1 TO N
    FOR L = 1 TO N
      U(K, L) = A(K, L)
    NEXT L
  NEXT K
  'Decompose matrix A
  CALL SVDCMP(U(), N, N, NP, NP, W(), V())
  'Find maximum singular value
  WMAX = 0!
  FOR K = 1 TO N
    IF W(K) > WMAX THEN WMAX = W(K)
  NEXT K
  'Define "small"
  WMIN = WMAX * (.000001)
  'Zero the "small" singular values
  FOR K = 1 TO N
    IF W(K) < WMIN THEN W(K) = 0!
  NEXT K
  'Backsubstitute for each right-hand side vector
  FOR L = 1 TO M
    PRINT "Vector number "; L
```

```
FOR K = 1 TO N
  C(K) = B(K, L)
NEXT K
CALL SVBKSB(U(), W(), V(), N, N, NP, NP, C(), X())
PRINT "    Solution vector is:"
FOR K = 1 TO N
  PRINT USING "#####.######"; X(K);
NEXT K
PRINT
PRINT "    Original right-hand side vector:"
FOR K = 1 TO N
  PRINT USING "#####.######"; C(K);
NEXT K
PRINT
PRINT "    Result of (matrix)*(sol'n vector):"
FOR K = 1 TO N
  C(K) = 0!
  FOR J = 1 TO N
    C(K) = C(K) + A(K, J) * X(J)
  NEXT J
NEXT K
FOR K = 1 TO N
  PRINT USING "#####.######"; C(K);
NEXT K
PRINT
NEXT L
PRINT "**************************************"
PRINT "Press RETURN for next problem"
LINE INPUT DUM$
LOOP
CLOSE #1
END
```

Companion driver D2R9 takes the same matrices from MATRX3.DAT and passes copies U to SVDCMP for singular value decomposition into U, W, and V. Then U, W, and the transpose of V are multiplied together. The result is compared to a saved copy of A.

Here is the recipe SVDCMP:

```
SUB SVDCMP (A(), M, N, MP, NP, W(), V())
```
Given a matrix A, with logical dimensions M by N and physical dimensions MP by NP, this routine computes its singular value decomposition, $A = U \cdot W \cdot V^T$. The matrix U replaces A on output. The diagonal matrix of singular values W is output as a vector W. The matrix V (not the transpose V^T) is output as V. M must be greater than or equal to N; if it is smaller, then A should be filled up to square with zero rows.
```
DIM RV1(N)
IF M < N THEN PRINT "You must augment A with extra zero rows.": EXIT SUB
```
   Householder reduction to bidiagonal form.
```
G = 0!
SCALE = 0!
ANORM = 0!
FOR I = 1 TO N
  L = I + 1
  RV1(I) = SCALE * G
```

```
G = 0!
S = 0!
SCALE = 0!
IF I <= M THEN
  FOR K = I TO M
    SCALE = SCALE + ABS(A(K, I))
  NEXT K
  IF SCALE <> 0! THEN
    FOR K = I TO M
      A(K, I) = A(K, I) / SCALE
      S = S + A(K, I) * A(K, I)
    NEXT K
    F = A(I, I)
    G = -ABS(SQR(S)) * SGN(F)
    H = F * G - S
    A(I, I) = F - G
    IF I <> N THEN
      FOR J = L TO N
        S = 0!
        FOR K = I TO M
          S = S + A(K, I) * A(K, J)
        NEXT K
        F = S / H
        FOR K = I TO M
          A(K, J) = A(K, J) + F * A(K, I)
        NEXT K
      NEXT J
    END IF
    FOR K = I TO M
      A(K, I) = SCALE * A(K, I)
    NEXT K
  END IF
END IF
W(I) = SCALE * G
G = 0!
S = 0!
SCALE = 0!
IF I <= M AND I <> N THEN
  FOR K = L TO N
    SCALE = SCALE + ABS(A(I, K))
  NEXT K
  IF SCALE <> 0! THEN
    FOR K = L TO N
      A(I, K) = A(I, K) / SCALE
      S = S + A(I, K) * A(I, K)
    NEXT K
    F = A(I, L)
    G = -ABS(SQR(S)) * SGN(F)
    H = F * G - S
    A(I, L) = F - G
    FOR K = L TO N
      RV1(K) = A(I, K) / H
    NEXT K
    IF I <> M THEN
      FOR J = L TO M
```

```
        S = 0!
        FOR K = L TO N
          S = S + A(J, K) * A(I, K)
        NEXT K
        FOR K = L TO N
          A(J, K) = A(J, K) + S * RV1(K)
        NEXT K
      NEXT J
    END IF
    FOR K = L TO N
      A(I, K) = SCALE * A(I, K)
    NEXT K
  END IF
 END IF
 IF ABS(W(I)) + ABS(RV1(I)) > ANORM THEN ANORM = ABS(W(I)) + ABS(RV1(I))
NEXT I
```
Accumulation of right-hand transformations.
```
FOR I = N TO 1 STEP -1
  IF I < N THEN
    IF G <> 0! THEN
      FOR J = L TO N                    Double division to avoid possible underflow:
        V(J, I) = (A(I, J) / A(I, L)) / G
      NEXT J
      FOR J = L TO N
        S = 0!
        FOR K = L TO N
          S = S + A(I, K) * V(K, J)
        NEXT K
        FOR K = L TO N
          V(K, J) = V(K, J) + S * V(K, I)
        NEXT K
      NEXT J
    END IF
    FOR J = L TO N
      V(I, J) = 0!
      V(J, I) = 0!
    NEXT J
  END IF
  V(I, I) = 1!
  G = RV1(I)
  L = I
NEXT I
```
Accumulation of left-hand transformations.
```
FOR I = N TO 1 STEP -1
  L = I + 1
  G = W(I)
  IF I < N THEN
    FOR J = L TO N
      A(I, J) = 0!
    NEXT J
  END IF
  IF G <> 0! THEN
    G = 1! / G
    IF I <> N THEN
```

```
        FOR J = L TO N
          S = 0!
          FOR K = L TO M
            S = S + A(K, I) * A(K, J)
          NEXT K
          F = (S / A(I, I)) * G
          FOR K = I TO M
            A(K, J) = A(K, J) + F * A(K, I)
          NEXT K
        NEXT J
      END IF
      FOR J = I TO M
        A(J, I) = A(J, I) * G
      NEXT J
    ELSE
      FOR J = I TO M
        A(J, I) = 0!
      NEXT J
    END IF
    A(I, I) = A(I, I) + 1!
NEXT I
        Diagonalization of the bidiagonal form.
FOR K = N TO 1 STEP -1            Loop over singular values.
  FOR ITS = 1 TO 30              Loop over allowed iterations.
    FOR L = K TO 1 STEP -1       Test for splitting:
      NM = L - 1                 Note that RV1(1) is always zero.
      IF ABS(RV1(L)) + ANORM = ANORM THEN EXIT FOR
      IF ABS(W(NM)) + ANORM = ANORM THEN EXIT FOR
    NEXT L
    IF ABS(RV1(L)) + ANORM <> ANORM THEN
      C = 0!                     Cancellation of RV1(L), if L> 1 :
      S = 1!
      FOR I = L TO K
        F = S * RV1(I)
        RV1(I) = C * RV1(I)
        IF ABS(F) + ANORM <> ANORM THEN
          G = W(I)
          H = SQR(F * F + G * G)
          W(I) = H
          H = 1! / H
          C = G * H
          S = -F * H
          FOR J = 1 TO M
            Y = A(J, NM)
            Z = A(J, I)
            A(J, NM) = Y * C + Z * S
            A(J, I) = -Y * S + Z * C
          NEXT J
        END IF
      NEXT I
    END IF
    Z = W(K)
    IF L = K THEN                Convergence.
      IF Z < 0! THEN             Singular value is made nonnegative.
```

```
      W(K) = -Z
      FOR J = 1 TO N
        V(J, K) = -V(J, K)
      NEXT J
    END IF
    EXIT FOR
  END IF
  IF ITS = 30 THEN PRINT "No convergence in 30 iterations": ERASE RV1: END
  X = W(L)                         Shift from bottom 2-by-2 minor:
  NM = K - 1
  Y = W(NM)
  G = RV1(NM)
  H = RV1(K)
  F = ((Y - Z) * (Y + Z) + (G - H) * (G + H)) / (2! * H * Y)
  G = SQR(F * F + 1!)
  F = ((X - Z) * (X + Z) + H * ((Y / (F + ABS(G) * SGN(F))) - H)) / X
  Next QR transformation:
  C = 1!
  S = 1!
  FOR J = L TO NM
    I = J + 1
    G = RV1(I)
    Y = W(I)
    H = S * G
    G = C * G
    Z = SQR(F * F + H * H)
    RV1(J) = Z
    C = F / Z
    S = H / Z
    F = X * C + G * S
    G = -X * S + G * C
    H = Y * S
    Y = Y * C
    FOR JJ = 1 TO N
      X = V(JJ, J)
      Z = V(JJ, I)
      V(JJ, J) = X * C + Z * S
      V(JJ, I) = -X * S + Z * C
    NEXT JJ
    Z = SQR(F * F + H * H)
    W(J) = Z                    Rotation can be arbitrary if Z=0.
    IF Z <> 0! THEN
      Z = 1! / Z
      C = F * Z
      S = H * Z
    END IF
    F = C * G + S * Y
    X = -S * G + C * Y
    FOR JJ = 1 TO M
      Y = A(JJ, J)
      Z = A(JJ, I)
      A(JJ, J) = Y * C + Z * S
      A(JJ, I) = -Y * S + Z * C
    NEXT JJ
```

```
      NEXT J
      RV1(L) = 0!
      RV1(K) = F
      W(K) = X
   NEXT ITS
NEXT K
ERASE RV1
END SUB
```

A sample program using SVDCMP is the following:

```
DECLARE SUB SVDCMP (A!(), M!, N!, MP!, NP!, W!(), V!())

'PROGRAM D2R9
'Driver for routine SVDCMP
CLS
NP = 20
DIM A(NP, NP), U(NP, NP), W(NP), V(NP, NP)
OPEN "MATRX3.DAT" FOR INPUT AS #1
DO
  LINE INPUT #1, DUM$
  IF DUM$ = "END" THEN EXIT DO
  LINE INPUT #1, DUM$
  LINE INPUT #1, DUM$
  M = VAL(MID$(DUM$, 1, 2))
  N = VAL(MID$(DUM$, 3, 2))
  LINE INPUT #1, DUM$
  'Copy original matrix into U
  FOR K = 1 TO M
    LINE INPUT #1, DUM$
    FOR L = 1 TO N
      A(K, L) = VAL(MID$(DUM$, 4 * L - 3, 4))
      U(K, L) = A(K, L)
    NEXT L
  NEXT K
  IF N > M THEN
    FOR K = M + 1 TO N
      FOR L = 1 TO N
        A(K, L) = 0!
        U(K, L) = 0!
      NEXT L
    NEXT K
    M = N
  END IF
  'Perform decomposition
  CALL SVDCMP(U(), M, N, NP, NP, W(), V())
  'Print results
  PRINT "Decomposition Matrices:"
  PRINT "Matrix U"
  FOR K = 1 TO M
    FOR L = 1 TO N
      PRINT USING "#####.######"; U(K, L);
    NEXT L
    PRINT
  NEXT K
```

```
PRINT "Diagonal of Matrix W"
FOR K = 1 TO N
  PRINT USING "#####.######"; W(K);
NEXT K
PRINT
PRINT "Matrix V-Transpose"
FOR K = 1 TO N
  FOR L = 1 TO N
    PRINT USING "#####.######"; V(L, K);
  NEXT L
  PRINT
NEXT K
PRINT "Check product against original matrix:"
PRINT "Original Matrix:"
FOR K = 1 TO M
  FOR L = 1 TO N
    PRINT USING "#####.######"; A(K, L);
  NEXT L
  PRINT
NEXT K
PRINT "Product U*W*(V-Transpose):"
FOR K = 1 TO M
  FOR L = 1 TO N
    A(K, L) = 0!
    FOR J = 1 TO N
      A(K, L) = A(K, L) + U(K, J) * W(J) * V(L, J)
    NEXT J
  NEXT L
  FOR L = 1 TO N
    PRINT USING "#####.######"; A(K, L);
  NEXT L
  PRINT
NEXT K
PRINT "***********************************"
PRINT "Press RETURN for next problem"
LINE INPUT DUM$
LOOP
CLOSE #1
END
```

Routine **SPARSE** solves linear systems $A \cdot x = b$ with a sparse matrix A. Rather than specifying the entire matrix A (most elements of which are zero), the program calls two subroutines **ASUB** and **ATSUB** which are, for any input vector **XIN**, supposed to return the result **XOUT** of applying A and its transpose to **XIN**, respectively. In our sample program we define these two subroutines to implement the 20×20 matrix

$$\begin{pmatrix} 1.0 & 2.0 & 0.0 & 0.0 & \dots \\ -2.0 & 1.0 & 2.0 & 0.0 & \dots \\ 0.0 & -2.0 & 1.0 & 2.0 & \dots \\ 0.0 & 0.0 & -2.0 & 1.0 & \dots \\ \vdots & \vdots & \vdots & \vdots & \ddots \end{pmatrix}$$

As a right-hand side vector **b** we have taken $(3.0, 1.0, 1.0, \dots, -1.0)$, and the solution is given as **x**. Notice that the components of **x** are all initialized to zero. You will set them to some initial guess of the solution to your own problem, but this guess will usually suffice. The solution in D2R10 is given the usual checks.

Here is the recipe SPARSE:

```
DECLARE SUB ASUB (XIN!(), XOUT!())
DECLARE SUB ATSUB (XIN!(), XOUT!())

SUB SPARSE (B(), N, DUM1, DUM2, X(), RSQ)
    Solves the linear system A · x = b for the vector X of length N, given the right-hand vector
    B, and given two subroutines, ASUB(XIN,XOUT) and ATSUB(XIN,XOUT), which respectively
    calculate A·x and A^T·x for x given as their first arguments, returning the result in their second
    arguments. These subroutines should take every advantage of the sparseness of the matrix A.
    On input, X should be set to a first guess of the desired solution (all zero components is fine).
    On output, X is the solution vector, and RSQ is the sum of the squares of the components of
    the residual vector A · x − b. If this is not small, then the matrix is numerically singular and
    the solution represents a least-squares best approximation.
EPS = .000001                               r.m.s. accuracy desired.
DIM G(N), H(N), XI(N), XJ(N)
EPS2 = N * EPS ^ 2                          Criterion for sum-squared residuals.
IRST = 0                                    Number of restarts attempted internally.
DO
  DONE% = -1
  IRST = IRST + 1
  CALL ASUB(X(), XI())                      Evaluate the starting gradient,
  RP = 0!
  BSQ = 0!
  FOR J = 1 TO N
    BSQ = BSQ + B(J) ^ 2                    and the magnitude of the right side.
    XI(J) = XI(J) - B(J)
    RP = RP + XI(J) ^ 2
  NEXT J
  CALL ATSUB(XI(), G())
  FOR J = 1 TO N
    G(J) = -G(J)
    H(J) = G(J)
  NEXT J
  FOR ITER = 1 TO 10 * N                    Main iteration loop.
    CALL ASUB(H(), XI())
    ANUM = 0!
    ADEN = 0!
    FOR J = 1 TO N
      ANUM = ANUM + G(J) * H(J)
      ADEN = ADEN + XI(J) ^ 2
    NEXT J
    IF ADEN = 0! THEN PRINT "very singular matrix": EXIT SUB
    ANUM = ANUM / ADEN                      Equation (2.10.21).
    FOR J = 1 TO N
      XI(J) = X(J)
      X(J) = X(J) + ANUM * H(J)
    NEXT J
    CALL ASUB(X(), XJ())
    RSQ = 0!
    FOR J = 1 TO N
      XJ(J) = XJ(J) - B(J)
```

```
      RSQ = RSQ + XJ(J) ^ 2
      NEXT J
      IF RSQ = RP OR RSQ <= BSQ * EPS2 THEN ERASE XJ, XI, H, G: EXIT SUB
      IF RSQ > RP THEN              Not improving. Do a restart.
        FOR J = 1 TO N
          X(J) = XI(J)
        NEXT J
        IF IRST >= 3 THEN ERASE XJ, XI, H, G: EXIT SUB      Return if too many restarts. This
        DONE% = 0                                           is the normal return when we
      END IF                                                run into roundoff error before
      IF NOT DONE% THEN EXIT FOR                            satisfying the return above.
      RP = RSQ
      CALL ATSUB(XJ(), XI())          Compute gradient for next iteration.
      GG = 0!
      DGG = 0!
      FOR J = 1 TO N
        GG = GG + G(J) ^ 2
        DGG = DGG + (XI(J) + G(J)) * XI(J)
      NEXT J
      IF GG = 0! THEN ERASE XJ, XI, H, G: EXIT SUB      A rare, but normal, return.
      GAM = DGG / GG
      FOR J = 1 TO N
        G(J) = -XI(J)
        H(J) = G(J) + GAM * H(J)
      NEXT J
    NEXT ITER
  LOOP WHILE NOT DONE%
  IF ITER = 10 * N THEN PRINT "too many iterations"
END SUB
```

A sample program using SPARSE is the following:

```
DECLARE SUB ASUB (XIN!(), XOUT!())
DECLARE SUB SPARSE (B!(), N!, ASUB!, ATSUB!, X!(), RSQ!)

'PROGRAM D2R10
'Driver for SPARSE
CLS
N = 20
DIM B(N), X(N), BCMP(N)
M = N
FOR I = 1 TO N
  X(I) = 0!
  B(I) = 1!
NEXT I
B(1) = 3!
B(N) = -1!
CALL SPARSE(B(), N, DUM, DUM, X(), RSQ)
PRINT "Sum-squared residual:",
PRINT USING "#.######^^^^"; RSQ
PRINT
PRINT "Solution vector:"
FOR I = 1 TO N
  PRINT USING "#####.######"; X(I);
  IF I MOD 5 = 0 THEN PRINT
```

```
NEXT I
PRINT
PRINT
CALL ASUB(X(), BCMP())
PRINT "press RETURN to continue..."
LINE INPUT DUM$
PRINT "Test of solution vector:"
PRINT "        a*x", "        b"
FOR I = 1 TO N
  PRINT USING "#####.######"; BCMP(I); B(I)
NEXT I
PRINT
END


SUB ASUB (XIN(), XOUT())
SHARED N
XOUT(1) = XIN(1) + 2! * XIN(2)
XOUT(N) = -2! * XIN(N - 1) + XIN(N)
FOR I = 2 TO N - 1
  XOUT(I) = -2! * XIN(I - 1) + XIN(I) + 2! * XIN(I + 1)
NEXT I
END SUB


SUB ATSUB (XIN(), XOUT())
SHARED N
XOUT(1) = XIN(1) - 2! * XIN(2)
XOUT(N) = 2! * XIN(N - 1) + XIN(N)
FOR I = 2 TO N - 1
  XOUT(I) = 2! * XIN(I - 1) + XIN(I) - 2! * XIN(I + 1)
NEXT I
END SUB
```

Appendix

File MATRX1.DAT:

```
MATRICES FOR INPUT TO TEST ROUTINES
Size of matrix (NxN), Number of solutions:
3,2
Matrix A:
1.0 0.0 0.0
0.0 2.0 0.0
0.0 0.0 3.0
Solution vectors:
1.0 0.0 0.0
1.0 1.0 1.0
NEXT PROBLEM
Size of matrix (NxN), Number of solutions:
3,2
Matrix A:
1.0 2.0 3.0
2.0 2.0 3.0
3.0 3.0 3.0
Solution vectors:
1.0 1.0 1.0
1.0 2.0 3.0
NEXT PROBLEM:
```

```
Size of matrix (NxN), Number of solutions:
5,2
Matrix A:
1.0 2.0 3.0 4.0 5.0
2.0 3.0 4.0 5.0 1.0
3.0 4.0 5.0 1.0 2.0
4.0 5.0 1.0 2.0 3.0
5.0 1.0 2.0 3.0 4.0
Solution vectors:
1.0 1.0 1.0 1.0 1.0
1.0 2.0 3.0 4.0 5.0
NEXT PROBLEM:
Size of matrix (NxN), Number of solutions:
5,2
Matrix A:
1.4 2.1 2.1 7.4 9.6
1.6 1.5 1.1 0.7 5.0
3.8 8.0 9.6 5.4 8.8
4.6 8.2 8.4 0.4 8.0
2.6 2.9 0.1 9.6 7.7
Solution vectors:
1.1 1.6 4.7 9.1 0.1
4.0 9.3 8.4 0.4 4.1
END
```

File MATRX2.DAT:

```
TRIDIAGONAL MATRICES FOR PROGRAM 'TRIDAG'
Dimension of matrix
3
Diagonal elements (N)
1.0 2.0 3.0
Super-diagonal elements (N-1)
2.0 3.0
Sub-diagonal elements (N-1)
2.0 3.0
Right-hand side vector (N)
1.0 2.0 3.0
NEXT PROBLEM:
Dimension of matrix
5
Diagonal elements (N)
1.0 1.0 1.0 1.0 1.0
Super-diagonal elements (N-1)
1.0 2.0 3.0 4.0
Sub-diagonal elements (N-1)
2.0 3.0 4.0 5.0
Right-hand side vector (N)
1.0 2.0 3.0 4.0 5.0
NEXT PROBLEM:
Dimension of matrix
5
Diagonal elements (N)
1.0 2.0 3.0 4.0 5.0
Super-diagonal elements (N-1)
2.0 3.0 4.0 5.0
Sub-diagonal elements (N-1)
2.0 3.0 4.0 5.0
Right-hand side vector (N)
1.0 1.0 1.0 1.0 1.0
```

NEXT PROBLEM:
Dimension of matrix
6
Diagonal elements (N)
9.7 9.5 5.2 3.5 5.1 6.0
Super-diagonal elements (N-1)
6.0 1.2 0.7 3.0 1.5
Sub-diagonal elements (N-1)
2.1 9.4 3.3 7.5 8.8
Right-hand side vector (N)
2.0 7.5 0.6 7.4 9.8 8.8
END

File MATRX3.DAT:

TEST MATRICES FOR SVDCMP:
Number of Rows, Columns
5,3
Matrix
1.0 2.0 3.0
2.0 3.0 4.0
3.0 4.0 5.0
4.0 5.0 6.0
5.0 6.0 7.0
NEXT PROBLEM:
Number of Rows, Columns
5,5
Matrix
1.0 2.0 3.0 4.0 5.0
2.0 2.0 3.0 4.0 5.0
3.0 3.0 3.0 4.0 5.0
4.0 4.0 4.0 4.0 5.0
5.0 5.0 5.0 5.0 5.0
NEXT PROBLEM:
Number of Rows, Columns
6,6
Matrix
3.0 5.3 5.6 3.5 6.8 5.7
0.4 8.2 6.7 1.9 2.2 5.3
7.8 8.3 7.7 3.3 1.9 4.8
5.5 8.8 3.0 1.0 5.1 6.4
5.1 5.1 3.6 5.8 5.7 4.9
3.5 2.7 5.7 8.2 9.6 2.9
END

Chapter 3: Interpolation and Extrapolation

Chapter 3 of *Numerical Recipes* deals with interpolation and extrapolation (the same routines are usable for both). Three fundamental interpolation methods are first discussed,

1. Polynomial interpolation (POLINT),
2. Rational function interpolation (RATINT), and
3. Cubic spline interpolation (SPLINE, SPLINT).

To find the place in an ordered table at which to perform an interpolation, two routines are given, LOCATQ and HUNT. Also, for cases in which the actual coefficients of a polynomial interpolation are desired, the routines POLCOE and POLCOF are provided (along with important warnings circumscribing their usefulness).

For higher-dimensional interpolations, *Numerical Recipes* treats only problems on a regularly spaced grid. Routine POLIN2 does a two-dimensional polynomial interpolation that aims at accuracy rather than smoothness. When smooth interpolation is desired, the methods shown in BCUCOF and BCUINT for bicubic interpolation are recommended. In the case of two-dimensional spline interpolations, the routines SPLIE2 and SPLIN2 are offered.

$$\star \quad \star \quad \star \quad \star$$

Program POLINT takes two arrays XA and YA of length N that express the known values of a function, and calculates the value, at a point x, of the unique polynomial of degree $N-1$ passing through all the given values. For the purpose of illustration, in D3R1 we have taken evenly spaced XA(I) and set YA(I) equal to simple functions (sines and exponentials) of these XA(I). For the sine we use an interval of length π, and for the exponential an interval of length 1.0. You may choose the number N of reference points and observe the improvement of the results as N increases. The test points x are slightly shifted from the reference points so that you can compare the estimated error DY with the actual error. By removing the shift, you may check that the polynomial actually hits all reference points.

Here is the recipe POLINT:

```
SUB POLINT (XA(), YA(), N, X, Y, DY)
          Given arrays XA and YA, each of length N, and given a value X, this routine returns a value Y,
          and an error estimate DY. If P(x) is the polynomial of degree N − 1 such that P(XA_i) =
          YA_i, i = 1,...,N, then the returned value Y = P(X).
DIM C(N), D(N)
NS = 1
DIF = ABS(X - XA(1))
FOR I = 1 TO N                    Here we find the index NS of the closest table entry,
  DIFT = ABS(X - XA(I))
  IF DIFT < DIF THEN
    NS = I
```

```
      DIF = DIFT
    END IF
    C(I) = YA(I)
    D(I) = YA(I)
  NEXT I
  Y = YA(NS)
  NS = NS - 1
  FOR M = 1 TO N - 1
    FOR I = 1 TO N - M
      HO = XA(I) - X
      HP = XA(I + M) - X
      W = C(I + 1) - D(I)
      DEN = HO - HP
      IF DEN = 0! THEN PRINT "Abnormal exit": EXIT SUB
      DEN = W / DEN
      D(I) = HP * DEN
      C(I) = HO * DEN
    NEXT I
    IF 2 * NS < N - M THEN
      DY = C(NS + 1)
    ELSE
      DY = D(NS)
      NS = NS - 1
    END IF
    Y = Y + DY
  NEXT M
  ERASE D, C
END SUB
```

and initialize the tableau of C's and D's.

This is the initial approximation to Y.

For each column of the tableau,
we loop over the current C's and D's and update them.

This error can occur only if two input XA's are (to within roundoff) identical.

Here the C's and D's are updated.

After each column in the tableau is completed, we decide which correction, C or D, we want to add to our accumulating value of Y, i.e. which path to take through the tableau—forking up or down. We do this in such a way as to take the most "straight line" route through the tableau to its apex, updating NS accordingly to keep track of where we are. This route keeps the partial approximations centered (insofar as possible) on the target X. The last DY added is thus the error indication.

A sample program using POLINT is the following:

```
DECLARE SUB POLINT (XA!(), YA!(), N!, X!, Y!, DY!)

'PROGRAM D3R1
'Driver for routine POLINT
CLS
NP = 10
PI = 3.1415926#
DIM XA(NP), YA(NP)
PRINT "Generation of interpolation tables"
PRINT " ... sin(x)     0<x<pi"
PRINT " ... exp(x)     0<x<1 "
PRINT "How many entries go in these tables? (note: N<10)"
INPUT N
FOR NFUNC = 1 TO 2
  IF NFUNC = 1 THEN
    PRINT "sine function from 0 to pi"
    FOR I = 1 TO N
      XA(I) = I * PI / N
      YA(I) = SIN(XA(I))
    NEXT I
  ELSEIF NFUNC = 2 THEN
    PRINT "exponential function from 0 to 1"
    FOR I = 1 TO N
      XA(I) = I * 1! / N
```

```
    YA(I) = EXP(XA(I))
  NEXT I
ELSE
  STOP
END IF
PRINT "         x          f(x)      interpolated       error"
FOR I = 1 TO 10
  IF NFUNC = 1 THEN
    X = (-.05 + I / 10!) * PI
    F = SIN(X)
  ELSEIF NFUNC = 2 THEN
    X = -.05 + I / 10!
    F = EXP(X)
  END IF
  CALL POLINT(XA(), YA(), N, X, Y, DY)
  PRINT USING "#####.######"; X; F; Y;
  PRINT "      ";
  PRINT USING "#.####^^^^"; DY
NEXT I
PRINT "**********************************"
PRINT "Press RETURN"
LINE INPUT DUM$
NEXT NFUNC
END
```

RATINT is functionally similar to POLINT in that it also returns a value y for the function at point x, and an error estimate DY as well. In this case the values are determined from the unique diagonal rational function that passes through all the reference points. If you inspect the driver closely, you will find that two of the test points fall directly on top of reference points and should give exact results. The remainder do not. You can compare the estimated error DYY to the actual error $|YY - YEXP|$ for these cases.

Here is the recipe RATINT:

```
SUB RATINT (XA(), YA(), N, X, Y, DY)
```
> Given arrays XA and YA, each of length N, and given a value of X, this routine returns a value of Y and an accuracy estimate DY. The value returned is that of the diagonal rational function, evaluated at X, which passes through the N points (XA_i, YA_i), $i = 1...N$.
```
TINY = 1E-25                    A small number.
DIM C(N), D(N)
NS = 1
HH = ABS(X - XA(1))
FOR I = 1 TO N
  H = ABS(X - XA(I))
  IF H = 0! THEN
    Y = YA(I)
    DY = 0!
    ERASE D, C
    EXIT SUB
  ELSEIF H < HH THEN
    NS = I
    HH = H
  END IF
  C(I) = YA(I)
```

```
    D(I) = YA(I) + TINY
  NEXT I
  Y = YA(NS)
  NS = NS - 1
  FOR M = 1 TO N - 1
    FOR I = 1 TO N - M
      W = C(I + 1) - D(I)
      H = XA(I + M) - X
      T = (XA(I) - X) * D(I) / H
      DD = T - C(I + 1)
      IF DD = 0! THEN PRINT "Abnormal exit": EXIT SUB
      DD = W / DD
      D(I) = C(I + 1) * DD
      C(I) = T * DD
    NEXT I
    IF 2 * NS < N - M THEN
      DY = C(NS + 1)
    ELSE
      DY = D(NS)
      NS = NS - 1
    END IF
    Y = Y + DY
  NEXT M
  ERASE D, C
  END SUB
```

The TINY part is needed to prevent a rare zero-over-zero condition.

H will never be zero, since this was tested in the initializing loop.

This error condition indicates that the interpolating function has a pole at the requested value of X.

A sample program using `RATINT` is the following:

```
DECLARE SUB RATINT (XA!(), YA!(), N!, X!, Y!, DY!)

'PROGRAM D3R2
'Driver for routine RATINT
CLS
NPT = 6
EPSSQ = 1!
DIM X(NPT), Y(NPT)
DEF FNF (Z) = Z * EXP(-Z) / ((Z - 1!) ^ 2 + EPSSQ)
FOR I = 1 TO NPT
  X(I) = I * 2! / NPT
  Y(I) = FNF(X(I))
NEXT I
PRINT "Diagonal rational function interpolation"
PRINT
PRINT "    x        interp.      accuracy      actual"
FOR I = 1 TO 10
  XX = .2 * I
  CALL RATINT(X(), Y(), NPT, XX, YY, DYY)
  YEXP = FNF(XX)
  PRINT USING "###.##"; XX;
  PRINT USING "#####.######"; YY;
  PRINT "        ";
  PRINT USING "#.####^^^^"; DYY;
  PRINT USING "#####.######"; YEXP
NEXT I
PRINT
```

END

Subroutine SPLINE generates a cubic spline. Given an array of x_i and $f(x_i)$, and given values of the first derivative of function f at the two endpoints of the tabulated region, it returns the second derivative of f at each of the tabulation points. As an example we chose the function $\sin x$ and evaluated it at evenly spaced points X(I). In this case the first derivatives at the endpoints are YP1 $= \cos x_1$ and YPN $= \cos x_N$. The output array of SPLINE is Y2(I), and this is listed along with $-\sin x_i$, the second derivative of $\sin x_i$, for comparison.

Here is the recipe SPLINE:

```
SUB SPLINE (X(), Y(), N, YP1, YPN, Y2())
```
Given arrays X and Y of length N containing a tabulated function, i.e. $Y_i = f(X_i)$, with $X_1 < X_2 < \ldots < X_N$, and given values YP1 and YPN for the first derivative of the interpolating function at points 1 and N, respectively, this routine returns an array Y2 of length N which contains the second derivatives of the interpolating function at the tabulated points X_i. If YP1 and/or YPN are equal to 1×10^{30} or larger, the routine is signaled to set the corresponding boundary condition for a natural spline, with zero second derivative on that boundary.

```
DIM U(N)
IF YP1 > 9.9E+29 THEN          The lower boundary condition is set either to be "natural"
  Y2(1) = 0!
  U(1) = 0!
ELSE                          or else to have a specified first derivative.
  Y2(1) = -.5
  U(1) = (3! / (X(2) - X(1))) * ((Y(2) - Y(1)) / (X(2) - X(1)) - YP1)
END IF
FOR I = 2 TO N - 1            This is the decomposition loop of
  SIG = (X(I) - X(I - 1)) / (X(I + 1) - X(I - 1))    the tridiagonal algorithm. Y2 and
  P = SIG * Y2(I - 1) + 2!    U are used for temporary storage of
  Y2(I) = (SIG - 1!) / P       the decomposed factors.
  DUM1 = (Y(I + 1) - Y(I)) / (X(I + 1) - X(I))
  DUM2 = (Y(I) - Y(I - 1)) / (X(I) - X(I - 1))
  U(I) = (6! * (DUM1 - DUM2) / (X(I + 1) - X(I - 1)) - SIG * U(I - 1)) / P
NEXT I
IF YPN > 9.9E+29 THEN         The upper boundary condition is set either to be "natural"
  QN = 0!
  UN = 0!
ELSE                         or else to have a specified first derivative.
  QN = .5
  UN = (3! / (X(N) - X(N - 1))) * (YPN - (Y(N) - Y(N - 1)) / (X(N) - X(N - 1)))
END IF
Y2(N) = (UN - QN * U(N - 1)) / (QN * Y2(N - 1) + 1!)
FOR K = N - 1 TO 1 STEP -1   This is the backsubstitution loop of
  Y2(K) = Y2(K) * Y2(K + 1) + U(K)    the tridiagonal algorithm.
NEXT K
ERASE U
END SUB
```

A sample program using SPLINE is the following:

```
DECLARE SUB SPLINE (X!(), Y!(), N!, YP1!, YPN!, Y2!())

'PROGRAM D3R3
'Driver for routine SPLINE
CLS
N = 20
PI = 3.141593
DIM X(N), Y(N), Y2(N)
PRINT "Second-derivatives for sin(x) from 0 to PI"
'Generate array for interpolation
FOR I = 1 TO 20
  X(I) = I * PI / N
  Y(I) = SIN(X(I))
NEXT I
'Calculate 2nd derivative with SPLINE
YP1 = COS(X(1))
YPN = COS(X(N))
CALL SPLINE(X(), Y(), N, YP1, YPN, Y2())
'Test result
PRINT "                  spline          actual"
PRINT "      angle      2nd deriv      2nd deriv"
FOR I = 1 TO N
  PRINT USING "#####.##"; X(I);
  PRINT USING "#########.######"; Y2(I), -SIN(X(I))
NEXT I
END
```

Actual cubic-spline interpolations, however, are carried out by SPLINT. This routine uses the output array from one call to SPLINE to service any subsequent number of spline interpolations with different x's. The demonstration program D3R4 tests this capability on both $\sin x$ and $\exp x$. The two are treated in succession according to whether NFUNC is one or two. In each case the function is tabulated at equally spaced points, and the derivatives are found at the first and last points. A call to SPLINE then produces an array of second derivatives Y2 which is fed to SPLINT. The interpolated values Y are compared with actual function values F at a different set of equally spaced points.

Here is the recipe SPLINT:

```
SUB SPLINT (XA(), YA(), Y2A(), N, X, Y)
```
Given the arrays XA and YA of length N, which tabulate a function (with the XA_i's in order), and given the array Y2A, which is the output from SPLINE above, and given a value of X, this routine returns a cubic-spline interpolated value Y.

```
KLO = 1                      We will find the right place in the table by means of bisection.
KHI = N                      This is optimal if sequential calls to this routine are at random
WHILE KHI - KLO > 1          values of X. If sequential calls are in order, and closely spaced,
  K = (KHI + KLO) / 2        one would do better to store previous values of KLO and KHI
  IF XA(K) > X THEN          and test if they remain appropriate on the next call.
    KHI = K
  ELSE
    KLO = K
  END IF
```

```
WEND                              KLO and KHI now bracket the input value of X.
H = XA(KHI) - XA(KLO)
IF H = 0! THEN PRINT "Bad XA input.": EXIT SUB    The XA's must be distinct.
A = (XA(KHI) - X) / H             Cubic spline polynomial is now evaluated.
B = (X - XA(KLO)) / H
Y = A * YA(KLO) + B * YA(KHI)
Y = Y + ((A ^ 3 - A) * Y2A(KLO) + (B ^ 3 - B) * Y2A(KHI)) * (H ^ 2) / 6!
END SUB
```

A sample program using SPLINT is the following:

```
DECLARE SUB SPLINE (X!(), Y!(), N!, YP1!, YPN!, Y2!())
DECLARE SUB SPLINT (XA!(), YA!(), Y2A!(), N!, X!, Y!)

'PROGRAM D3R4
'Driver for routine SPLINT, which calls SPLINE
CLS
NP = 10
PI = 3.141593
DIM XA(NP), YA(NP), Y2(NP)
FOR NFUNC = 1 TO 2
  IF NFUNC = 1 THEN
    PRINT "Sine function from 0 to pi"
    FOR I = 1 TO NP
      XA(I) = I * PI / NP
      YA(I) = SIN(XA(I))
    NEXT I
    YP1 = COS(XA(1))
    YPN = COS(XA(NP))
  ELSEIF NFUNC = 2 THEN
    PRINT "Exponential function from 0 to 1"
    FOR I = 1 TO NP
      XA(I) = 1! * I / NP
      YA(I) = EXP(XA(I))
    NEXT I
    YP1 = EXP(XA(1))
    YPN = EXP(XA(NP))
  ELSE
    STOP
  END IF
'Call SPLINE to get second derivatives
CALL SPLINE(XA(), YA(), NP, YP1, YPN, Y2())
'Call SPLINT for interpolations
PRINT "        x        f(x)      interpolation"
FOR I = 1 TO 10
  IF NFUNC = 1 THEN
    X = (-.05 + I / 10!) * PI
    F = SIN(X)
  ELSEIF NFUNC = 2 THEN
    X = -.05 + I / 10!
    F = EXP(X)
  END IF
  CALL SPLINT(XA(), YA(), Y2(), NP, X, Y)
  PRINT USING "#####.######"; X; F; Y
NEXT I
```

```
PRINT "**********************************"
PRINT "Press RETURN"
LINE INPUT DUM$
NEXT NFUNC
END
```

The next program, LOCATQ, may be used in conjunction with any interpolation method to bracket the x-position for which $f(x)$ is sought by two adjacent tabulated positions. That is, given a monotonic array of x_i, and given a value of x, it finds the two values x_i, x_{i+1} that surround x. In D3R5 we chose the array x_i to be nonuniform, varying exponentially with i. Then we took a uniform series of x-values and sought their position in the array using LOCATQ. For each X, LOCATQ finds the value J for which X(J) is nearest below X. Then the driver shows J, and the two bracketing values XX(J) and XX(J+1). If J is 0 or N, then X is not within the tabulated range. The program thereby flags 'lower lim' if X is below X(1) or 'upper lim' if X is above X(N).

Here is the recipe LOCATQ:

```
SUB LOCATQ (XX(), N, X, J)
```
Given an array XX of length N, and given a value X, returns a value J such that X is between XX(J) and XX(J+1). XX must be monotonic, either increasing or decreasing. J=0 or J=N is returned to indicate that X is out of range.

```
JL = 0                              Initialize lower
JU = N + 1                          and upper limits.
WHILE JU - JL > 1                   If we are not yet done,
  JM = INT((JU + JL) / 2)           compute a midpoint,
  IF XX(N) > XX(1) EQV X > XX(JM) THEN
    JL = JM                         and replace either the lower limit
  ELSE
    JU = JM                         or the upper limit, as appropriate.
  END IF
WEND                                Repeat until the test condition is satisfied.
J = JL                              Then set the output and return.
END SUB
```

A sample program using LOCATQ is the following:

```
DECLARE SUB LOCATQ (XX!(), N!, X!, J!)

'PROGRAM D3R5
'Driver for routine LOCATQ
CLS
N = 100
DIM XX(N)
'Create array to be searched
FOR I = 1 TO N
  XX(I) = EXP(I / 20!) - 74!
NEXT I
PRINT "Result of:    j=0 indicates x too small"
PRINT "             j=100 indicates x too large"
PRINT "    locate    j      xx(j)      xx(j+1)"
'Do test
FOR I = 1 TO 19
```

```
X = -100! + 200! * I / 20!
CALL LOCATQ(XX(), N, X, J)
IF J = 0 THEN
   PRINT USING "#####.####"; X;
   PRINT USING "######"; J;
   PRINT "    lower lim";
   PRINT USING "#####.######"; XX(J + 1)
ELSEIF J = N THEN
   PRINT USING "#####.####"; X;
   PRINT USING "######"; J;
   PRINT USING "#####.######"; XX(J);
   PRINT "    upper lim"
ELSE
   PRINT USING "#####.####"; X;
   PRINT USING "######"; J;
   PRINT USING "#####.######"; XX(J);
   PRINT USING "#####.######"; XX(J + 1)
END IF
NEXT I
END
```

Routine HUNT serves the same function as LOCATQ, but is used when the table is to be searched many times and the abscissa each time is close to its value on the previous search. D3R6 sets up the array XX(I) and then a series X of points to locate. The hunt begins with a trial value JI (which is fed to HUNT through variable J) and HUNT returns solution J such that X lies between XX(J) and XX(J+1). The two cases J=0 and J=N have the same meaning as in D3R5 and are treated in the same way.

Here is the recipe HUNT:

```
SUB HUNT (XX(), N, X, JLO)
        Given an array XX of length N, and given a value X, returns a value JLO such that X is between
        XX(JLO) and XX(JLO+1). XX must be monotonic, either increasing or decreasing. JLO=0
        or JLO=N is returned to indicate that X is out of range. JLO on input is taken as the initial
        guess for JLO on output.
ASCND% = XX(N) > XX(1)                   True if ascending order of table, false otherwise.
IF JLO <= 0 OR JLO > N THEN              Input guess not useful. Go immediately to bisection.
   JLO = 0
   JHI = N + 1
ELSE
   INC = 1                              Set the hunting increment.
   IF X >= XX(JLO) EQV ASCND% THEN            Hunt up:
1     JHI = JLO + INC
      IF JHI > N THEN                   Done hunting, since off end of table.
         JHI = N + 1
      ELSEIF X >= XX(JHI) EQV ASCND% THEN     Not done hunting,
         JLO = JHI
         INC = INC + INC                so double the increment
         GOTO 1                         and try again.
      END IF                           Done hunting, value bracketed.
   ELSE                                 Hunt down:
      JHI = JLO
2     JLO = JHI - INC
      IF JLO < 1 THEN                   Done hunting, since off end of table.
```

```
      JLO = 0
      ELSEIF X < XX(JLO) EQV ASCND% THEN        Not done hunting,
        JHI = JLO
        INC = INC + INC                      so double the increment
        GOTO 2                               and try again.
      END IF                                 Done hunting, value bracketed.
    END IF
  END IF
  END IF                                     Hunt is done, so begin the final bisection phase:
  DO
    IF JHI - JLO = 1 THEN EXIT SUB
    JM = INT((JHI + JLO) / 2)
    IF X > XX(JM) EQV ASCND% THEN
      JLO = JM
    ELSE
      JHI = JM
    END IF
  LOOP
END SUB
```

A sample program using HUNT is the following:

```
DECLARE SUB HUNT (XX!(), N!, X!, JLO!)

'PROGRAM D3R6
'Driver for routine HUNT
CLS
N = 100
DIM XX(N)
'Create array to be searched
FOR I = 1 TO N
  XX(I) = EXP(I / 20!) - 74!
NEXT I
PRINT "Result of:    j=0 indicates x too small"
PRINT "              j=100 indicates x too large"
PRINT "    locate:   guess   j      xx(j)      xx(j+1)"
'Do test
FOR I = 1 TO 19
  X = -100! + 200! * I / 20!
  'Trial parameter
  JI = 5 * I
  J = JI
  'Begin search
  CALL HUNT(XX(), N, X, J)
  IF J = 0 THEN
    PRINT USING "#####.######"; X;
    PRINT USING "######"; JI;
    PRINT USING "######"; J;
    PRINT "    lower lim";
    PRINT USING "#####.######"; XX(J + 1)
  ELSEIF J = N THEN
    PRINT USING "#####.######"; X;
    PRINT USING "######"; JI;
    PRINT USING "######"; J;
    PRINT USING "#####.######"; XX(J);
    PRINT "    upper lim"
```

```
ELSE
  PRINT USING "####.######"; X;
  PRINT USING "######"; JI;
  PRINT USING "######"; J;
  PRINT USING "####.######"; XX(J);
  PRINT USING "####.######"; XX(J + 1)
  END IF
NEXT I
END
```

The next two demonstration programs, D3R7 and D3R8, are so nearly identical that they may be discussed together. POLCOE and POLCOF themselves both find coefficients of interpolating polynomials. In the present instance we have tried both a sine function and an exponential function for YA(I), each tabulated at uniformly spaced points XA(I). The validity of the array of polynomial coefficients COEFF is tested by calculating the value SUM of the polynomials at a series of test points and listing these alongside the functions F that they represent.

Here is the recipe POLCOE:

```
SUB POLCOE (X(), Y(), N, COF())
```
   Given arrays X and Y of length N containing a tabulated function $Y_i = f(X_i)$, this routine returns an array of coefficients COF, also of length N, such that $Y_i = \sum_j COF_j X_i^{j-1}$

```
DIM S(N)
FOR I = 1 TO N
  S(I) = 0!
  COF(I) = 0!
NEXT I
S(N) = -X(1)
FOR I = 2 TO N                          Coefficients $S_i$ of the master polynomial $P(x)$ are found
  FOR J = N + 1 - I TO N - 1            by recurrence.
    S(J) = S(J) - X(I) * S(J + 1)
  NEXT J
  S(N) = S(N) - X(I)
NEXT I
FOR J = 1 TO N
  PHI = N
  FOR K = N - 1 TO 1 STEP -1           The quantity PHI $= \prod_{j \neq k}(x_j - x_k)$ is found as a deriva-
    PHI = K * S(K + 1) + X(J) * PHI    tive of $P(x_j)$.
  NEXT K
  FF = Y(J) / PHI
  B = 1!                               Coefficients of polynomials in each term of the Lagrange
  FOR K = N TO 1 STEP -1               formula are found by synthetic division of $P(x)$ by $(x -$
    COF(K) = COF(K) + B * FF           $x_j)$. The solution $C_k$ is accumulated.
    B = S(K) + X(J) * B
  NEXT K
NEXT J
ERASE S
END SUB
```

A sample program using POLCOE is the following:

```
DECLARE SUB POLCOE (X!(), Y!(), N!, COF!())

'PROGRAM D3R7
'Driver for routine POLCOE
CLS
NP = 5
PI = 3.1415926#
DIM XA(NP), YA(NP), COEFF(NP)
FOR NFUNC = 1 TO 2
  IF NFUNC = 1 THEN
    PRINT "Sine function from 0 to PI"
    FOR I = 1 TO NP
      XA(I) = I * PI / NP
      YA(I) = SIN(XA(I))
    NEXT I
  ELSEIF NFUNC = 2 THEN
    PRINT "Exponential function from 0 to 1"
    FOR I = 1 TO NP
      XA(I) = 1! * I / NP
      YA(I) = EXP(XA(I))
    NEXT I
  ELSE
    STOP
  END IF
  CALL POLCOE(XA(), YA(), NP, COEFF())
  PRINT "    coefficients"
  FOR I = 1 TO NP
    PRINT USING "#####.######"; COEFF(I);
  NEXT I
  PRINT
  PRINT "         x           f(x)       polynomial"
  FOR I = 1 TO 10
    IF NFUNC = 1 THEN
      X = (-.05 + I / 10!) * PI
      F = SIN(X)
    ELSEIF NFUNC = 2 THEN
      X = -.05 + I / 10!
      F = EXP(X)
    END IF
    SUM = COEFF(NP)
    FOR J = NP - 1 TO 1 STEP -1
      SUM = COEFF(J) + SUM * X
    NEXT J
    PRINT USING "#####.######"; X; F; SUM
  NEXT I
  PRINT "**********************************"
  PRINT "Press RETURN"
  LINE INPUT DUM$
NEXT NFUNC
END
```

Here is the recipe POLCOF:

```
DECLARE SUB POLINT (XA!(), YA!(), N!, X!, Y!, DY!)

SUB POLCOF (XA(), YA(), N, COF())
```

Given arrays XA and YA of length N containing a tabulated function $YA_i = f(XA_i)$, this routine returns an array of coefficients COF, also of length N, such that $YA_i = \sum_j COF_j XA_i^{j-1}$

```
DIM X(N), Y(N)
FOR J = 1 TO N
  X(J) = XA(J)
  Y(J) = YA(J)
NEXT J
FOR J = 1 TO N
  CALL POLINT(X(), Y(), N + 1 - J, 0!, COF(J), DY)      This is the polynomial interpo-
  XMIN = 1E+38                                          lation routine of §3.1.
  K = 0                                      We extrapolate to x = 0.
  FOR I = 1 TO N + 1 - J                     Find the remaining Xᵢ of smallest absolute value,
    IF ABS(X(I)) < XMIN THEN
      XMIN = ABS(X(I))
      K = I
    END IF
    IF X(I) <> 0! THEN Y(I) = (Y(I) - COF(J)) / X(I)   (meanwhile reducing all the terms)
  NEXT I
  FOR I = K + 1 TO N + 1 - J          and eliminate it.
    Y(I - 1) = Y(I)
    X(I - 1) = X(I)
  NEXT I
NEXT J
ERASE Y, X
END SUB
```

A sample program using POLCOF is the following:

```
DECLARE SUB POLCOF (XA!(), YA!(), N!, COF!())

'PROGRAM D3R8
'Driver for routine POLCOF
CLS
NP = 5
PI = 3.141593
DIM XA(NP), YA(NP), COEFF(NP)
FOR NFUNC = 1 TO 2
  IF NFUNC = 1 THEN
    PRINT "Sine function from 0 to PI"
    FOR I = 1 TO NP
      XA(I) = I * PI / NP
      YA(I) = SIN(XA(I))
    NEXT I
  ELSEIF NFUNC = 2 THEN
    PRINT "Exponential function from 0 to 1"
    FOR I = 1 TO NP
      XA(I) = 1! * I / NP
      YA(I) = EXP(XA(I))
    NEXT I
```

```
ELSE
  STOP
END IF
CALL POLCOF(XA(), YA(), NP, COEFF())
PRINT "     coefficients"
FOR I = 1 TO NP
  PRINT USING "#####.######"; COEFF(I);
NEXT I
PRINT
PRINT "        x          f(x)       polynomial"
FOR I = 1 TO 10
  IF NFUNC = 1 THEN
    X = (-.05 + I / 10!) * PI
    F = SIN(X)
  ELSEIF NFUNC = 2 THEN
    X = -.05 + I / 10!
    F = EXP(X)
  END IF
  SUM = COEFF(NP)
  FOR J = NP - 1 TO 1 STEP -1
    SUM = COEFF(J) + SUM * X
  NEXT J
  PRINT USING "#####.######"; X; F; SUM
NEXT I
PRINT "**********************************"
PRINT "Press RETURN"
LINE INPUT DUM$
NEXT NFUNC
END
```

For two-dimensional interpolation, POLIN2 implements a bilinear interpolation. We feed it coordinates X1A,X2A for an $M \times N$ array of gridpoints as well as the function value at each gridpoint. In return it gives the value Y of the interpolated function at a given point X1,X2, and the estimated accuracy DY of the interpolation. D3R9 runs the test on a uniform grid for the function $f(x, y) = \sin x \exp y$. Then, for an offset grid of test points, the interpolated value Y is compared to the actual function value F, and the actual error is compared to the estimated error DY.

Here is the recipe POLIN2:

```
DECLARE SUB POLINT (XA!(), YA!(), N!, X!, Y!, DY!)

SUB POLIN2 (X1A(), X2A(), YA(), M, N, X1, X2, Y, DY)
```
Given arrays X1A (length M) and X2A (length N) of independent variables, and an M by N array of function values YA, tabulated at the grid points defined by X1A and X2A; and given values X1 and X2 of the independent variables; this routine returns an interpolated function value Y, and an accuracy indication DY (based only on the interpolation in the X1 direction, however).
```
DIM YNTMP(N), YMTMP(M)
FOR J = 1 TO M             Loop over rows.
  FOR K = 1 TO N           Copy the row into temporary storage.
    YNTMP(K) = YA(J, K)
  NEXT K
  CALL POLINT(X2A(), YNTMP(), N, X2, YMTMP(J), DY)    Interpolate answer into tempo-
NEXT J                                                 rary storage.
```

```
CALL POLINT(X1A(), YMTMP(), M, X1, Y, DY)          Do the final interpolation.
ERASE YMTMP, YNTMP
END SUB
```

A sample program using POLIN2 is the following:

```
DECLARE SUB POLIN2 (X1A!(), X2A!(), YA!(), M!, N!, X1!, X2!, Y!, DY!)

'PROGRAM D3R9
'Driver for routine POLIN2
CLS
N = 5
PI = 3.141593
DIM X1A(N), X2A(N), YA(N, N)
FOR I = 1 TO N
  X1A(I) = I * PI / N
  FOR J = 1 TO N
    X2A(J) = 1! * J / N
    YA(I, J) = SIN(X1A(I)) * EXP(X2A(J))
  NEXT J
NEXT I
'Test 2-dimensional interpolation
PRINT "        x1          x2        f(x)      interpolated      error"
FOR I = 1 TO 4
  X1 = (-.1 + I / 5!) * PI
  FOR J = 1 TO 4
    X2 = -.1 + J / 5!
    F = SIN(X1) * EXP(X2)
    CALL POLIN2(X1A(), X2A(), YA(), N, N, X1, X2, Y, DY)
    PRINT USING "#####.######"; X1; X2; F; Y; DY
  NEXT J
  PRINT "***********************************"
NEXT I
END
```

Bicubic interpolation in two dimensions is carried out with BCUCOF and BCUINT. The first supplies interpolating coefficients within a grid square and the second calculates interpolated values. The calculation provides not only interpolated function values, but also interpolated values of two partial derivatives, all of which are guaranteed to be smooth. To get this, we are required to supply more information than we have needed in previous interpolation routines.

Demonstration program D3R10 works with the function $f(x,y) = xy \exp(-xy)$. You may compare the two first derivatives and the cross derivative of this function with what you find computed in the routine. The function and derivatives are calculated at the four corners of a rectangular grid cell, in this case a 2×2 unit square with one corner at the origin. The points are supplied counterclockwise around the cell. D1 and D2 are the dimensions of the cell. A call to BCUCOF provides sixteen coefficients which are listed below for your reference.

```
Coefficients for bicubic interpolation
  0.000000E+00    0.000000E+00    0.000000E+00     0.000000E+00
  0.000000E+00    0.400000E+01    0.000000E+00     0.000000E+00
  0.000000E+00    0.000000E+00   -0.136556E+02     0.609517E+01
  0.000000E+00    0.000000E+00    0.609517E+01    -0.246149E+01
```

Here is the recipe BCUCOF:

```
DATA 1.,0.,-3.,2.,0.,0.,0.,0.,-3.,0.,9.,-6.,2.,0.,-6.,4.,0.,0.,0.,0.,0.,0.,
DATA 0.,3.,0.,-9.,6.,-2.,0.,6.,-4.,0.,0.,0.,0.,0.,0.,0.,0.,0.,0.,9.,-6.,0.,
DATA -6.,4.,0.,0.,3.,-2.,0.,0.,0.,0.,0.,0.,-9.,6.,0.,0.,6.,-4.,0.,0.,0.,0.,
DATA 0.,-3.,2.,-2.,0.,6.,-4.,1.,0.,-3.,2.,0.,0.,0.,0.,0.,0.,0.,0.,-1.,0.,3.
DATA -2.,1.,0.,-3.,2.,0.,0.,0.,0.,0.,0.,0.,0.,0.,0.,-3.,2.,0.,0.,3.,-2.,0.,
DATA 0.,0.,0.,0.,3.,-2.,0.,0.,-6.,4.,0.,0.,3.,-2.,0.,1.,-2.,1.,0.,0.,0.,0.,
DATA -3.,6.,-3.,0.,2.,-4.,2.,0.,0.,0.,0.,0.,0.,0.,0.,3.,-6.,3.,0.,-2.,4.
DATA -2.,0.,0.,0.,0.,0.,0.,0.,0.,0.,0.,-3.,3.,0.,0.,2.,-2.,0.,0.,-1.,1.,0.,
DATA 0.,0.,0.,0.,3.,-3.,0.,0.,-2.,2.,0.,0.,0.,0.,1.,-2.,1.,0.,-2.,4.,-2.
DATA 1.,-2.,1.,0.,0.,0.,0.,0.,0.,0.,0.,0.,-1.,2.,-1.,0.,1.,-2.,1.,0.,0.,0.,
DATA 0.,0.,0.,0.,0.,0.,1.,-1.,0.,0.,-1.,1.,0.,0.,0.,0.,0.,0.,-1.,1.,0.,0.,2
DATA -2.,0.,0.,-1.,1.
```

SUB BCUCOF (Y(), Y1(), Y2(), Y12(), D1, D2, C())
> Given arrays Y,Y1,Y2, and Y12, each of length 4, containing the function, gradients, and cross derivative at the four grid points of a rectangular grid cell (numbered counterclockwise from the lower left), and given D1 and D2, the length of the grid cell in the 1- and 2-directions, this routine returns the table C that is used by routine BCUINT for bicubic interpolation.

```
DIM CL(16), X(16), WT(16, 16)
RESTORE
FOR J = 1 TO 16
  FOR I = 1 TO 16
    READ WT(I, J)
  NEXT I
NEXT J
D1D2 = D1 * D2
FOR I = 1 TO 4                    Pack a temporary vector X.
  X(I) = Y(I)
  X(I + 4) = Y1(I) * D1
  X(I + 8) = Y2(I) * D2
  X(I + 12) = Y12(I) * D1D2
NEXT I
FOR I = 1 TO 16                   Matrix multiply by the stored table.
  XX = 0!
  FOR K = 1 TO 16
    XX = XX + WT(I, K) * X(K)
  NEXT K
  CL(I) = XX
NEXT I
L = 0
FOR I = 1 TO 4                    Unpack the result into the output table.
  FOR J = 1 TO 4
    L = L + 1
    C(I, J) = CL(L)
  NEXT J
NEXT I
ERASE WT, X, CL
END SUB
```

A sample program using BCUCOF is the following:

```
DECLARE SUB BCUCOF (Y!(), Y1!(), Y2!(), Y12!(), D1!, D2!, C!())

'PROGRAM D3R10
'Driver for routine BCUCOF
CLS
DIM C(4, 4), Y(4), Y1(4), Y2(4)
DIM Y12(4), X1(4), X2(4)
FOR I = 1 TO 4
  READ X1(I)
NEXT I
DATA 0.0,2.0,2.0,0.0
FOR I = 1 TO 4
  READ X2(I)
NEXT I
DATA 0.0,0.0,2.0,2.0
D1 = X1(2) - X1(1)
D2 = X2(4) - X2(1)
FOR I = 1 TO 4
  X1X2 = X1(I) * X2(I)
  EE = EXP(-X1X2)
  Y(I) = X1X2 * EE
  Y1(I) = X2(I) * (1! - X1X2) * EE
  Y2(I) = X1(I) * (1! - X1X2) * EE
  Y12(I) = (1! - 3! * X1X2 + X1X2 ^ 2) * EE
NEXT I
CALL BCUCOF(Y(), Y1(), Y2(), Y12(), D1, D2, C())
PRINT "Coefficients for bicubic interpolation"
FOR I = 1 TO 4
  FOR J = 1 TO 4
    PRINT "  ";
    PRINT USING "#.######^^^^"; C(I, J);
  NEXT J
  PRINT
NEXT I
END
```

Program D3R11 works with the function $f(x,y) = (xy)^2$, which has derivatives $\partial f/\partial x = 2xy^2$, $\partial f/\partial y = 2yx^2$, and $\partial^2 f/\partial x \partial y = 4xy$. These are supplied to BCUINT along with the locations of the grid points. BCUINT calls BCUCOF internally to determine coefficients, and then calculates ANSY, ANSY1, ANSY2, the interpolated values of f, $\partial f/\partial x$ and $\partial f/\partial y$ at the specified test point (X1,X2). These are compared by the demonstration program to expected values for the three quantities, which are called EY, EY1, and EY2. The test points run along the diagonal of the grid square.

Here is the recipe BCUINT:

```
DECLARE SUB BCUCOF (Y!(), Y1!(), Y2!(), Y12!(), D1!, D2!, C!())

SUB BCUINT (Y(), Y1(), Y2(), Y12(), X1L, X1U, X2L, X2U, X1, X2, ANSY, ANSY1,
ANSY2)
```

Bicubic interpolation within a grid square. Input quantities are Y,Y1,Y2,Y12 (as described in BCUCOF); X1L and X1U, the lower and upper coordinates of the grid square in the 1-direction; X2L and X2U likewise for the 2-direction; and X1,X2, the coordinates of the desired point for the interpolation. The interpolated function value is returned as ANSY, and the interpolated gradient values as ANSY1 and ANSY2. This routine calls BCUCOF.

```
DIM C(4, 4)
CALL BCUCOF(Y(), Y1(), Y2(), Y12(), X1U - X1L, X2U - X2L, C()) Get the c's.
IF X1U = X1L OR X2U = X2L THEN PRINT "bad input": EXIT SUB
T = (X1 - X1L) / (X1U - X1L)         Equation 3.6.4.
U = (X2 - X2L) / (X2U - X2L)
ANSY = 0!
ANSY2 = 0!
ANSY1 = 0!
FOR I = 4 TO 1 STEP -1               Equation 3.6.6.
  ANSY = T * ANSY + ((C(I, 4) * U + C(I, 3)) * U + C(I, 2)) * U + C(I, 1)
  ANSY2 = T * ANSY2 + (3! * C(I, 4) * U + 2! * C(I, 3)) * U + C(I, 2)
  ANSY1 = U * ANSY1 + (3! * C(4, I) * T + 2! * C(3, I)) * T + C(2, I)
NEXT I
ANSY1 = ANSY1 / (X1U - X1L)
ANSY2 = ANSY2 / (X2U - X2L)
ERASE C
END SUB
```

A sample program using BCUINT is the following:

```
DECLARE SUB BCUINT (Y!(), Y1!(), Y2!(), Y12!(), X1L!, X1U!, X2L!, X2U!, X1!, X2!,
ANSY!, ANSY1!, ANSY2!)

'PROGRAM D3R11
'Driver for routine BCUINT
CLS
DIM Y(4), Y1(4), Y2(4), Y12(4), XX(4), YY(4)
FOR I = 1 TO 4
  READ XX(I)
NEXT I
DATA 0.0,2.0,2.0,0.0
FOR I = 1 TO 4
  READ YY(I)
NEXT I
DATA 0.0,0.0,2.0,2.0
X1L = XX(1)
X1U = XX(2)
X2L = YY(1)
X2U = YY(4)
FOR I = 1 TO 4
  XXYY = XX(I) * YY(I)
  Y(I) = XXYY ^ 2
  Y1(I) = 2! * YY(I) * XXYY
  Y2(I) = 2! * XX(I) * XXYY
  Y12(I) = 4! * XXYY
NEXT I
```

```
PRINT "    X1        X2        Y     EXPECT   Y1    EXPECT   Y2    EXPECT"
PRINT
FOR I = 1 TO 10
  X1 = .2 * I
  X2 = X1
  CALL BCUINT(Y(), Y1(), Y2(), Y12(), X1L, X1U, X2L, X2U, X1, X2, ANSY, ANSY1, ANSY2)
  X1X2 = X1 * X2
  EY = X1X2 ^ 2
  EY1 = 2! * X2 * X1X2
  EY2 = 2! * X1 * X1X2
  PRINT USING "###.####"; X1; X2; ANSY; EY; ANSY1; EY1; ANSY2; EY2
NEXT I
END
```

Routines SPLIE2 and SPLIN2 work as a pair to perform bicubic spline interpolations. SPLIE2 takes a function tabulated on an $M \times N$ grid and performs one dimensional natural cubic splines along the rows of the grid to generate an array of second derivatives. These are fodder for SPLIN2 which takes the grid points, function values, and second derivative values and returns the interpolated function value for a desired point in the grid region.

Demonstration program D3R12 exercises SPLIE2 on a regular 10×10 grid of points with coordinates X1 and X2, for the function $y = (x_1 x_2)^2$. The calculated second derivative array is compared with the actual second derivative $2x_1 x_2$ of the function. Keep in mind that a natural spline is assumed, so that agreement will not be so good near the boundaries of the grid. (This shows that you should *not* assume a natural spline if you have better derivative information at the endpoints.)

Here is the recipe SPLIE2:

```
DECLARE SUB SPLINE (X!(), Y!(), N!, YP1!, YPN!, Y2!())

SUB SPLIE2 (X1A(), X2A(), YA(), M, N, Y2A())
        Given an M by N tabulated function YA, and tabulated independent variables X1A (M values)
        and X2A (N values), this routine constructs one-dimensional natural cubic splines of the rows
        of YA and returns the second-derivatives in the array Y2A.
DIM YTMP(N), Y2TMP(N)
FOR J = 1 TO M
  FOR K = 1 TO N
    YTMP(K) = YA(J, K)
  NEXT K
  CALL SPLINE(X2A(), YTMP(), N, 1E+30, 1E+30, Y2TMP())    Values 1 × 10³⁰ signal a
  FOR K = 1 TO N                                          natural spline.
    Y2A(J, K) = Y2TMP(K)
  NEXT K
NEXT J
ERASE Y2TMP, YTMP
END SUB
```

A sample program using `SPLIE2` is the following:

```
DECLARE SUB SPLIE2 (X1A!(), X2A!(), YA!(), M!, N!, Y2A!())

'PROGRAM D3R12
'Driver for routine SPLIE2
CLS
M = 10
N = 10
DIM X1(M), X2(N), Y(M, N), Y2(M, N)
FOR I = 1 TO M
  X1(I) = .2 * I
NEXT I
FOR I = 1 TO N
  X2(I) = .2 * I
NEXT I
FOR I = 1 TO M
  FOR J = 1 TO N
    X1X2 = X1(I) * X2(J)
    Y(I, J) = X1X2 ^ 2
  NEXT J
NEXT I
CALL SPLIE2(X1(), X2(), Y(), M, N, Y2())
PRINT "Second derivatives from SPLIE2"
PRINT "Natural spline assumed"
PRINT
FOR I = 1 TO 5
  FOR J = 1 TO 5
    PRINT USING "#####.######"; Y2(I, J);
  NEXT J
  PRINT
NEXT I
PRINT
PRINT "Actual second derivatives"
PRINT
FOR I = 1 TO 5
  FOR J = 1 TO 5
    Y2(I, J) = 2! * X1(I) ^ 2
  NEXT J
  FOR J = 1 TO 5
    PRINT USING "#####.######"; Y2(I, J);
  NEXT J
  PRINT
NEXT I
END
```

The demonstration program `D3R13` establishes a similar 10×10 grid for the function $y = x_1 x_2 \exp(-x_1 x_2)$. It makes a single call to `SPLIE2` to produce second derivatives `Y2`, and then finds function values `F` through calls to `SPLIN2`, comparing them to actual function values `FF`. These values are determined, for no reason better than perversity, along a quadratic path $x_2 = x_1^2$ through the grid region.

Here is the recipe `SPLIN2`:

```
DECLARE SUB SPLINE (X!(), Y!(), N!, YP1!, YPN!, Y2!())
DECLARE SUB SPLINT (XA!(), YA!(), Y2A!(), N!, X!, Y!)

SUB SPLIN2 (X1A(), X2A(), YA(), Y2A(), M, N, X1, X2, Y)
```
Given X1A, X2A, YA, M, N as described in SPLIE2 and Y2A as produced by that routine; and given a desired interpolating point X1, X2; this routine returns an interpolated function value Y by bicubic spline interpolation.

```
DIM YTMP(N), Y2TMP(N), YYTMP(N)
FOR J = 1 TO M                          Perform M evaluations of the row splines constructed by SPLIE2,
  FOR K = 1 TO N                        using the one-dimensional spline evaluator SPLINT.
    YTMP(K) = YA(J, K)
    Y2TMP(K) = Y2A(J, K)
  NEXT K
  CALL SPLINT(X2A(), YTMP(), Y2TMP(), N, X2, YYTMP(J))
NEXT J
CALL SPLINE(X1A(), YYTMP(), M, 1E+30, 1E+30, Y2TMP())    Construct the 1-dim. spline
CALL SPLINT(X1A(), YYTMP(), Y2TMP(), M, X1, Y)           and evaluate it.
ERASE YYTMP, Y2TMP, YTMP
END SUB
```

A sample program using SPLIN2 is the following:

```
DECLARE SUB SPLIE2 (X1A!(), X2A!(), YA!(), M!, N!, Y2A!())
DECLARE SUB SPLIN2 (X1A!(), X2A!(), YA!(), Y2A!(), M!, N!, X1!, X2!, Y!)

'PROGRAM D3R13
'Driver for routine SPLIN2
CLS
M = 10
N = 10
DIM X1(M), X2(N), Y(M, N), Y2(M, N)
FOR I = 1 TO M
  X1(I) = .2 * I
NEXT I
FOR I = 1 TO N
  X2(I) = .2 * I
NEXT I
FOR I = 1 TO M
  FOR J = 1 TO N
    X1X2 = X1(I) * X2(J)
    Y(I, J) = X1X2 * EXP(-X1X2)
  NEXT J
NEXT I
CALL SPLIE2(X1(), X2(), Y(), M, N, Y2())
PRINT "       x1          x2         splin2       actual"
FOR I = 1 TO 10
  XX1 = .1 * I
  XX2 = XX1 ^ 2
  CALL SPLIN2(X1(), X2(), Y(), Y2(), M, N, XX1, XX2, F)
  X1X2 = XX1 * XX2
  FF = X1X2 * EXP(-X1X2)
  PRINT USING "#####.######"; XX1; XX2; F; FF
NEXT I
END
```

Chapter 4: Integration of Functions

Numerical integration, or "quadrature", has been treated with some degree of detail in *Numerical Recipes*. Chapter 4 begins with TRAPZD, a subroutine for applying the extended trapezoidal rule. It can be used in successive calls for sequentially improving accuracy, and is used as a foundation for several other programs. For example QTRAP is an integrating routine that makes repeated calls to TRAPZD until a certain fractional accuracy is achieved. QSIMP also calls TRAPZD, and in this case performs integration by Simpson's rule. Romberg integration, a generalization of Simpson's rule to successively higher orders, is performed with QROMB—this one also calls TRAPZD. For improper integrals a different "workhorse" is used, the subroutine MIDPNT. This routine applies the extended midpoint rule to avoid function evaluations at an endpoint of the region of integration. It can be used in QTRAP or QSIMP in place of TRAPZD. Routine QROMB can be generalized similarly, and we have implemented this idea in QROMO, a Romberg integrator for open intervals. The chapter also offers a number of exact replacements for MIDPNT, to be used for various types of singularity in the integrand:

1. MIDINF - if one or the other of the limits of integration is infinite.
2. MIDSQL - if there is an inverse square root singularity of the integrand at the lower limit of integration.
3. MIDSQU - if there is an inverse square root singularity of the integrand at the upper limit of integration.
4. MIDEXP - when the upper limit of integration is infinite and the integrand decreases exponentially at infinity.

The somewhat more subtle method of Gaussian quadrature uses unequally spaced abscissas, and weighting coefficients which can be read from tables. Routine QGAUS computes integrals with a ten-point Gauss-Legendre weighting using such coefficients. GAULEG calculates the tables of abscissas and weights that would apply to an N-point Gauss-Legendre quadrature.

$$\star \quad \star \quad \star \quad \star$$

TRAPZD applies the extended trapezoidal rule for integration. It is called sequentially for higher and higher stages of refinement of the integral. The sample program D4R1 uses TRAPZD to perform a numerical integration of the function

$$\text{FUNC} = x^2(x^2 - 2)\sin x$$

whose definite integral is

$$\text{FINT} = 4x(x^2 - 7)\sin x - (x^4 - 14x^2 + 28)\cos x.$$

The integral is performed from $A = 0.0$ to $B = \pi/2$. To demonstrate the increasing accuracy on sequential calls, TRAPZD is called 14 times with the index I increasing by one each time. The improving values of the integral are listed for comparison to the actual value $\text{FINT}(B) - \text{FINT}(A)$.

Here is the recipe TRAPZD:

```
DECLARE FUNCTION FUNC! (X!)

SUB TRAPZD (DUM, A, B, S, N) STATIC
```

This routine computes the N'th stage of refinement of an extended trapezoidal rule. FUNC is the name of a user-supplied function to be integrated between limits A and B, also input. When called with N=1, the routine returns as S the crudest estimate of $\int_a^b f(x)dx$. Subsequent calls with N=2,3,... (in that sequential order) will improve the accuracy of S by adding 2^{N-2} additional interior points. S should not be modified between sequential calls.

```
IF N = 1 THEN
  S = .5 * (B - A) * (FUNC(A) + FUNC(B))
  IT = 1                          IT is the number of points to be added on the next
ELSE                              call.
  TNM = IT
  DEL = (B - A) / TNM             This is the spacing of the points to be added.
  X = A + .5 * DEL
  SUM = 0!
  FOR J = 1 TO IT
    SUM = SUM + FUNC(X)
    X = X + DEL
  NEXT J
  S = .5 * (S + (B - A) * SUM / TNM)    This replaces S by its refined value.
  IT = 2 * IT
END IF
END SUB
```

A sample program using TRAPZD is the following:

```
DECLARE FUNCTION FUNC! (X!)
DECLARE FUNCTION FINT! (X!)
DECLARE SUB TRAPZD (DUM!, A!, B!, S!, N!)

'PROGRAM D4R1
'Driver for routine TRAPZD
CLS
NMAX = 14
PIO2 = 1.5707963#
A = 0!
B = PIO2
PRINT "Integral of FUNC with 2^(n-1) points"
PRINT "Actual value of integral is  ";
PRINT USING "#.######"; FINT(B) - FINT(A)
PRINT "     n          Approx. Integral"
FOR I = 1 TO NMAX
  CALL TRAPZD(DUM, A, B, S, I)
  PRINT USING "######"; I;
  PRINT USING "#############.######"; S
NEXT I
END

FUNCTION FINT (X)
FINT = 4! * X * (X ^ 2 - 7!) * SIN(X) - (X ^ 4 - 14! * X ^ 2 + 28!) * COS(X)
END FUNCTION
```

```
FUNCTION FUNC (X)
FUNC = X ^ 2 * (X ^ 2 - 2!) * SIN(X)
END FUNCTION
```

QTRAP carries out the same integration algorithm but allows us to specify the accuracy with which we wish the integration done. (It is specified within QTRAP as EPS=1.0E-6.) QTRAP itself makes the sequential calls to TRAPZD until the desired accuracy is reached. Then QTRAP issues a single result. In sample program D4R2 we compare this result to the exact value of the integral.

Here is the recipe QTRAP:

```
DECLARE SUB TRAPZD (DUM!, A!, B!, S!, N!)

SUB QTRAP (DUM, A, B, S)
```
    Returns as S the integral of the user-supplied function FUNC from A to B. The parameters EPS can be set to the desired fractional accuracy and JMAX so that 2^{JMAX-1} is the maximum allowed number of steps. Integration is performed by the trapezoidal rule.
```
EPS = .000001
JMAX = 20
OLDS = -1E+30                    Any number that is unlikely to be the average of the function
FOR J = 1 TO JMAX               at its endpoints will do here.
  CALL TRAPZD(DUM, A, B, S, J)
  IF ABS(S - OLDS) < EPS * ABS(OLDS) THEN EXIT SUB
  OLDS = S
NEXT J
PRINT "Too many steps."
END SUB
```

A sample program using QTRAP is the following:

```
DECLARE FUNCTION FUNC! (X!)
DECLARE FUNCTION FINT! (X!)
DECLARE SUB QTRAP (DUM!, A!, B!, S!)

'PROGRAM D4R2
'Driver for routine QTRAP
CLS
PIO2 = 1.5707963#
A = 0!
B = PIO2
PRINT "Integral of FUNC computed with QTRAP"
PRINT "Actual value of integral is  ";
PRINT USING "#.######"; FINT(B) - FINT(A)
CALL QTRAP(DUM, A, B, S)
PRINT "Result from routine QTRAP is  ";
PRINT USING "#.######"; S
END

FUNCTION FINT (X)
FINT = 4! * X * (X ^ 2 - 7!) * SIN(X) - (X ^ 4 - 14! * X ^ 2 + 28!) * COS(X)
END FUNCTION

FUNCTION FUNC (X)
```

```
FUNC = X ^ 2 * (X ^ 2 - 2!) * SIN(X)
END FUNCTION
```

Alternatively, the integral may be handled by QSIMP which applies Simpson's rule. Sample program D4R3 carries out the same integration as the previous program, and reports the result in the same way as well.

Here is the recipe QSIMP:

```
DECLARE SUB TRAPZD (DUM!, A!, B!, S!, N!)

SUB QSIMP (DUM, A, B, S)
```
> Returns as S the integral of the user-supplied function FUNC from A to B. The parameters EPS can be set to the desired fractional accuracy and JMAX so that 2^{JMAX-1} is the maximum allowed number of steps. Integration is performed by Simpson's rule.

```
EPS = .000001
JMAX = 20
OST = -1E+30
OS = -1E+30
FOR J = 1 TO JMAX
  CALL TRAPZD(DUM, A, B, ST, J)
  S = (4! * ST - OST) / 3!          Compare equation (4.2.4).
  IF ABS(S - OS) < EPS * ABS(OS) THEN EXIT SUB
  OS = S
  OST = ST
NEXT J
PRINT "Too many steps."
END SUB
```

A sample program using QSIMP is the following:

```
DECLARE FUNCTION FUNC! (X!)
DECLARE FUNCTION FINT! (X!)
DECLARE SUB QSIMP (DUM!, A!, B!, S!)

'PROGRAM D4R3
'Driver for routine QSIMP
CLS
PIO2 = 1.5707963#
A = 0!
B = PIO2
PRINT "Integral of FUNC computed with QSIMP"
PRINT "Actual value of integral is   ";
PRINT USING "#.######"; FINT(B) - FINT(A)
CALL QSIMP(DUM, A, B, S)
PRINT "Result from routine QSIMP is   ";
PRINT USING "#.######"; S
END

FUNCTION FINT (X)
'Integral of FUNC
FINT = 4! * X * (X ^ 2 - 7!) * SIN(X) - (X ^ 4 - 14! * X ^ 2 + 28!) * COS(X)
END FUNCTION
```

```
FUNCTION FUNC (X)
FUNC = X ^ 2 * (X ^ 2 - 2!) * SIN(X)
END FUNCTION
```

QROMB generalizes Simpson's rule to higher orders. It makes successive calls to TRAPZD and stores the results. Then it uses POLINT, the polynomial interpolater/extrapolator, to project the value of integral that would be obtained were we to continue indefinitely with TRAPZD. Sample program D4R4 is essentially identical to the sample programs for QTRAP and QSIMP.

Here is the recipe QROMB:

```
DECLARE SUB TRAPZD (DUM!, A!, B!, S!, N!)
DECLARE SUB POLINT (XA!(), YA!(), N!, X!, Y!, DY!)

SUB QROMB (DUM, A, B, SS)
```
> Returns as SS the integral of the user-supplied function FUNC from A to B. Integration is performed by Romberg's method of order 2K, where, e.g., K=2 is Simpson's rule.
```
EPS = .000001
JMAX = 20
JMAXP = JMAX + 1
K = 5
```
> Here EPS is the fractional accuracy desired, as determined by the extrapolation error estimate; JMAX limits the total number of steps; K is the number of points used in the extrapolation.
```
KM = K - 1
DIM S(JMAXP), H(JMAXP)        These store the successive trapezoidal approximations and
H(1) = 1!                     their relative step-sizes.
FOR J = 1 TO JMAX
  CALL TRAPZD(DUM, A, B, S(J), J)
  IF J >= K THEN
    CALL POLINT(H(), S(), K, 0!, SS, DSS)
    IF ABS(DSS) < EPS * ABS(SS) THEN ERASE H, S: EXIT SUB
  END IF
  S(J + 1) = S(J)
  H(J + 1) = .25 * H(J)       This is a key step: The factor is 0.25 even though the step-
NEXT J                        size is decreased by only 0.5. This makes the extrapolation
PRINT "Too many steps."       a polynomial in h² as allowed by equation (4.2.1), not just a
END SUB                       polynomial in h.
```

A sample program using QROMB is the following:

```
DECLARE FUNCTION FUNC! (X!)
DECLARE FUNCTION FINT! (X!)
DECLARE SUB QROMB (DUM!, A!, B!, SS!)

'PROGRAM D4R4
'Driver for routine QROMB
CLS
PIO2 = 1.5707963#
A = 0!
B = PIO2
PRINT "Integral of FUNC computed with QROMB"
PRINT "Actual value of integral is  ";
PRINT USING "#.######"; FINT(B) - FINT(A)
CALL QROMB(DUM, A, B, S)
```

```
PRINT "Result from routine QROMB is   ";
PRINT USING "#.######"; S
END


FUNCTION FINT (X)
'Integral of FUNC
FINT = 4! * X * (X ^ 2 - 7!) * SIN(X) - (X ^ 4 - 14! * X ^ 2 + 28!) * COS(X)
END FUNCTION


FUNCTION FUNC (X)
FUNC = X ^ 2 * (X ^ 2 - 2!) * SIN(X)
END FUNCTION
```

Sample program D4R5 uses the function FUNC $= 1/\sqrt{x}$ which is singular at the origin. Limits of integration are set at $A = 0.0$ and $B = 1.0$. MIDPNT, however, implements an open formula and does not evaluate the function exactly at $x = 0$. In this case the integral is compared to FINT(B) $-$ FINT(A) where FINT $= 2\sqrt{x}$, the integral of FUNC.

Here is the recipe MIDPNT:

```
DECLARE FUNCTION FUNC! (X!)

SUB MIDPNT (DUM, A, B, S, N) STATIC
```
> This routine computes the Nth stage of refinement of an extended midpoint rule. The user-supplied routine FUNC is a function to be integrated between limits A and B. When called with N=1, the routine returns as S the crudest estimate of $\int_a^b f(x)dx$. Subsequent calls with N=2,3,... (in that sequential order) will improve the accuracy of S by adding $(2/3) \times 3^{N-1}$ additional interior points. S should not be modified between sequential calls.

```
IF N = 1 THEN
  S = (B - A) * FUNC(.5 * (A + B))
  IT = 1                          2×IT points will be added on the next refinement.
ELSE
  TNM = IT
  DEL = (B - A) / (3! * TNM)
  DDEL = DEL + DEL                The added points alternate in spacing between DEL and DDEL.
  X = A + .5 * DEL
  SUM = 0!
  FOR J = 1 TO IT
    SUM = SUM + FUNC(X)
    X = X + DDEL
    SUM = SUM + FUNC(X)
    X = X + DEL
  NEXT J
  S = (S + (B - A) * SUM / TNM) / 3!   The new sum is combined with the old integral to
  IT = 3 * IT                          give a refined integral.
END IF
END SUB
```

A sample program using MIDPNT is the following:

```
DECLARE FUNCTION FUNC! (X!)
DECLARE FUNCTION FINT! (X!)
DECLARE SUB MIDPNT (DUM!, A!, B!, S!, N!)

'PROGRAM D4R5
'Driver for routine MIDPNT
CLS
NMAX = 10
A = 0!
B = 1!
PRINT "Integral of FUNC computed with MIDPNT"
PRINT "Actual value of integral is  ";
PRINT USING "#####.######"; FINT(B) - FINT(A)
PRINT "    n            Approx. Integral"
FOR I = 1 TO NMAX
  CALL MIDPNT(DUM, A, B, S, I)
  PRINT USING "######"; I;
  PRINT USING "################.######"; S
NEXT I
END


FUNCTION FINT (X)
'Integral of FUNC
FINT = 2! * SQR(X)
END FUNCTION


FUNCTION FUNC (X)
FUNC = 1! / SQR(X)
END FUNCTION
```

Various special forms of MIDPNT (i.e. MIDSQL, MIDSQU, MIDINF) are demonstrated by sample program D4R6. For those tests that integrate to infinity, we take infinity to be 1.0E20. The following integrations are performed:

1. Integral of $\sqrt{x}/\sin x$ from 0.0 to $\pi/2$. (This has a $1/\sqrt{x}$ singularity at $x = 0$, and uses MIDSQL.)

2. Integral of $\sqrt{(\pi - x)}/\sin x$ from $\pi/2$ to π. (This has the $1/\sqrt{x}$ singularity at the upper limit $x = \pi$, and uses MIDSQU.)

3. Integral of $(\sin x)/x^2$ from $\pi/2$ to ∞. (This has a region of integration extending to ∞ and uses MIDINF. It is quite slowly convergent, as is the next integral.)

4. Integral of $(\sin x)/x^2$ from $-\infty$ to $-\pi/2$. (Region of integration goes to $-\infty$; uses MIDINF.)

5. Integral of $\exp(-x)/\sqrt{x}$ from 0.0 to ∞. (This has a singularity at $x = 0.0$ and also integrates to ∞. It is performed in two pieces, (0.0 to $\pi/2$) and ($\pi/2$ to ∞) using MIDSQL and MIDINF. The two calculations give results RES1 and RES2 respectively, which are added to give the entire integral.)

Here is the recipe QROMO:

```
DECLARE SUB MIDINF (DUM!, AA!, BB!, S!, N!)
DECLARE SUB MIDSQL (DUM!, AA!, BB!, S!, N!)
DECLARE SUB MIDSQU (DUM!, AA!, BB!, S!, N!)
DECLARE SUB MIDPNT (DUM!, A!, B!, S!, N!)
DECLARE SUB POLINT (XA!(), YA!(), N!, X!, Y!, DY!)


SUB QROMO (DUM, A, B, SS, PICK$)
```
Romberg integration on an open interval. Returns as SS the integral of the function FUNC from A to B, using any specified integrating subroutine PICK$ and Romberg's method. Normally PICK$ will be an open formula, not evaluating the function at the endpoints. It is assumed that PICK$ triples the number of steps on each call, and that its error series contains only even powers of the number of steps. The routines MIDPNT, MIDINF, MIDSQL, MIDSQU, are possible choices for PICK$.
```
EPS = .00003
JMAX = 14
JMAXP = JMAX + 1
K = 5
KM = K - 1
```
   The parameters have the same meaning as in QROMB.
```
DIM S(JMAXP), H(JMAXP)
H(1) = 1!
FOR J = 1 TO JMAX
  IF PICK$ = "MIDPNT" THEN CALL MIDPNT(DUM, A, B, S(J), J)
  IF PICK$ = "MIDINF" THEN CALL MIDINF(DUM, A, B, S(J), J)
  IF PICK$ = "MIDSQL" THEN CALL MIDSQL(DUM, A, B, S(J), J)
  IF PICK$ = "MIDSQU" THEN CALL MIDSQU(DUM, A, B, S(J), J)
  IF J >= K THEN
     CALL POLINT(H(), S(), K, 0!, SS, DSS)
     IF ABS(DSS) < EPS * ABS(SS) THEN ERASE H, S: EXIT SUB
  END IF
  S(J + 1) = S(J)
  H(J + 1) = H(J) / 9!      This is where the assumption of step tripling and an even
NEXT J                      error series is used.
PRINT "Too many steps."
END SUB
```

Here is the recipe MIDSQL:

```
DECLARE FUNCTION SQL! (X!, AA!)
DECLARE FUNCTION FUNC! (X!)

SUB MIDSQL (DUM, AA, BB, S, N) STATIC
B = SQR(BB - AA)
A = 0!
IF N = 1 THEN
  S = (B - A) * SQL(.5 * (A + B), AA)
  IT = 1
ELSE
  TNM = IT
  DEL = (B - A) / (3! * TNM)
  DDEL = DEL + DEL
  X = A + .5 * DEL
  SUM = 0!
  FOR J = 1 TO IT
```

```
      SUM = SUM + SQL(X, AA)
      X = X + DDEL
      SUM = SUM + SQL(X, AA)
      X = X + DEL
   NEXT J
   S = (S + (B - A) * SUM / TNM) / 3!
   IT = 3 * IT
END IF
END SUB

FUNCTION SQL (X, AA)
SQL = 2 * X * FUNC(AA + X ^ 2)
END FUNCTION
```

A sample program using QROMO and MIDSQL is the following:

```
DECLARE FUNCTION FUNCL! (X!)
DECLARE FUNCTION FUNCU! (X!)
DECLARE FUNCTION FUNCINF! (X!)
DECLARE FUNCTION FUNCEND! (X!)
DECLARE FUNCTION FUNC! (X!)
DECLARE SUB QROMO (DUM!, A!, B!, SS!, CHOOSE$)

'PROGRAM D4R6
'Driver for QROMO
CLS
X1 = 0!
X2 = 1.5707963#
X3 = 3.1415926#
AINF = 1E+20
PRINT "Improper integrals:"
PRINT
CHOOSE$ = "FUNCL"
CALL QROMO(DUM, X1, X2, RESULT, "MIDSQL")
PRINT "Function: SQR(x)/SIN(x)        Interval: (0,pi/2)"
PRINT "Using: MIDSQL                  Result:";
PRINT USING "###.####"; RESULT
PRINT
CHOOSE$ = "FUNCU"
CALL QROMO(DUM, X2, X3, RESULT, "MIDSQU")
PRINT "Function: SQR(pi-x)/SIN(x)     Interval: (pi/2,pi)"
PRINT "Using: MIDSQU                  Result:";
PRINT USING "###.####"; RESULT
PRINT
CHOOSE$ = "FUNCINF"
CALL QROMO(DUM, X2, AINF, RESULT, "MIDINF")
PRINT "Function: SIN(x)/x^2           Interval: (pi/2,infty)"
PRINT "Using: MIDINF                  Result:";
PRINT USING "###.####"; RESULT
PRINT
CHOOSE$ = "FUNCINF"
CALL QROMO(DUM, -AINF, -X2, RESULT, "MIDINF")
PRINT "Function: SIN(x)/x^2           Interval: (-infty,-pi/2)"
PRINT "Using: MIDINF                  Result:";
PRINT USING "###.####"; RESULT
```

```
PRINT
CHOOSE$ = "FUNCEND"
CALL QROMO(DUM, X1, X2, RES1, "MIDSQL")
CALL QROMO(DUM, X2, AINF, RES2, "MIDINF")
PRINT "Function: EXP(-x)/SQR(x)          Interval: (0.0,infty)"
PRINT "Using: MIDSQL,MIDINF             Result:";
PRINT USING "###.####"; RES1 + RES2
END

FUNCTION FUNC (X)
SHARED CHOOSE$
IF CHOOSE$ = "FUNCL" THEN FUNC = FUNCL(X)
IF CHOOSE$ = "FUNCU" THEN FUNC = FUNCU(X)
IF CHOOSE$ = "FUNCINF" THEN FUNC = FUNCINF(X)
IF CHOOSE$ = "FUNCEND" THEN FUNC = FUNCEND(X)
END FUNCTION

FUNCTION FUNCEND (X)
IF X > 80 THEN
  FUNCEND = 0!
ELSE
  FUNCEND = EXP(-X) / SQR(X)
END IF
END FUNCTION

FUNCTION FUNCINF (X)
FUNCINF = SIN(X) / X ^ 2
END FUNCTION

FUNCTION FUNCL (X)
FUNCL = SQR(X) / SIN(X)
END FUNCTION

FUNCTION FUNCU (X)
PI = 3.1415926#
FUNCU = SQR(PI - X) / SIN(X)
END FUNCTION
```

Here is the recipe MIDEXP:

```
DECLARE FUNCTION FEXP! (X!)
DECLARE FUNCTION FUNC! (X!)

FUNCTION FEXP (X)
FEXP = FUNC(1! / X) / X ^ 2
END FUNCTION

SUB MIDEXP (DUM, AA, BB, S, N) STATIC
B = EXP(-AA)
A = 0!
IF N = 1 THEN
  S = (B - A) * FEXP(.5 * (A + B))
  IT = 1
ELSE
  TNM = IT
```

```
DEL = (B - A) / (3! * TNM)
DDEL = DEL + DEL
X = A + .5 * DEL
SUM = 0!
FOR J = 1 TO IT
  SUM = SUM + FEXP(X)
  X = X + DDEL
  SUM = SUM + FEXP(X)
  X = X + DEL
NEXT J
S = (S + (B - A) * SUM / TNM) / 3!
IT = 3 * IT
END IF
END SUB
```

Here is the recipe MIDINF:

```
DECLARE FUNCTION INF! (X!)
DECLARE FUNCTION FUNC! (X!)

FUNCTION INF (X)
INF = FUNC(1! / X) / X ^ 2
END FUNCTION

SUB MIDINF (DUM, AA, BB, S, N) STATIC
```
      This routine is an exact replacement for MIDPNT, i.e. returns as S the Nth stage of refinement of the integral of FUNC from AA to BB, except that the function is evaluated at evenly spaced points in $1/x$ rather than in x. This allows the upper limit BB to be as large and positive as the computer allows, or the lower limit AA to be as large and negative, but not both. AA and BB must have the same sign.

```
B = 1! / AA
```
These two statements change the limits of integration as required by the change of variable.
```
A = 1! / BB
IF N = 1 THEN
```
From this point on, the routine is identical to MIDPNT.
```
  X = .5 * (A + B)
  S = (B - A) * INF(.5 * (A + B))
  IT = 1
ELSE
  TNM = IT
  DEL = (B - A) / (3! * TNM)
  DDEL = DEL + DEL
  X = A + .5 * DEL
  SUM = 0!
  FOR J = 1 TO IT
    SUM = SUM + INF(X)
    X = X + DDEL
    SUM = SUM + INF(X)
    X = X + DEL
  NEXT J
  S = (S + (B - A) * SUM / TNM) / 3!
  IT = 3 * IT
END IF
END SUB
```

Here is the recipe MIDSQU:

```
DECLARE FUNCTION SQU! (X!, BB!)
DECLARE FUNCTION FUNC! (X!)

SUB MIDSQU (DUM, AA, BB, S, N) STATIC
B = SQR(BB - AA)
A = 0!
IF N = 1 THEN
  S = (B - A) * SQU(.5 * (A + B), BB)
  IT = 1
ELSE
  TNM = IT
  DEL = (B - A) / (3! * TNM)
  DDEL = DEL + DEL
  X = A + .5 * DEL
  SUM = 0!
  FOR J = 1 TO IT
    SUM = SUM + SQU(X, BB)
    X = X + DDEL
    SUM = SUM + SQU(X, BB)
    X = X + DEL
  NEXT J
  S = (S + (B - A) * SUM / TNM) / 3!
  IT = 3 * IT
END IF
END SUB

FUNCTION SQU (X, BB)
SQU = 2! * X * FUNC(BB - X ^ 2)
END FUNCTION
```

Subroutine QGAUS performs a Gauss-Legendre integration, using only ten function evaluations. Sample program D4R8 applies it to the function $x \exp(-x)$ whose integral from x_1 to x is $(1 + x_1)\exp(-x_1) - (1 + x)\exp(-x)$. QGAUS returns this integral as parameter SS. The method is used for a series of intervals, as short as $(0.0 - 0.5)$ and as long as $(0.0 - 5.0)$. You may observe how the accuracy depends on the interval.

Here is the recipe QGAUS:

```
DECLARE FUNCTION FUNC! (X!)
DATA 0.1488743389,0.4333953941,0.6794095682,0.8650633666,0.9739065285
DATA 0.2955242247,0.2692667193,0.2190863625,0.1494513491,0.0666713443

SUB QGAUS (DUM, A, B, SS)
```
        Returns as SS the integral of the function FUNC between A and B, by ten-point Gauss-Legendre
        integration: the function is evaluated exactly ten times at interior points in the range of inte-
        gration.
```
DIM X(5), W(5)                    The abscissas and weights.
RESTORE
FOR I = 1 TO 5
  READ X(I)
NEXT I
FOR I = 1 TO 5
```

```
  READ W(I)
NEXT I
XM = .5 * (B + A)
XR = .5 * (B - A)
SS = 0
FOR J = 1 TO 5
  DX = XR * X(J)
  SS = SS + W(J) * (FUNC(XM + DX) + FUNC(XM - DX))
NEXT J
SS = XR * SS
ERASE W, X
END SUB
```

Will be twice the average value of the function, since the ten weights (five numbers above each used twice) sum to 2.

Scale the answer to the range of integration.

A sample program using QGAUS is the following:

```
DECLARE FUNCTION FUNC! (X!)
DECLARE SUB QGAUS (DUM!, A!, B!, SS!)

'PROGRAM D4R8
'Driver for routine QGAUS
CLS
X1 = 0!
X2 = 5!
NVAL = 10
DX = (X2 - X1) / NVAL
PRINT "0.0 to     QGAUS       Expected"
PRINT
FOR I = 1 TO NVAL
  X = X1 + I * DX
  CALL QGAUS(DUM, X1, X, SS)
  PRINT USING "##.##"; X;
  PRINT USING "#####.######"; SS;
  PRINT USING "#####.######"; -(1! + X) * EXP(-X) + (1! + X1) * EXP(-X1)
NEXT I
END

FUNCTION FUNC (X)
FUNC = X * EXP(-X)
END FUNCTION
```

Sample program D4R9, which drives GAULEG, performs the same method of quadrature, and on the same function. However, it chooses its own abscissas and weights for the Gauss-Legendre calculation, and is not restricted to a ten-point formula; it can do an N-point calculation for any N. The N abscissas and weights appropriate to an interval $x = 0.0$ to 1.0 are found by sample program D4R9 for the case $N = 10$. The results you should find are listed below. Next the program applies these values to a quadrature and compares the result to that from a formal integration.

position	weight
.013047	.033336
.067468	.074726
.160295	.109543
.283302	.134633
.425563	.147762
.574437	.147762

.716698	.134633
.839705	.109543
.932532	.074726
.986953	.033336

Here is the recipe GAULEG:

```
SUB GAULEG (X1, X2, X(), W(), N)
```
Given the lower and upper limits of integration X1 and X2, and given N, this routine returns arrays X and W of length N, containing the abscissas and weights of the Gauss-Legendre N-point quadrature formula. High precision is a good idea for this routine.

```
EPS = .00000000000003#              Increase if you don't have this floating precision.
M = INT((N + 1) / 2)                The roots are symmetric in the interval, so we only have to
XM = .5# * (X2 + X1)                find half of them.
XL = .5# * (X2 - X1)
FOR I = 1 TO M                      Loop over the desired roots.
  Z = COS(3.141592654# * (I - .25#) / (N + .5#))
```
Starting with the above approximation to the Ith root, we enter the main loop of refinement by Newton's method.
```
  DO
    P1 = 1#
    P2 = 0#
    FOR J = 1 TO N                  Loop up the recurrence relation to get the Legendre polyno-
      P3 = P2                       mial evaluated at Z.
      P2 = P1
      P1 = ((2# * J - 1#) * Z * P2 - (J - 1#) * P3) / J
    NEXT J
```
P1 is now the desired Legendre polynomial. We next compute PP, its derivative, by a standard relation involving also P2, the polynomial of one lower order.
```
    PP = N * (Z * P1 - P2) / (Z * Z - 1#)
    Z1 = Z
    Z = Z1 - P1 / PP                Newton's method.
  LOOP WHILE ABS(Z - Z1) > EPS
  X(I) = XM - XL * Z                Scale the root to the desired interval,
  X(N + 1 - I) = XM + XL * Z        and put in its symmetric counterpart.
  W(I) = 2# * XL / ((1# - Z * Z) * PP * PP)     Compute the weight
  W(N + 1 - I) = W(I)              and its symmetric counterpart.
NEXT I
END SUB
```

A sample program using GAULEG is the following:

```
DECLARE FUNCTION FUNC! (X!)
DECLARE SUB GAULEG (X1!, X2!, X!(), W!(), N!)

'PROGRAM D4R9
'Driver for routine GAULEG
CLS
NPOINT = 10
X1 = 0!
X2 = 1!
X3 = 10!
DIM X(NPOINT), W(NPOINT)
CALL GAULEG(X1, X2, X(), W(), NPOINT)
PRINT " #        X(I)           W(I)"
```

```
PRINT
FOR I = 1 TO NPOINT
  PRINT USING "##"; I;
  PRINT USING "####.######"; X(I);
  PRINT USING "####.######"; W(I)
NEXT I
PRINT
'Demonstrate the use of GAULEG for an integral
CALL GAULEG(X1, X3, X(), W(), NPOINT)
XX = 0!
FOR I = 1 TO NPOINT
  XX = XX + W(I) * FUNC(X(I))
NEXT I
PRINT "Integral from GAULEG:";
PRINT USING "#####.######"; XX
PRINT "Actual value:";
PRINT USING "#####.######"; 1! - (1! + X3) * EXP(-X3)
END


FUNCTION FUNC (X)
FUNC = X * EXP(-X)
END FUNCTION
```

Chapter 4 of *Numerical Recipes* ends with a short discussion of multidimensional integration, exemplified by routine QUAD3D which does a 3-dimensional integration by repeated 1-dimensional integration. Sample program D4R10 applies the method to the integration of FUNC $= x^2 + y^2 + z^2$ over a spherical volume with a radius XMAX which is taken successively as $0.1, 0.2, ..., 1.0$. The integral is done in Cartesian rather than spherical coordinates, but the result is compared to that easily found in spherical coordinates, $4\pi(\text{XMAX})^5/5$. The concept is quite simple, but you may find the zoo of subroutines somewhat confusing. Subroutine FUNC generates the function. Subroutines Y1 and Y2 supply the two limits of the y-integration for each value of x. Similarly Z1 and Z2 give the limits of z-integration for given x and y. Routines QGAUSX, QGAUSY, and QGAUSZ are all identical except for their names, and are used to do Gauss-Legendre integration along the three coordinate directions.

Here is the recipe QUAD3D:

```
DECLARE SUB QGAUSX (DUM!, A!, B!, SS!)
DECLARE SUB QGAUSY (DUM!, A!, B!, SS!)
DECLARE SUB QGAUSZ (DUM!, A!, B!, SS!)
DECLARE FUNCTION F! (ZZ!)
DECLARE FUNCTION G! (YY!)
DECLARE FUNCTION H! (XX!)
DECLARE FUNCTION Y1! (X!)
DECLARE FUNCTION Y2! (X!)
DECLARE FUNCTION Z1! (X!, Y!)
DECLARE FUNCTION Z2! (X!, Y!)
DECLARE FUNCTION FUNC! (X!, Y!, Z!)
COMMON SHARED X, Y, Z

FUNCTION F (ZZ)                    Called by QGAUSZ. Calls FUNC.
Z = ZZ
F = FUNC(X, Y, Z)
```

```
END FUNCTION

FUNCTION G (YY)                    Called by QGAUSY. Calls QGAUSZ.
Y = YY
CALL QGAUSZ(DUM, Z1(X, Y), Z2(X, Y), SS)
G = SS
END FUNCTION

FUNCTION H (XX)                    Called by QGAUSX. Calls QGAUSY.
X = XX
CALL QGAUSY(DUM, Y1(X), Y2(X), SS)
H = SS
END FUNCTION

SUB QUAD3D (X1, X2, SS)
      Returns as SS the integral of a user-supplied function FUNC over a three-dimensional region
      specified by the limits X1, X2, and by the user-supplied functions Y1, Y2, Z1, and Z2, as
      defined in (4.6.2).
CALL QGAUSX(DUM, X1, X2, SS)
END SUB
```

A sample program using QUAD3D is the following:

```
DECLARE FUNCTION FUNC! (X!, Y!, Z!)
DECLARE FUNCTION Z1! (X!, Y!)
DECLARE FUNCTION Z2! (X!, Y!)
DECLARE FUNCTION Y1! (X!)
DECLARE FUNCTION Y2! (X!)
DECLARE FUNCTION F! (X!)
DECLARE FUNCTION G! (X!)
DECLARE FUNCTION H! (X!)
DECLARE SUB QUAD3D (X1!, X2!, SS!)
COMMON SHARED X, Y, Z

'PROGRAM D4R10
'Driver for routine QUAD3D
CLS
PI = 3.1415926#
NVAL = 10
PRINT "Integral of r^2 over a spherical volume"
PRINT
PRINT "  Radius    QUAD3D    Actual"
FOR I = 1 TO NVAL
  XMAX = .1 * I
  XMIN = -XMAX
  CALL QUAD3D(XMIN, XMAX, S)
  PRINT USING "#####.##"; XMAX;
  PRINT USING "#####.####"; S;
  PRINT USING "#####.####"; 4! * PI * (XMAX ^ 5) / 5!
NEXT I
END
DATA .1488743389,.4333953941,.6794095682,.8650633666,.9739065285
DATA .2955242247,.2692667193,.2190863625,.1494513491,.0666713443
```

```
FUNCTION FUNC (X, Y, Z)
FUNC = X ^ 2 + Y ^ 2 + Z ^ 2
END FUNCTION


SUB QGAUSX (DUM, A, B, SS)
DIM X(5), W(5)
RESTORE
FOR J = 1 TO 5
  READ X(J)
NEXT J
FOR J = 1 TO 5
  READ W(J)
NEXT J
XM = .5 * (B + A)
XR = .5 * (B - A)
SS = 0!
FOR J = 1 TO 5
  DX = XR * X(J)
  SS = SS + W(J) * (H(XM + DX) + H(XM - DX))
NEXT J
SS = XR * SS
ERASE W, X
END SUB


SUB QGAUSY (DUM, A, B, SS)
DIM X(5), W(5)
RESTORE
FOR J = 1 TO 5
  READ X(J)
NEXT J
FOR J = 1 TO 5
  READ W(J)
NEXT J
XM = .5 * (B + A)
XR = .5 * (B - A)
SS = 0!
FOR J = 1 TO 5
  DX = XR * X(J)
  SS = SS + W(J) * (G(XM + DX) + G(XM - DX))
NEXT J
SS = XR * SS
ERASE W, X
END SUB


SUB QGAUSZ (DUM, A, B, SS)
DIM X(5), W(5)
RESTORE
FOR J = 1 TO 5
  READ X(J)
NEXT J
FOR J = 1 TO 5
  READ W(J)
NEXT J
XM = .5 * (B + A)
XR = .5 * (B - A)
```

```
SS = 0!
FOR J = 1 TO 5
  DX = XR * X(J)
  SS = SS + W(J) * (F(XM + DX) + F(XM - DX))
NEXT J
SS = XR * SS
ERASE W, X
END SUB

FUNCTION Y1 (X)
SHARED XMAX
Y1 = -SQR(ABS(XMAX ^ 2 - X ^ 2))
END FUNCTION

FUNCTION Y2 (X)
SHARED XMAX
Y2 = SQR(ABS(XMAX ^ 2 - X ^ 2))
END FUNCTION

FUNCTION Z1 (X, Y)
SHARED XMAX
Z1 = -SQR(ABS(XMAX ^ 2 - X ^ 2 - Y ^ 2))
END FUNCTION

FUNCTION Z2 (X, Y)
SHARED XMAX
Z2 = SQR(ABS(XMAX ^ 2 - X ^ 2 - Y ^ 2))
END FUNCTION
```

Chapter 5: Evaluation of Functions

Chapter 5 of *Numerical Recipes* treats the approximation and evaluation of functions. The methods, along with a few others, are applied in Chapter 6 to the calculation of a collection of "special" functions. Polynomial or power series expansions are perhaps the most often used approximations, and a few tips are given for accelerating the convergence of some series. In the case of alternating series, Euler's transformation is popular, and is implemented in program EUL-SUM. For general polynomials, DDPOLY demonstrates the evaluation of both the polynomial and its derivatives from a list of its coefficients. The division of one polynomial into another, giving a quotient and remainder polynomial, is done by POLDIV.

The approximation of functions by Chebyshev polynomial series is presented as a method of arriving at the approximation of nearly smallest deviation from the true function over a given region for a specified order of approximation. The coefficients for such polynomials are given by CHEBFT, and function approximations are subsequently carried out by CHEBEV. To generate the derivative or integral of a function from its Chebyshev coefficients, use CHDER or CHINT respectively. Finally, to convert Chebyshev coefficients into coefficients of a polynomial for the same function (a dangerous procedure about which we offer due warning in the text) use CHEBPC and PCSHFT in succession.

Chapter 5 also treats several methods for which we supply no programs. These are continued fractions, rational functions, recurrence relations, and the solution of quadratic and cubic equations.

$$\star \quad \star \quad \star \quad \star$$

Subroutine EULSUM applies Euler's transformation to the summation of an alternating series. It is called successively for each term to be summed. Our sample program D5R1 evaluates the approximation

$$\ln(1 + x) = x - \frac{x^2}{2} + \frac{x^3}{3} - \frac{x^4}{4} + \cdots \qquad -1 < x < 1$$

It asks how many terms MVAL are to be included in the approximation and then makes MVAL calls to EULSUM. Each time, index j increases and TERM takes the value $(-1)^{j+1}x^j/j$. Both this approximation and the function $\ln(1 + x)$ itself are evaluated across the region -1 to 1 for comparison. If MVAL is set less than 1 or more than 40, the program terminates.

Here is the recipe EULSUM:

```
SUB EULSUM (SUM, TERM, JTERM, WKSP()) STATIC
     Incorporates into SUM the JTERMth term, with value TERM, of an alternating series. SUM is
     input as the previous partial sum, and is output as the new partial sum. The first call to this
     routine, with the first TERM in the series, should be with JTERM=1. On the second call, TERM
```

should be set to the second term of the series, with sign opposite to that of the first call, and JTERM should be 2. And so on.

```
IF JTERM = 1 THEN             Initialize:
  NTERM = 1                   Number of saved differences in WKSP.
  WKSP(1) = TERM
  SUM = .5 * TERM             Return first estimate.
ELSE
  TMP = WKSP(1)
  WKSP(1) = TERM
  FOR J = 1 TO NTERM - 1      Update saved quantities by van Wijngaarden's algorithm.
    DUM = WKSP(J + 1)
    WKSP(J + 1) = .5 * (WKSP(J) + TMP)
    TMP = DUM
  NEXT J
  WKSP(NTERM + 1) = .5 * (WKSP(NTERM) + TMP)
  IF ABS(WKSP(NTERM + 1)) <= ABS(WKSP(NTERM)) THEN      Favorable to increase p,
    SUM = SUM + .5 * WKSP(NTERM + 1)
    NTERM = NTERM + 1         and the table becomes longer.
  ELSE                        Favorable to increase n,
    SUM = SUM + WKSP(NTERM + 1)   the table doesn't become longer.
  END IF
END IF
END SUB
```

A sample program using **EULSUM** is the following:

```
DECLARE SUB EULSUM (SUM!, TERM!, JTERM!, WKSP!())

'PROGRAM D5R1
'Driver for routine EULSUM
CLS
NVAL = 40
DIM WKSP(NVAL)
'Evaluate ln(1+x)=x-x^2/2+x^3/3-x^4/4... for -1<x<1
DO
  PRINT "How many terms in polynomial?"
  PRINT "Enter n between 1 and"; STR$(NVAL); ". Enter n=0 to end."
  LINE INPUT MVAL$
  MVAL = VAL(MVAL$)
  IF MVAL <= 0 OR MVAL > NVAL THEN EXIT DO
  PRINT "        X          Actual      Polynomial"
  FOR I = -8 TO 8 STEP 1
    X = I / 10!
    SUM = 0!
    XPOWER = -1
    FOR J = 1 TO MVAL
      XPOWER = -X * XPOWER
      TERM = XPOWER / J
      CALL EULSUM(SUM, TERM, J, WKSP())
    NEXT J
    PRINT USING "#####.######"; X; LOG(1! + X); SUM
  NEXT I
LOOP
END
```

DDPOLY evaluates a polynomial and its derivatives, given the coefficients of the polynomial in the form of an input vector. Sample program D5R2 illustrates this for the polynomial:

$$(x-1)^5 = -1 + 5x - 10x^2 + 10x^3 - 5x^4 + x^5$$

(This is a foolish example, of course. No one would knowingly evaluate $(x-1)^5$ by multiplying it out and evaluating terms individually—but it gives us a convenient way to check the result!). Since this is a fifth order polynomial, we set NC, the number of coeffients, to 6, and initialize the array C of coefficients, with C(1) being the constant coefficient and C(6) the highest-order coefficient. There are two loops, one of which evaluates for x values from 0.0 to 2.0, and the other of which stores the value of the function and NC-1 derivatives. D(J,I) keeps the entire array of values for printing. In the second part of the program, the polynomial evaluations are compared with

$$f^{(n-1)}(x) = \frac{5!}{(6-n)!}(x-1.0)^{6-n} \qquad n = 1,\ldots,5$$

Here is the recipe DDPOLY:

```
SUB DDPOLY (C(), NC, X, PD(), ND)
     Given the NC coefficients of a polynomial of degree NC-1 as an array C with C(1) being the
     constant term, and given a value X, and given a value ND>1, this routine returns the polynomial
     evaluated at X as PD(1) and ND-1 derivatives as PD(2)...PD(ND).
PD(1) = C(NC)
FOR J = 2 TO ND
  PD(J) = 0!
NEXT J
FOR I = NC - 1 TO 1 STEP -1
  IF NC + 1 - I < ND THEN NND = NC + 1 - I ELSE NND = ND
  FOR J = NND TO 2 STEP -1
    PD(J) = PD(J) * X + PD(J - 1)
  NEXT J
  PD(1) = PD(1) * X + C(I)
NEXT I
CONSQ = 2!                        After the first derivative, factorial constants come in.
FOR I = 3 TO ND
  PD(I) = CONSQ * PD(I)
  CONSQ = CONSQ * I
NEXT I
END SUB
```

A sample program using DDPOLY is the following:

```
DECLARE SUB DDPOLY (C!(), NC!, X!, PD!(), ND!)
DECLARE FUNCTION FACTRL! (X!)

'PROGRAM D5R2
'Driver for routine DDPOLY
'Polynomial (X-1)^5
CLS
NC = 6
NCM1 = 5
NP = 20
```

```
DIM C(NC), PD(NCM1), D(NCM1, NP)
DIM A$(NCM1)
FOR I = 1 TO NCM1
  READ A$(I)
NEXT I
DATA "polynomial:","first deriv:","second deriv:","third deriv:"
DATA "fourth deriv:"
FOR I = 1 TO NC
  READ C(I)
NEXT I
DATA -1.0,5.0,-10.0,10.0,-5.0,1.0
FOR I = 1 TO NP
  X = .1 * I
  CALL DDPOLY(C(), NC, X, PD(), NC - 1)
  FOR J = 1 TO NC - 1
    D(J, I) = PD(J)
  NEXT J
NEXT I
FOR I = 1 TO NC - 1
  PRINT "      "; A$(I)
  PRINT "          X          DDPOLY        actual"
  FOR J = 1 TO NP
    X = .1 * J
    PRINT USING "########.######"; X; D(I, J);
    PRINT USING "########.######"; FACTRL(NC - 1) / FACTRL(NC - I) * (X - 1!) ^
(NC - I)
  NEXT J
  PRINT "press ENTER to continue..."
  LINE INPUT DUM$
NEXT I
END
```

POLDIV divides polynomials. Given the coefficients of a numerator and denominator polynomial, POLDIV returns the coefficients of a quotient and a remainder polynomial. Sample program D5R3 takes

$$\text{Numerator} = U = -1 + 5x - 10x^2 + 10x^3 - 5x^4 + x^5 = (x - 1)^5$$

$$\text{Denominator} = V = 1 + 3x + 3x^2 + x^3 = (x + 1)^3$$

for which we expect

$$\text{Quotient} = Q = 31 - 8x + x^2$$

$$\text{Remainder} = R = -32 - 80x - 80x^2$$

The program compares these with the output of POLDIV.

Here is the recipe POLDIV:

```
SUB POLDIV (U(), N, V(), NV, Q(), R())
```
Given the N coefficients of a polynomial in U, and the NV coefficients of another polynomial in V, divide the polynomial U by the polynomial V ("U"/"V") giving a quotient polynomial whose coefficients are returned in Q, and a remainder polynomial whose coefficients are returned in R. The arrays Q and R are dimensioned with lengths N, but the elements R(NV)...R(N) and Q(N-NV+2)...Q(N) will be returned as zero.

```
FOR J = 1 TO N
  R(J) = U(J)
  Q(J) = 0!
NEXT J
FOR K = N - NV TO 0 STEP -1
  Q(K + 1) = R(NV + K) / V(NV)
  FOR J = NV + K - 1 TO K + 1 STEP -1
    R(J) = R(J) - Q(K + 1) * V(J - K)
  NEXT J
NEXT K
FOR J = NV TO N
  R(J) = 0!
NEXT J
END SUB
```

A sample program using POLDIV is the following:

```
DECLARE SUB POLDIV (U!(), N!, V!(), NV!, Q!(), R!())

'PROGRAM D5R3
'Driver for routine POLDIV
' (X-1)^5/(X+1)^3
CLS
N = 6
NV = 4
DIM U(N), V(NV), Q(N), R(N)
FOR I = 1 TO N
  READ U(I)
NEXT I
DATA -1.0,5.0,-10.0,10.0,-5.0,1.0
FOR I = 1 TO NV
  READ V(I)
NEXT I
DATA 1.0,3.0,3.0,1.0
CALL POLDIV(U(), N, V(), NV, Q(), R())
PRINT "      X^0      X^1      X^2      X^3      X^4      X^5"
PRINT
PRINT "Quotient polynomial coefficients:"
FOR I = 1 TO 6
  PRINT USING "#######.##"; Q(I);
NEXT I
PRINT
PRINT
PRINT "Expected quotient coefficients:"
PRINT USING "#######.##"; 31!; -8!; 1!; 0!; 0!; 0!
PRINT
PRINT
PRINT
PRINT "Remainder polynomial coefficients:"
```

```
FOR I = 1 TO 4
  PRINT USING "#######.##"; R(I);
NEXT I
PRINT
PRINT
PRINT "Expected remainder coefficients:"
PRINT USING "#######.##"; -32!; -80!; -80!; 0!
PRINT
END
```

The remaining six programs all deal with Chebyshev polynomials. CHEBFT evaluates the coefficients for a Chebyshev polynomial approximation of a function on a specified interval and for a maximum degree N of polynomial. Demonstration program D5R4 uses the function FUNC $= x^2(x^2 - 2)\sin x$ on the interval $(-\pi/2, \pi/2)$ with the maximum degree of NVAL=40. Notice that CHEBFT is called with this maximum degree specified, even though subsequent evaluations may truncate the Chebyshev series at much lower terms. After we choose the number MVAL of terms in the evaluation, the Chebyshev polynomial is evaluated term by term, for x values between -0.8π and 0.8π, and the result F is compared to the actual function value.

Here is the recipe CHEBFT:

```
DECLARE FUNCTION FUNC! (X!)

SUB CHEBFT (A, B, C(), N, DUM)
```
> Chebyshev fit: Given a function FUNC, lower and upper limits of the interval [A,B], and a maximum degree N, this routine computes the N coefficients C_k such that FUNC$(x) \approx [\sum_{k=1}^{N} C_k T_{k-1}(y)] - c_1/2$, where y and x are related by (5.6.10). This routine is to be used with moderately large N (e.g. 30 or 50), the array of C's subsequently to be truncated at the smaller value m such that C_{m+1} and subsequent elements are negligible.

```
PI# = 3.141592653589793#
DIM F(N)
BMA = .5 * (B - A)
BPA = .5 * (B + A)
FOR K = 1 TO N                        We evaluate the function at the N points required by (5.6.7).
  Y = COS(PI# * (K - .5) / N)
  F(K) = FUNC(Y * BMA + BPA)
NEXT K
FAC = 2! / N
FOR J = 1 TO N
  SUM# = 0#                           We will accumulate the sum in double precision, a nicety
  FOR K = 1 TO N                      which you can ignore.
    SUM# = SUM# + F(K) * COS((PI# * (J - 1)) * ((K - .5#) / N))
  NEXT K
  C(J) = FAC * SUM#
NEXT J
ERASE F
END SUB
```

A sample program using **CHEBFT** is the following:

```
DECLARE SUB CHEBFT (A!, B!, C! (), N!, DUM!)
DECLARE FUNCTION FUNC! (X!)

'PROGRAM D5R4
'Driver for routine CHEBFT
CLS
NVAL = 40
PIO2 = 1.5707963#
EPS = .000001
DIM C(NVAL)
A = -PIO2
B = PIO2
CALL CHEBFT(A, B, C(), NVAL, DUM)
'Test result
DO
  PRINT "How many terms in Chebyshev evaluation?"
  PRINT "Enter n between 6 and"; STR$(NVAL); ". Enter n=0 to end."
  LINE INPUT MVAL$
  MVAL = VAL(MVAL$)
  IF MVAL <= 0 OR MVAL > NVAL THEN EXIT DO
  PRINT "        X         Actual    Chebyshev fit"
  FOR I = -8 TO 8 STEP 1
    X = I * PIO2 / 10!
    Y = (X - .5 * (B + A)) / (.5 * (B - A))
    'Evaluate Chebyshev polynomial without using routine CHEBEV
    T0 = 1!
    T1 = Y
    F = C(2) * T1 + C(1) * .5
    FOR J = 3 TO MVAL
      DUM = T1
      T1 = 2! * Y * T1 - T0
      T0 = DUM
      TERM = C(J) * T1
      F = F + TERM
    NEXT J
    PRINT USING "#####.######"; X; FUNC(X); F
  NEXT I
LOOP
END

FUNCTION FUNC (X)
FUNC = (X ^ 2) * (X ^ 2 - 2!) * SIN(X)
END FUNCTION
```

CHEBEV is the Chebyshev polynomial evaluator, and the next sample program **D5R5** uses it for the same problem just discussed. In fact, the program is identical except that it replaces the internal polynomial summation with **CHEBEV**, which applies Clenshaw's recurrence to find the polynomial values.

Here is the recipe **CHEBEV**:

```
DECLARE FUNCTION CHEBEV! (A!, B!, DUM!, M!, X!)
COMMON SHARED C()

FUNCTION CHEBEV (A, B, DUM, M, X)
```

Chebyshev evaluation: All arguments are input. C is an array of Chebyshev coefficients, of length M, the first M elements of C output from CHEBFT (which must have been called with the same A and B). The Chebyshev polynomial is evaluated at a point Y determined from X, A, and B, and the result is returned as the function value.

```
IF (X - A) * (X - B) > 0! THEN PRINT "X not in range.": EXIT FUNCTION
D = 0!
DD = 0!
Y = (2! * X - A - B) / (B - A)      Change of variable.
Y2 = 2! * Y
FOR J = M TO 2 STEP -1              Clenshaw's recurrence.
  SV = D
  D = Y2 * D - DD + C(J)
  DD = SV
NEXT J
CHEBEV = Y * D - DD + .5 * C(1)     Last step is different.
END FUNCTION
```

A sample program using CHEBEV is the following:

```
DECLARE SUB CHEBFT (A!, B!, C!(), N!, DUM!)
DECLARE FUNCTION FUNC! (X!)
DECLARE FUNCTION CHEBEV! (A!, B!, DUM!, MVAL!, X!)
COMMON SHARED C()

'PROGRAM D5R5
'Driver for routine CHEBEV
CLS
NVAL = 40
PIO2 = 1.5707963#
DIM C(NVAL)
A = -PIO2
B = PIO2
CALL CHEBFT(A, B, C(), NVAL, DUM)
'Test Chebyshev evaluation routine
DO
  PRINT "How many terms in Chebyshev evaluation?"
  PRINT "Enter n between 6 and"; STR$(NVAL); ". Enter n=0 to end."
  LINE INPUT MVAL$
  MVAL = VAL(MVAL$)
  IF MVAL <= 0 OR MVAL > NVAL THEN EXIT DO
  PRINT "        X        Actual    Chebyshev fit"
  FOR I = -8 TO 8 STEP 1
    X = I * PIO2 / 10!
    PRINT USING "#####.######"; X; FUNC(X); CHEBEV(A, B, DUM, MVAL, X)
  NEXT I
LOOP
END

FUNCTION FUNC (X)
FUNC = (X ^ 2) * (X ^ 2 - 2!) * SIN(X)
END FUNCTION
```

By the same token, the tests for CHINT and CHDER needn't be much different. CHINT determines Chebyshev coefficients for the integral of the function, and CHDER for the derivative of the function, given the Chebyshev coefficients for the function itself (from CHEBFT) and the interval (A, B) of evaluation. When applied to the function above, the true integral is

$$\text{FINT} = 4x(x^2 - 7)\sin x - (x^4 - 14x^2 + 28)\cos x$$

and the true derivative is

$$\text{FDER} = 4x(x^2 - 1)\sin x + x^2(x^2 - 2)\cos x$$

The code in sample programs D5R6 and D5R7 compares the true and Chebyshev-derived integral and derivative values for a range of x in the interval of evaluation. Since CHINT and CHDER return Chebyshev coefficients, and not the integral and derivative values themselves, calls to CHEBEV are required for the comparison.

Here is the recipe CHINT:

```
SUB CHINT (A, B, C(), CINQ(), N)
```
> Given A,B,C, as output from routine CHEBFT §5.6, and given N, the desired degree of approximation (length of C to be used), this routine returns the array CINQ, the Chebyshev coefficients of the integral of the function whose coefficients are C. The constant of integration is set so that the integral vanishes at A.

```
CON = .25 * (B - A)                              Factor which normalizes to the interval B − A.
SUM = 0!                                         Accumulates the constant of integration.
FAC = 1!                                         Will equal ±1.
FOR J = 2 TO N - 1
  CINQ(J) = CON * (C(J - 1) - C(J + 1)) / (J - 1)      Equation 5.7.1.
  SUM = SUM + FAC * CINQ(J)
  FAC = -FAC
NEXT J
CINQ(N) = CON * C(N - 1) / (N - 1)       Special case of 5.7.1 for N.
SUM = SUM + FAC * CINQ(N)
CINQ(1) = 2! * SUM                        Set the constant of integration.
END SUB
```

A sample program using CHINT is the following:

```
DECLARE FUNCTION FUNC! (X!)
DECLARE SUB CHEBFT (A!, B!, C!(), N!, DUM!)
DECLARE SUB CHINT (A!, B!, C!(), CINQ!(), N!)
DECLARE FUNCTION FINT! (X!)
DECLARE FUNCTION CHEBEV! (A!, B!, DUM!, MVAL!, X!)
COMMON SHARED C()

'PROGRAM D5R6
'Driver for routine CHINT
CLS
NVAL = 40
PIO2 = 1.5707963#
DIM C(NVAL), CINQ(NVAL)
A = -PIO2
B = PIO2
```

```
CALL CHEBFT (A, B, C (), NVAL, DUM)
'Test integral
DO
   PRINT "How many terms in Chebyshev evaluation?"
   PRINT "Enter n between 6 and"; STR$ (NVAL); ". Enter n=0 to end."
   LINE INPUT MVAL$
   MVAL = VAL (MVAL$)
   IF MVAL <= 0 OR MVAL > NVAL THEN EXIT DO
   CALL CHINT (A, B, C (), CINQ (), MVAL)
   FOR I = 1 TO NVAL
     C (I) = CINQ (I)
   NEXT I
   PRINT "        X        Actual     Cheby. Integ."
   FOR I = -8 TO 8
     X = I * PIO2 / 10!
     PRINT USING "#####.######"; X; FINT (X) - FINT (-PIO2); CHEBEV (A, B, DUM, MVAL,
X)
   NEXT I
LOOP
END


FUNCTION FINT (X)
'Integral of FUNC
FINT = 4! * X * (X ^ 2 - 7!) * SIN (X) - (X ^ 4 - 14! * X ^ 2 + 28!) * COS (X)
END FUNCTION


FUNCTION FUNC (X)
FUNC = (X ^ 2) * (X ^ 2 - 2!) * SIN (X)
END FUNCTION
```

Here is the recipe CHDER:

```
SUB CHDER (A, B, C (), CDER (), N)
```

Given A, B, C, as output from routine CHEBFT §5.6, and given N, the desired degree of approximation (length of C to be used), this routine returns the array CDER, the Chebyshev coefficients of the derivative of the function whose coefficients are C.

```
CDER (N) = 0!                                    N and N-1 are special cases.
CDER (N - 1) = 2 * (N - 1) * C (N)
IF N >= 3 THEN
  FOR J = N - 2 TO 1 STEP -1
    CDER (J) = CDER (J + 2) + 2 * J * C (J + 1)   Equation 5.7.2.
  NEXT J
END IF
CON = 2! / (B - A)
FOR J = 1 TO N                                    Normalize to the interval B − A.
  CDER (J) = CDER (J) * CON
NEXT J
END SUB
```

A sample program using CHDER is the following:

```
DECLARE FUNCTION FUNC! (X!)
DECLARE SUB CHEBFT (A!, B!, C!(), N!, DUM!)
DECLARE SUB CHDER (A!, B!, C!(), CDER!(), N!)
DECLARE FUNCTION FDER! (X!)
DECLARE FUNCTION CHEBEV! (A!, B!, DUM!, MVAL!, X!)
COMMON SHARED C()

'PROGRAM D5R7
'Driver for routine CHDER
CLS
NVAL = 40
PIO2 = 1.5707963#
DIM C(NVAL), CDER(NVAL)
A = -PIO2
B = PIO2
CALL CHEBFT(A, B, C(), NVAL, DUM)
'Test derivative
DO
   PRINT "How many terms in Chebyshev evaluation?"
   PRINT "Enter n between 6 and"; STR$(NVAL); ". Enter n=0 to end."
   LINE INPUT MVAL$
   MVAL = VAL(MVAL$)
   IF MVAL <= 0 OR MVAL > NVAL THEN EXIT DO
   CALL CHDER(A, B, C(), CDER(), MVAL)
   PRINT "        X          Actual    Cheby. Deriv."
   FOR I = 1 TO NVAL
     C(I) = CDER(I)
   NEXT I
   FOR I = -8 TO 8 STEP 1
     X = I * PIO2 / 10!
     PRINT USING "#####.######"; X; FDER(X); CHEBEV(A, B, DUM, MVAL, X)
   NEXT I
LOOP
END

FUNCTION FDER (X)
'Derivative of FUNC
FDER = 4! * X * ((X ^ 2) - 1!) * SIN(X) + (X ^ 2) * (X ^ 2 - 2!) * COS(X)
END FUNCTION

FUNCTION FUNC (X)
FUNC = (X ^ 2) * (X ^ 2 - 2!) * SIN(X)
END FUNCTION
```

The final two programs of this chapter turn the coefficients of a Chebyshev approximation into those of a polynomial approximation in the variable

$$y = \frac{x - \frac{1}{2}(B + A)}{\frac{1}{2}(B - A)}$$

(routine CHEBPC), or of a polynomial approximation in x itself (routine CHEBPC followed by PCSHFT). These procedures are discouraged for reasons discussed in *Numerical Recipes*, but should they serve some special purpose for you, we have at least warned that you will be sacrificing accuracy, particularly for polynomials above order 7 or 8. Sample program D5R8 calls CHEBFT and CHEBPC to find polynomial coefficients in y for a truncated series. For a set of x values between $-\pi$ and π it calculates y and then the terms of the y-polynomial, which are summed in variable POLY. Finally, POLY is compared to the true function value. (The function FUNC is the same used before.)

Here is the recipe CHEBPC:

SUB CHEBPC (C(), D(), N)

> Chebyshev polynomial coefficients. Given a coefficient array C of length N, this routine generates a coefficient array D of length N such that $\sum_{k=1}^{N} D_k y^{k-1} = \sum_{k=1}^{N} c_k T_{k-1}(y)$. The method is Clenshaw's recurrence (5.6.11), but now applied algebraically rather than arithmetically.

```
DIM DD(N)
FOR J = 1 TO N
  D(J) = 0!
  DD(J) = 0!
NEXT J
D(1) = C(N)
FOR J = N - 1 TO 2 STEP -1
  FOR K = N - J + 1 TO 2 STEP -1
    SV = D(K)
    D(K) = 2! * D(K - 1) - DD(K)
    DD(K) = SV
  NEXT K
  SV = D(1)
  D(1) = -DD(1) + C(J)
  DD(1) = SV
NEXT J
FOR J = N TO 2 STEP -1
  D(J) = D(J - 1) - DD(J)
NEXT J
D(1) = -DD(1) + .5 * C(1)
ERASE DD
END SUB
```

A sample program using CHEBPC is the following:

```
DECLARE FUNCTION FUNC! (X!)
DECLARE SUB CHEBFT (A!, B!, C!(), N!, DUM!)
DECLARE SUB CHEBPC (C!(), D!(), N!)

'PROGRAM D5R8
'Driver for routine CHEBPC
CLS
NVAL = 40
PIO2 = 1.5707963#
DIM C(NVAL), D(NVAL)
A = -PIO2
B = PIO2
CALL CHEBFT(A, B, C(), NVAL, DUM)
DO
  PRINT "How many terms in Chebyshev evaluation?"
```

```
PRINT "Enter n between 6 and"; STR$(NVAL); ". Enter n=0 to end."
LINE INPUT MVAL$
MVAL = VAL(MVAL$)
IF MVAL <= 0 OR MVAL > NVAL THEN EXIT DO
CALL CHEBPC(C(), D(), MVAL)
'Test polynomial
PRINT "        X        Actual     Polynomial"
FOR I = -8 TO 8 STEP 1
  X = I * PIO2 / 10!
  Y = (X - (.5 * (B + A))) / (.5 * (B - A))
  POLY = D(MVAL)
  FOR J = MVAL - 1 TO 1 STEP -1
    POLY = POLY * Y + D(J)
  NEXT J
  PRINT USING "#####.######"; X; FUNC(X); POLY
NEXT I
LOOP
END


FUNCTION FUNC (X)
FUNC = (X ^ 2) * (X ^ 2 - 2!) * SIN(X)
END FUNCTION
```

PCSHFT shifts the polynomial to be one in variable x. Sample program D5R9 is like the previous program except that it follows the call to CHEBPC with a call to PCSHFT.

Here is the recipe PCSHFT:

```
SUB PCSHFT (A, B, D(), N)
```
Polynomial coefficient shift. Given a coefficient array D of length N, this routine generates a coefficient array g of length N such that $\sum_{k=1}^{N} D_k y^{k-1} = \sum_{k=1}^{N} g_k x^{k-1}$, where x and y are related by (5.6.10), i.e. the interval $-1 < y < 1$ is mapped to the interval $A < x < B$. The array g is returned in D.
```
CONSQ = 2! / (B - A)
FAC = CONSQ
FOR J = 2 TO N                 First we rescale by the factor CONSQ...
  D(J) = D(J) * FAC
  FAC = FAC * CONSQ
NEXT J
CONSQ = .5 * (A + B)           ...which is then redefined as the desired shift.
FOR J = 1 TO N - 1             We accomplish the shift by synthetic division. Synthetic
  FOR K = N - 1 TO J STEP -1   division is a miracle of high-school algebra. If you never
    D(K) = D(K) - CONSQ * D(K + 1)   learned it, go do so. You won't be sorry.
  NEXT K
NEXT J
END SUB
```

A sample program using `PCSHFT` is the following:

```
DECLARE FUNCTION FUNC! (X!)
DECLARE SUB CHEBFT (A!, B!, C!(), N!, DUM!)
DECLARE SUB CHEBPC (C!(), D!(), N!)
DECLARE SUB PCSHFT (A!, B!, D!(), N!)

'PROGRAM D5R9
'Driver for routine PCSHFT
CLS
NVAL = 40
PIO2 = 1.5707963#
DIM C(NVAL), D(NVAL)
A = -PIO2
B = PIO2
CALL CHEBFT(A, B, C(), NVAL, DUM)
DO
  PRINT "How many terms in Chebyshev evaluation?"
  PRINT "Enter n between 6 and"; STR$(NVAL); ". Enter n=0 to end."
  LINE INPUT MVAL$
  MVAL = VAL(MVAL$)
  IF MVAL <= 0 OR MVAL > NVAL THEN EXIT DO
  CALL CHEBPC(C(), D(), MVAL)
  CALL PCSHFT(A, B, D(), MVAL)
  'Test shifted polynomial
  PRINT "         X         Actual     Polynomial"
  FOR I = -8 TO 8
    X = I * PIO2 / 10!
    POLY = D(MVAL)
    FOR J = MVAL - 1 TO 1 STEP -1
      POLY = POLY * X + D(J)
    NEXT J
    PRINT USING "#####.######"; X; FUNC(X); POLY
  NEXT I
LOOP
END

FUNCTION FUNC (X)
FUNC = (X ^ 2) * (X ^ 2 - 2!) * SIN(X)
END FUNCTION
```

Chapter 6: Special Functions

This chapter on special functions provides illustrations of techniques developed in Chapter 5. At the same time, it offers routines for calculating many of the functions that arise frequently in analytical work, but which are not so common as to be included, for example, as a single keystroke on your pocket calculator. In terms of demonstration programs, they represent a simple bunch. The test routines are all virtually identical, all making reference to a single file of function values called FNCVAL.DAT which is listed in the Appendix at the end of this chapter. In this file are accurate values for the individual functions for a variety of values for each argument. We have aimed to "stress" the routines a bit by throwing in some extreme values for the arguments.

Many of the function values came from Abramowitz and Stegun's *Handbook of Mathematical Functions*. Some others, however, came from our library of dusty volumes from past masters. There is an implicit danger in a comparison test like this—namely, that our source has used the same algorithms as ours to construct the tables. In that case, we test only our mutual competence at computing, not the correctness of the result. Nevertheless, there is some assurance in knowing that the values we calculate are the ones that have been used and scrutinized for many years. Moreover, the expressions for the functions themselves can be worked out in certain special or limiting cases without computer aid, and in these instances the results have proven correct.

★ ★ ★ ★

With few exceptions, the routines that follow work in this fashion:
1. Open file FNCVAL.DAT.
2. Find the appropriate data table according to its title.
3. Read the argument list for each table entry and pass them to the routine to be tested.
4. Print the arguments along with the expected and actual results.

For the routines in this list, therefore, we forgo any further comment, but simply identify them by the special function that they evaluate.

Natural logarithm of the gamma function for positive arguments:

Here is the recipe GAMMLN:

```
DECLARE FUNCTION GAMMLN! (XX!)
DECLARE FUNCTION RAN1! (IDUM&)
DATA 76.18009173D0,-86.50532033D0,24.01409822D0
DATA -1.231739516D0,.120858003D-2,-.536382D-5,2.50662827465D0
DATA 0.5D0,1.0D0,5.5D0

FUNCTION GAMMLN (XX)
```

Returns the value $\ln[\Gamma(\mathbf{XX})]$ for $\mathbf{XX} > 0$. Full accuracy is obtained for $\mathbf{XX} > 1$. For $0 < \mathbf{XX} < 1$, the reflection formula (6.1.4) can be used first.

```
DIM COF#(6)
```

Internal arithmetic will be done in double precision, a nicety that you can omit if five-figure accuracy is good enough.

```
RESTORE
FOR J = 1 TO 6
  READ COF#(J)
NEXT J
READ STP#
READ HALF#, ONE#, FPF#
X# = XX - ONE#
TMP# = X# + FPF#
TMP# = (X# + HALF#) * LOG(TMP#) - TMP#
SER# = ONE#
FOR J = 1 TO 6
  X# = X# + ONE#
  SER# = SER# + COF#(J) / X#
NEXT J
GAMMLN = TMP# + LOG(STP# * SER#)
ERASE COF#
END FUNCTION
```

A sample program using GAMMLN is the following:

```
DECLARE FUNCTION GAMMLN! (X!)

'PROGRAM D6R1
'Driver for routine GAMMLN
CLS
PI = 3.1415926#
OPEN "FNCVAL.DAT" FOR INPUT AS #1
DO
  LINE INPUT #1, TEXT$
LOOP WHILE TEXT$ <> "Gamma Function"
LINE INPUT #1, NVAL$
NVAL = VAL(NVAL$)
PRINT "Log of gamma function:"
PRINT "        X              Actual            GAMMLN(X)"
FOR I = 1 TO NVAL
  LINE INPUT #1, DUM$
  X = VAL(MID$(DUM$, 1, 4))
  ACTUAL = VAL(MID$(DUM$, 8))
  IF X > 0! THEN
    IF X >= 1! THEN
      CALC = GAMMLN(X)
    ELSE
      CALC = GAMMLN(X + 1!) - LOG(X)
    END IF
    PRINT USING "########.##"; X;
    PRINT USING "###########.######"; LOG(ACTUAL); CALC
  END IF
NEXT I
CLOSE #1
```

END

Factorial function N!:

Here is the recipe **FACTRL**:

```
DECLARE FUNCTION FACTRL! (N!)
DECLARE FUNCTION GAMMLN! (X!)

FUNCTION FACTRL (N)
        Returns the value N! as a floating point number.
DIM A(33)                          Table to be filled in only as required
NTOP = 0
A(1) = 1!                          Table initialized with 0! only.
IF N < 0 THEN
  PRINT "negative factorial"
  EXIT FUNCTION
ELSEIF N <= NTOP THEN              Already in table.
  FACTRL = A(N + 1)
ELSEIF N <= 32 THEN                Fill in table up to desired value.
  FOR J = NTOP + 1 TO N
    A(J + 1) = J * A(J)
  NEXT J
  NTOP = N
  FACTRL = A(N + 1)
ELSE                               Larger value than size of table is required.  Actually, this big
  FACTRL = EXP(GAMMLN(N + 1!))     a value is going to overflow on many computers, but no harm
END IF                             in trying.
END FUNCTION
```

A sample program using **FACTRL** is the following:

```
DECLARE FUNCTION FACTRL! (N!)

'PROGRAM D6R2
'Driver for routine FACTRL
CLS
OPEN "FNCVAL.DAT" FOR INPUT AS #1
DO
  LINE INPUT #1, TEXT$
LOOP WHILE TEXT$ <> "N-factorial"
LINE INPUT #1, NVAL$
NVAL = VAL(NVAL$)
PRINT TEXT$
PRINT "    N                  Actual                FACTRL(N)"
FOR I = 1 TO NVAL
  LINE INPUT #1, DUM$
  N = VAL(MID$(DUM$, 1, 2))
  ACTUAL = VAL(MID$(DUM$, 8))
  IF ACTUAL < 1E+10 THEN
    PRINT USING "#####"; N;
    PRINT "          ";
    PRINT USING "############."; ACTUAL;
    PRINT "          ";
```

```
    PRINT USING "###########."; FACTRL(N)
  ELSE
    PRINT USING "#####"; N;
    PRINT "       ";
    PRINT USING "#.#######^^^^"; ACTUAL;
    PRINT "       ";
    PRINT USING "#.#######^^^^"; FACTRL(N)
  END IF
NEXT I
CLOSE #1
END
```

Binomial coefficients:

Here is the recipe BICO:

```
DECLARE FUNCTION BICO! (N!, K!)
DECLARE FUNCTION FACTLN! (N!)

FUNCTION BICO (N, K)
```
Returns the binomial coefficient $\binom{n}{k}$ as a floating-point number.
```
BICO = CLNG(EXP(FACTLN(N) - FACTLN(K) - FACTLN(N - K)))
END FUNCTION
```
The nearest-integer function cleans up round-off error for smaller values of N and K.

A sample program using BICO is the following:

```
DECLARE FUNCTION BICO! (N!, K!)

'PROGRAM D6R3
'Driver for routine BICO
CLS
OPEN "FNCVAL.DAT" FOR INPUT AS #1
DO
  LINE INPUT #1, TEXT$
LOOP WHILE TEXT$ <> "Binomial Coefficients"
LINE INPUT #1, NVAL$
NVAL = VAL(NVAL$)
PRINT TEXT$
PRINT "    N     K      Actual    BICO(N,K)"
FOR I = 1 TO NVAL
  LINE INPUT #1, DUM$
  N = VAL(MID$(DUM$, 1, 4))
  K = VAL(MID$(DUM$, 8, 6))
  BINCO = VAL(MID$(DUM$, 16))
  PRINT USING "#####"; N;
  PRINT " ";
  PRINT USING "#####"; K;
  PRINT USING "##########."; BINCO; BICO(N, K)
NEXT I
CLOSE #1
END
```

Natural logarithm of N!:

Here is the recipe `FACTLN`:

```
DECLARE FUNCTION FACTLN! (N!)
DECLARE FUNCTION GAMMLN! (X!)

FUNCTION FACTLN (N)
      Returns ln(N!).
DIM A(100)
FOR I = 1 TO 100                     Initialize the table to negative values.
  A(I) = -1
NEXT I
IF N < 0 THEN PRINT "negative factorial": EXIT FUNCTION
IF N <= 99 THEN                      In range of the table.
  IF A(N + 1) < 0! THEN A(N + 1) = GAMMLN(N + 1!)    If not already in the table, put
  FACTLN = A(N + 1)                                   it in.
ELSE
  FACTLN = GAMMLN(N + 1!)           Out of range of the table.
END IF
ERASE A
END FUNCTION
```

A sample program using `FACTLN` is the following:

```
DECLARE FUNCTION FACTLN! (N!)

'PROGRAM D6R4
'Driver for routine FACTLN
CLS
OPEN "FNCVAL.DAT" FOR INPUT AS #1
DO
   LINE INPUT #1, TEXT$
LOOP WHILE TEXT$ <> "N-factorial"
LINE INPUT #1, NVAL$
NVAL = VAL(NVAL$)
PRINT "Log of N-factorial"
PRINT "    N             Actual          FACTLN(N)"
FOR I = 1 TO NVAL
   LINE INPUT #1, DUM$
   N = VAL(MID$(DUM$, 1, 6))
   VALUE = VAL(MID$(DUM$, 8))
   PRINT USING "#####"; N;
   PRINT USING "##########.######"; LOG(VALUE); FACTLN(N)
NEXT I
CLOSE #1
END
```

Beta function:

Here is the recipe BETA:

```
DECLARE FUNCTION BETA! (Z!, W!)
DECLARE FUNCTION GAMMLN! (X!)

FUNCTION BETA (Z, W)
```
Returns the value of the beta function $B(z,w)$.
```
BETA = EXP(GAMMLN(Z) + GAMMLN(W) - GAMMLN(Z + W))
END FUNCTION
```

A sample program using BETA is the following:

```
DECLARE FUNCTION BETA! (Z!, W!)

'PROGRAM D6R5
'Driver for routine BETA
CLS
OPEN "FNCVAL.DAT" FOR INPUT AS #1
DO
   LINE INPUT #1, TEXT$
LOOP WHILE TEXT$ <> "Beta Function"
LINE INPUT #1, NVAL$
NVAL = VAL(NVAL$)
PRINT TEXT$
PRINT "   W      Z              Actual              BETA(W,Z)"
FOR I = 1 TO NVAL
   LINE INPUT #1, DUM$
   W = VAL(MID$(DUM$, 1, 6))
   Z = VAL(MID$(DUM$, 7, 6))
   VALUE = VAL(MID$(DUM$, 13))
   PRINT USING "##.##"; W;
   PRINT USING "###.##"; Z;
   PRINT "        ";
   PRINT USING "#.######^^^^"; VALUE;
   PRINT "        ";
   PRINT USING "#.######^^^^"; BETA(W, Z)
NEXT I
CLOSE #1
END
```

Incomplete gamma function $P(a,x)$:

Here is the recipe GAMMP:

```
DECLARE SUB GCF (GAMMCF!, A!, X!, GLN!)
DECLARE FUNCTION GAMMP! (A!, X!)
DECLARE SUB GSER (G!, A!, X!, L!)
```

```
FUNCTION GAMMP (A, X)
```
        Returns the incomplete gamma function $P(a,x)$
```
IF X < 0! OR A <= 0! THEN PRINT "Abnormal exit": EXIT FUNCTION
IF X < A + 1! THEN                Use the series representation.
  CALL GSER(GAMSER, A, X, GLN)
  GAMMP = GAMSER
ELSE                             Use the continued fraction representation
  CALL GCF(GAMMCF, A, X, GLN)
  GAMMP = 1! - GAMMCF            and take its complement.
END IF
END FUNCTION
```

A sample program using GAMMP is the following:

```
DECLARE FUNCTION GAMMP! (A!, X!)

'PROGRAM D6R6
'Driver for routine GAMMP
CLS
OPEN "FNCVAL.DAT" FOR INPUT AS #1
DO
  LINE INPUT #1, TEXT$
LOOP WHILE TEXT$ <> "Incomplete Gamma Function"
LINE INPUT #1, NVAL$
NVAL = VAL(NVAL$)
PRINT TEXT$
PRINT "    A          X        Actual     GAMMP(A,X)"
FOR I = 1 TO NVAL
  LINE INPUT #1, DUM$
  A = VAL(MID$(DUM$, 1, 6))
  X = VAL(MID$(DUM$, 7, 16))
  VALUE = VAL(MID$(DUM$, 23))
  PRINT USING "###.##"; A;
  PRINT USING "#####.######"; X;
  PRINT USING "#####.######"; VALUE;
  PRINT USING "#####.######"; GAMMP(A, X)
NEXT I
CLOSE #1
END
```

Incomplete gamma function $Q(a,x) = 1 - P(a,x)$:

Here is the recipe GAMMQ:

```
DECLARE FUNCTION GAMMQ! (A!, X!)
DECLARE SUB GSER (GAMSER!, A!, X!, GLN!)
DECLARE SUB GCF (GAMMCF!, A!, X!, GLN!)

FUNCTION GAMMQ (A, X)
     Returns the incomplete gamma function Q(a,x) ≡ 1 − P(a,x).
IF X < 0! OR A <= 0! THEN PRINT "Abnormal exit": EXIT FUNCTION
IF X < A + 1! THEN                    Use the series representation
   CALL GSER(GAMSER, A, X, GLN)
   GAMMQ = 1! - GAMSER                and take its complement.
ELSE                                  Use the continued fraction representation.
   CALL GCF(GAMMCF, A, X, GLN)
   GAMMQ = GAMMCF
END IF
END FUNCTION
```

A sample program using GAMMQ is the following:

```
DECLARE FUNCTION GAMMQ! (A!, X!)

'PROGRAM D6R7
'Driver for routine GAMMQ
CLS
OPEN "FNCVAL.DAT" FOR INPUT AS #1
DO
  LINE INPUT #1, TEXT$
LOOP WHILE TEXT$ <> "Incomplete Gamma Function"
LINE INPUT #1, NVAL$
NVAL = VAL(NVAL$)
PRINT TEXT$
PRINT "    A            X        Actual      GAMMQ(A,X)"
FOR I = 1 TO NVAL
  LINE INPUT #1, DUM$
  A = VAL(MID$(DUM$, 1, 6))
  X = VAL(MID$(DUM$, 7, 16))
  VALUE = VAL(MID$(DUM$, 23))
  PRINT USING "####.##"; A;
  PRINT USING "#####.######"; X;
  PRINT USING "#####.######"; 1! - VALUE;
  PRINT USING "#####.######"; GAMMQ(A, X)
NEXT I
CLOSE #1
END
```

Incomplete gamma function $P(a,x)$ evaluated from series representation:

Here is the recipe GSER:

```
DECLARE FUNCTION GAMMLN! (X!)

SUB GSER (GAMSER, A, X, GLN)
```
Returns the incomplete gamma function $P(a, x)$ evaluated by its series representation as GAMSER. Also returns $\ln \Gamma(a)$ as GLN.
```
ITMAX = 100
EPS = .0000003
GLN = GAMMLN(A)
IF X <= 0! THEN
  IF X < 0! THEN PRINT "Abnormal exit": EXIT SUB
  GAMSER = 0!
  EXIT SUB
END IF
AP = A
SUM = 1! / A
DEL = SUM
FOR N = 1 TO ITMAX
  AP = AP + 1!
  DEL = DEL * X / AP
  SUM = SUM + DEL
  IF ABS(DEL) < ABS(SUM) * EPS THEN EXIT FOR
NEXT N
IF ABS(DEL) >= ABS(SUM) * EPS THEN
  PRINT "A too large, ITMAX too small"
  EXIT SUB
END IF
GAMSER = SUM * EXP(-X + A * LOG(X) - GLN)
END SUB
```

A sample program using GSER is the following:

```
DECLARE SUB GSER (GAMSER!, A!, X!, GLN!)
DECLARE FUNCTION GAMMLN! (X!)

'PROGRAM D6R8
'Driver for routine GSER
CLS
OPEN "FNCVAL.DAT" FOR INPUT AS #1
DO
  LINE INPUT #1, TEXT$
LOOP WHILE TEXT$ <> "Incomplete Gamma Function"
LINE INPUT #1, NVAL$
NVAL = VAL(NVAL$)
PRINT TEXT$
PRINT "    A          X         Actual     GSER(A,X)   GAMMLN(A)        GLN"
FOR I = 1 TO NVAL
  LINE INPUT #1, DUM$
  A = VAL(MID$(DUM$, 1, 6))
  X = VAL(MID$(DUM$, 7, 16))
  VALUE = VAL(MID$(DUM$, 23))
  CALL GSER(GAMSER, A, X, GLN)
  PRINT USING "####.##"; A;
  PRINT USING "#####.######"; X;
  PRINT USING "#####.######"; VALUE;
```

```
PRINT USING "#####.######"; GAMSER;
PRINT USING "#####.######"; GAMMLN(A);
PRINT USING "#####.######"; GLN
NEXT I
CLOSE #1
END
```

Incomplete gamma function $Q(a, x)$ evaluated by continued fraction representation:

Here is the recipe GCF:

```
DECLARE FUNCTION GAMMLN! (X!)

SUB GCF (GAMMCF, A, X, GLN)
```
Returns the incomplete gamma function $Q(a, x)$ evaluated by its continued fraction representation as GAMMCF. Also returns $\ln \Gamma(a)$ as GLN.
```
ITMAX = 100
EPS = .0000003
GLN = GAMMLN(A)
GOLD = 0!                              This is the previous value, tested against for convergence.
A0 = 1!
A1 = X                                 We are here setting up the A's and B's of equation (5.2.4)
B0 = 0!                                for evaluating the continued fraction.
B1 = 1!
FAC = 1!                               FAC is the renormalization factor for preventing overflow of
FOR N = 1 TO ITMAX                     the partial numerators and denominators.
  AN = CSNG(N)
  ANA = AN - A
  A0 = (A1 + A0 * ANA) * FAC           One step of the recurrence (5.2.5).
  B0 = (B1 + B0 * ANA) * FAC
  ANF = AN * FAC
  A1 = X * A0 + ANF * A1               The next step of the recurrence (5.2.5).
  B1 = X * B0 + ANF * B1
  IF A1 <> 0! THEN                     Shall we renormalize?
    FAC = 1! / A1                      Yes. Set FAC so it happens.
    G = B1 * FAC                       New value of answer.
    IF ABS((G - GOLD) / G) < EPS THEN EXIT FOR    Converged? If so, exit.
    GOLD = G                           If not, save value.
  END IF
NEXT N
IF ABS((G - GOLD) / G) >= EPS THEN PRINT "A too large, ITMAX too small": EXIT
SUB
IF -X + A * LOG(X) - GLN < -100 THEN
  GAMMCF = 0!
ELSE
  GAMMCF = EXP(-X + A * LOG(X) - GLN) * G        Put factors in front.
END IF
END SUB
```

A sample program using **GCF** is the following:

```
DECLARE FUNCTION GAMMLN! (X!)
DECLARE SUB GCF (G!, A!, X!, L!)

'PROGRAM D6R9
'Driver for routine GCF
CLS
OPEN "FNCVAL.DAT" FOR INPUT AS #1
DO
  LINE INPUT #1, TEXT$
LOOP WHILE TEXT$ <> "Incomplete Gamma Function"
LINE INPUT #1, NVAL$
NVAL = VAL(NVAL$)
PRINT TEXT$
PRINT "    A            X         Actual     GCF(A,X)     GAMMLN(A)        GLN"
FOR I = 1 TO NVAL
  LINE INPUT #1, DUM$
  A = VAL(MID$(DUM$, 1, 6))
  X = VAL(MID$(DUM$, 7, 16))
  VALUE = VAL(MID$(DUM$, 23))
  IF X >= A + 1! THEN
    CALL GCF(GAMMCF, A, X, GLN)
    PRINT USING "###.##"; A;
    PRINT USING "#####.######"; X;
    PRINT USING "#####.######"; 1! - VALUE;
    PRINT USING "#####.######"; GAMMCF;
    PRINT USING "#####.######"; GAMMLN(A);
    PRINT USING "#####.######"; GLN
  END IF
NEXT I
CLOSE #1
END
```

Error function:

Here is the recipe **ERF**:

```
DECLARE FUNCTION ERF! (X!)
DECLARE FUNCTION GAMMP! (A!, X!)

FUNCTION ERF (X)
      Returns the error function erf(X).
IF X < 0! THEN
  ERF = -GAMMP(.5, X ^ 2)
ELSE
  ERF = GAMMP(.5, X ^ 2)
END IF
END FUNCTION
```

A sample program using ERF is the following:

```
DECLARE FUNCTION ERF! (X!)

'PROGRAM D6R10
'Driver for routine ERF
CLS
OPEN "FNCVAL.DAT" FOR INPUT AS #1
DO
  LINE INPUT #1, TEXT$
LOOP WHILE TEXT$ <> "Error Function"
LINE INPUT #1, NVAL$
NVAL = VAL(NVAL$)
PRINT TEXT$
PRINT "   X         Actual        ERF(X)"
FOR I = 1 TO NVAL
  LINE INPUT #1, DUM$
  X = VAL(MID$(DUM$, 1, 6))
  VALUE = VAL(MID$(DUM$, 7))
  PRINT USING "##.##"; X;
  PRINT USING "####.#######"; VALUE;
  PRINT USING "####.#######"; ERF(X)
NEXT I
CLOSE #1
END
```

Complementary error function:

Here is the recipe ERFC:

```
DECLARE FUNCTION ERFC! (X!)
DECLARE FUNCTION GAMMP! (A!, X!)
DECLARE FUNCTION GAMMQ! (A!, X!)

FUNCTION ERFC (X)
        Returns the complementary error function erfc(X).
IF X < 0! THEN
  ERFC = 1! + GAMMP(.5, X ^ 2)
ELSE
  ERFC = GAMMQ(.5, X ^ 2)
END IF
END FUNCTION
```

A sample program using ERFC is the following:

```
DECLARE FUNCTION ERFC! (X!)

'PROGRAM D6R11
'Driver for routine ERFC
CLS
OPEN "FNCVAL.DAT" FOR INPUT AS #1
DO
   LINE INPUT #1, TEXT$
LOOP WHILE TEXT$ <> "Error Function"
LINE INPUT #1, NVAL$
NVAL = VAL(NVAL$)
PRINT "Complementary error function"
PRINT "   X        Actual        ERFC(X)"
FOR I = 1 TO NVAL
   LINE INPUT #1, DUM$
   X = VAL(MID$(DUM$, 1, 6))
   VALUE = VAL(MID$(DUM$, 7))
   VALUE = 1! - VALUE
   PRINT USING "##.##"; X;
   PRINT USING "####.#######"; VALUE;
   PRINT USING "####.#######"; ERFC(X)
NEXT I
CLOSE #1
END
```

Complementary error function from a Chebyshev fit to a guessed functional form:

Here is the recipe ERFCC:

```
DECLARE FUNCTION ERFCC! (X!)

FUNCTION ERFCC (X)
```
   Returns the complementary error function $\text{erfc}(X)$ with fractional error everywhere less than 1.2×10^{-7}.
```
Z = ABS(X)
T = 1! / (1! + .5 * Z)
DUM = T * (-.82215223# + T * .17087277#)
DUM = T * (-1.13520398# + T * (1.48851587# + DUM))
DUM = T * (.09678418# + T * (-.18628806# + T * (.27886807# + DUM)))
DUM = -Z * Z - 1.26551223# + T * (1.00002368# + T * (.37409196# + DUM))
DUM = T * EXP(DUM)
IF X < 0! THEN
   ERFCC = 2! - DUM
ELSE
   ERFCC = DUM
END IF
END FUNCTION
```

A sample program using **ERFCC** is the following:

```
DECLARE FUNCTION ERFCC! (X!)

'PROGRAM D6R12
'Driver for routine ERFCC
CLS
OPEN "FNCVAL.DAT" FOR INPUT AS #1
DO
  LINE INPUT #1, TEXT$
LOOP WHILE TEXT$ <> "Error Function"
LINE INPUT #1, NVAL$
NVAL = VAL(NVAL$)
PRINT "Complementary error function"
PRINT "   X        Actual      ERFCC(X)"
FOR I = 1 TO NVAL
  LINE INPUT #1, DUM$
  X = VAL(MID$(DUM$, 1, 6))
  VALUE = VAL(MID$(DUM$, 7))
  VALUE = 1! - VALUE
  PRINT USING "##.##"; X;
  PRINT USING "####.#######"; VALUE;
  PRINT USING "####.#######"; ERFCC(X)
NEXT I
CLOSE #1
END
```

Incomplete Beta function:

Here is the recipe **BETAI**:

```
DECLARE FUNCTION BETAI! (A!, B!, X!)
DECLARE FUNCTION BETACF! (A!, B!, X!)
DECLARE FUNCTION GAMMLN! (X!)

FUNCTION BETAI (A, B, X)
      Returns the incomplete beta function I_X(A,B).
IF X < 0! OR X > 1! THEN PRINT "bad argument X in BETAI": EXIT FUNCTION
IF X = 0! OR X = 1! THEN
  BT = 0!
ELSE                                    Factors in front of the continued fraction.
  BT = EXP(GAMMLN(A + B) - GAMMLN(A) - GAMMLN(B) + A * LOG(X) + B * LOG(1! -
X))
END IF
IF X < (A + 1!) / (A + B + 2!) THEN
  BETAI = BT * BETACF(A, B, X) / A          Use continued fraction directly.
ELSE
  BETAI = 1! - BT * BETACF(B, A, 1! - X) / B   Use continued fraction after making the
END IF                                        symmetry transformation.
END FUNCTION
```

A sample program using **BETAI** is the following:

```
DECLARE FUNCTION BETAI! (A!, B!, X!)

'PROGRAM D6R13
'Driver for routine BETAI,BETACF
CLS
OPEN "FNCVAL.DAT" FOR INPUT AS #1
DO
  LINE INPUT #1, TEXT$
LOOP WHILE TEXT$ <> "Incomplete Beta Function"
LINE INPUT #1, NVAL$
NVAL = VAL(NVAL$)
PRINT TEXT$
PRINT "    A          B           X         Actual      BETAI(X)"
FOR I = 1 TO NVAL
  LINE INPUT #1, DUM$
  A = VAL(MID$(DUM$, 1, 7))
  B = VAL(MID$(DUM$, 8, 7))
  X = VAL(MID$(DUM$, 15, 7))
  VALUE = VAL(MID$(DUM$, 22))
  PRINT USING "##.##"; A;
  PRINT USING "#####.######"; B;
  PRINT USING "#####.######"; X;
  PRINT USING "#####.######"; VALUE;
  PRINT USING "#####.######"; BETAI(A, B, X)
NEXT I
CLOSE #1
END
```

Here is the recipe **BETACF**:

```
DECLARE FUNCTION BETACF! (A!, B!, X!)

FUNCTION BETACF (A, B, X)
```
        Continued fraction for incomplete beta function, used by **BETAI**.
```
ITMAX = 100
EPS = .0000003
AM = 1!
BM = 1!
AZ = 1!
QAB = A + B                    These Q's will be used in factors that occur in the coefficients
QAP = A + 1!                   (6.3.6).
QAM = A - 1!
BZ = 1! - QAB * X / QAP
FOR M = 1 TO ITMAX             Continued fraction evaluation by the recurrence method (5.2.5).
  EM = INT(M)
  TEM = EM + EM
  D = EM * (B - M) * X / ((QAM + TEM) * (A + TEM))
  AP = AZ + D * AM             One step (the even one) of the recurrence.
  BP = BZ + D * BM
  D = -(A + EM) * (QAB + EM) * X / ((A + TEM) * (QAP + TEM))
  APP = AP + D * AZ            Next step of the recurrence (the odd one).
  BPP = BP + D * BZ
```

```
AOLD = AZ                              Save the old answer.
AM = AP / BPP                          Renormalize to prevent overflows.
BM = BP / BPP
AZ = APP / BPP
BZ = 1!
IF ABS(AZ - AOLD) < EPS * ABS(AZ) THEN EXIT FOR     Are we done?
NEXT M
IF ABS(AZ - AOLD) >= EPS * ABS(AZ) THEN
  PRINT "A or B too big, or ITMAX too small"
ELSE
  BETACF = AZ
END IF
END FUNCTION
```

Bessel function J_0:
  Here is the recipe BESSJ0:

```
DECLARE FUNCTION BESSJ0! (X!)
DATA 1.0D0,-0.1098628627D-2,0.2734510407D-4,-0.2073370639D-5
DATA 0.2093887211D-6,-0.1562499995D-1,0.1430488765D-3
DATA -0.6911147651D-5,0.7621095161D-6,-0.934945152D-7
DATA 57568490574.0D0,-13362590354.0D0,651619640.7D0
DATA -11214424.18D0,77392.33017D0,-184.9052456D0
DATA 57568490411.0D0,1029532985.0D0,9494680.718D0,59272.64853D0
DATA 267.8532712D0,1.D0

FUNCTION BESSJ0 (X)
        Returns the Bessel function $J_0(X)$ for any real X.
RESTORE
READ P1#, P2#, P3#, P4#, P5#           We'll accumulate polynomials in double precision.
READ Q1#, Q2#, Q3#, Q4#, Q5#
READ R1#, R2#, R3#, R4#, R5#, R6#
READ S1#, S2#, S3#, S4#, S5#, S6#
IF ABS(X) < 8! THEN                    Direct rational function fit.
  Y# = X ^ 2
  DUM# = R1# + Y# * (R2# + Y# * (R3# + Y# * (R4# + Y# * (R5# + Y# * R6#))))
  BESSJ0 = DUM# / (S1# + Y# * (S2# + Y# * (S3# + Y# * (S4# + Y# * (S5# + Y#
* S6#)))))
ELSE                                   Fitting function (6.4.9).
  AX = ABS(X)
  Z = 8! / AX
  Y# = Z ^ 2
  XX = AX - .785398164#
  DUM# = COS(XX) * (P1# + Y# * (P2# + Y# * (P3# + Y# * (P4# + Y# * P5#))))
  DUM# = DUM# - Z * SIN(XX) * (Q1# + Y# * (Q2# + Y# * (Q3# + Y# * (Q4# + Y#
* Q5#))))
  BESSJ0 = SQR(.636619772# / AX) * DUM#
END IF
END FUNCTION
```

A sample program using BESSJ0 is the following:

```
DECLARE FUNCTION BESSJ0! (X!)

'PROGRAM D6R15
'Driver for routine BESSJ0
CLS
OPEN "FNCVAL.DAT" FOR INPUT AS #1
DO
  LINE INPUT #1, TEXT$
LOOP WHILE TEXT$ <> "Bessel Function J0"
LINE INPUT #1, NVAL$
NVAL = VAL(NVAL$)
PRINT TEXT$
PRINT "    X       Actual    BESSJ0(X)"
FOR I = 1 TO NVAL
  LINE INPUT #1, DUM$
  X = VAL(MID$(DUM$, 1, 6))
  VALUE = VAL(MID$(DUM$, 7))
  PRINT USING "##.##"; X;
  PRINT USING "####.#######"; VALUE;
  PRINT USING "####.#######"; BESSJ0(X)
NEXT I
CLOSE #1
END
```

Bessel function Y_0:

Here is the recipe BESSY0:

```
DECLARE FUNCTION BESSY0! (X!)
DECLARE FUNCTION BESSJ0! (X!)
DATA 1.0D0,-0.1098628627D-2,0.2734510407D-4,-0.2073370639D-5
DATA 0.2093887211D-6,-0.1562499995D-1,0.1430488765D-3
DATA -0.6911147651D-5,0.7621095161D-6,-0.934945152D-7
DATA -2957821389.0D0,7062834065.0D0,-512359803.6D0,10879881.29D0
DATA -86327.92757D0,228.4622733D0
DATA 40076544269.0D0,745249964.8D0,7189466.438D0,47447.26470D0
DATA 226.1030244D0,1.D0

FUNCTION BESSY0 (X)
        Returns the Bessel function $Y_0(X)$ for positive X.
RESTORE
READ P1#, P2#, P3#, P4#, P5#        We'll accumulate polynomials in double precision.
READ Q1#, Q2#, Q3#, Q4#, Q5#
READ R1#, R2#, R3#, R4#, R5#, R6#
READ S1#, S2#, S3#, S4#, S5#, S6#
IF X < 8! THEN                      Rational function approximation of (6.4.8).
  Y# = X ^ 2
  DUM# = R1# + Y# * (R2# + Y# * (R3# + Y# * (R4# + Y# * (R5# + Y# * R6#))))
  DUM# = DUM# / (S1# + Y# * (S2# + Y# * (S3# + Y# * (S4# + Y# * (S5# + Y# *
S6#)))))
  BESSY0 = DUM# + .636619772# * BESSJ0(X) * LOG(X)
ELSE                                Fitting function 6.4.10.
```

```
Z = 8! / X
Y# = Z ^ 2
XX = X - .785398164#
DUM# = SIN(XX) * (P1# + Y# * (P2# + Y# * (P3# + Y# * (P4# + Y# * P5#))))
DUM# = DUM# + Z * COS(XX) * (Q1# + Y# * (Q2# + Y# * (Q3# + Y# * (Q4# + Y#
* Q5#))))
BESSYO = SQR(.636619772# / X) * DUM#
END IF
END FUNCTION
```

A sample program using BESSY0 is the following:

```
DECLARE FUNCTION BESSY0! (X!)

'PROGRAM D6R16
'Driver for routine BESSY0
CLS
OPEN "FNCVAL.DAT" FOR INPUT AS #1
DO
  LINE INPUT #1, TEXT$
LOOP WHILE TEXT$ <> "Bessel Function Y0"
LINE INPUT #1, NVAL$
NVAL = VAL(NVAL$)
PRINT TEXT$
PRINT "   X       Actual      BESSY0(X)"
FOR I = 1 TO NVAL
  LINE INPUT #1, DUM$
  X = VAL(MID$(DUM$, 1, 6))
  VALUE = VAL(MID$(DUM$, 7))
  PRINT USING "##.##"; X;
  PRINT USING "####.#######"; VALUE;
  PRINT USING "####.#######"; BESSY0(X)
NEXT I
CLOSE #1
END
```

Bessel function J_1:

Here is the recipe BESSJ1:

```
DECLARE FUNCTION BESSJ1! (X!)
DATA 72362614232.0D0,-7895059235.0D0,242396853.1D0,-2972611.439D0
DATA 15704.48260D0,-30.16036606D0
DATA 144725228442.0D0,2300535178.0D0,18583304.74D0,99447.43394D0
DATA 376.9991397D0,1.0D0
DATA 1.0D0,0.183105D-2,-0.3516396496D-4,0.2457520174D-5,-0.240337019D-6
DATA 0.04687499995D0,-0.2002690873D-3,0.8449199096D-5,-0.88228987D-6
DATA 0.105787412D-6

FUNCTION BESSJ1 (X)
      Returns the Bessel function $J_1(X)$ for any real X.
RESTORE
READ R1#, R2#, R3#, R4#, R5#, R6#     We'll accumulate polynomials in double precision.
READ S1#, S2#, S3#, S4#, S5#, S6#
```

```
READ P1#, P2#, P3#, P4#, P5#
READ Q1#, Q2#, Q3#, Q4#, Q5#
IF ABS(X) < 8! THEN              Direct rational approximation.
  Y# = X ^ 2
  NUM# = R1# + Y# * (R2# + Y# * (R3# + Y# * (R4# + Y# * (R5# + Y# * R6#))))
  DUM# = S1# + Y# * (S2# + Y# * (S3# + Y# * (S4# + Y# * (S5# + Y# * S6#))))
  BESSJ1 = X * NUM# / DUM#
ELSE                             Fitting function (6.4.9).
  AX = ABS(X)
  Z = 8! / AX
  Y# = Z ^ 2
  XX = AX - 2.356194491#
  DUM# = COS(XX) * (P1# + Y# * (P2# + Y# * (P3# + Y# * (P4# + Y# * P5#))))
  DUM# = DUM# - Z * SIN(XX) * (Q1# + Y# * (Q2# + Y# * (Q3# + Y# * (Q4# + Y#
* Q5#))))
  DUM# = DUM# * SGN(X)
  BESSJ1 = SQR(.636619772# / AX) * DUM#
END IF
END FUNCTION
```

A sample program using BESSJ1 is the following:

```
DECLARE FUNCTION BESSJ1! (X!)

'PROGRAM D6R17
'Driver for routine BESSJ1
CLS
OPEN "FNCVAL.DAT" FOR INPUT AS #1
DO
  LINE INPUT #1, TEXT$
LOOP WHILE TEXT$ <> "Bessel Function J1"
LINE INPUT #1, NVAL$
NVAL = VAL(NVAL$)
PRINT TEXT$
PRINT "   X        Actual      BESSJ1(X)"
FOR I = 1 TO NVAL
  LINE INPUT #1, DUM$
  X = VAL(MID$(DUM$, 1, 6))
  VALUE = VAL(MID$(DUM$, 7))
  PRINT USING "##.##"; X;
  PRINT USING "####.#######"; VALUE;
  PRINT USING "####.#######"; BESSJ1(X)
NEXT I
CLOSE #1
END
```

Bessel function Y_1:

Here is the recipe BESSY1:

```
DECLARE FUNCTION BESSY1! (X!)
DECLARE FUNCTION BESSJ1! (X!)
DATA 1.0D0,0.183105D-2,-0.3516396496D-4,0.2457520174D-5,-0.240337019D-6
DATA 0.04687499995D0,-0.2002690873D-3,0.8449199096D-5,-0.88228987D-6
DATA 0.105787412D-6
DATA -0.4900604943D13,0.1275274390D13,-0.5153438139D11,0.7349264551D9
DATA -0.4237922726D7,0.8511937935D4
DATA 0.2499580570D14,0.4244419664D12,0.3733650367D10,0.2245904002D8
DATA 0.1020426050D6,0.3549632885D3,1.0D0


FUNCTION BESSY1 (X)
     Returns the Bessel function Y₁(X) for positive X.
RESTORE
READ P1#, P2#, P3#, P4#, P5#          We'll accumulate polynomials in double precision.
READ Q1#, Q2#, Q3#, Q4#, Q5#
READ R1#, R2#, R3#, R4#, R5#, R6#
READ S1#, S2#, S3#, S4#, S5#, S6#, S7#
IF X < 8! THEN                        Rational function approximation of (6.4.8).
   Y# = X ^ 2
   NUM# = R1# + Y# * (R2# + Y# * (R3# + Y# * (R4# + Y# * (R5# + Y# * R6#))))
   DUM# = S4# + Y# * (S5# + Y# * (S6# + Y# * S7#))
   DUM# = S1# + Y# * (S2# + Y# * (S3# + Y# * DUM#))
   BESSY1 = X * NUM# / DUM# + .636619772# * (BESSJ1(X) * LOG(X) - 1! / X)
ELSE                                  Fitting function (6.4.10).
   Z = 8! / X
   Y# = Z ^ 2
   XX = X - 2.356194491#
   NUM# = SIN(XX) * (P1# + Y# * (P2# + Y# * (P3# + Y# * (P4# + Y# * P5#))))
   DUM# = Q1# + Y# * (Q2# + Y# * (Q3# + Y# * (Q4# + Y# * Q5#)))
   BESSY1 = SQR(.636619772# / X) * (NUM# + Z * COS(XX) * DUM#)
END IF
END FUNCTION
```

A sample program using BESSY1 is the following:

```
DECLARE FUNCTION BESSY1! (X!)

'PROGRAM D6R18
'Driver for routine BESSY1
CLS
OPEN "FNCVAL.DAT" FOR INPUT AS #1
DO
  LINE INPUT #1, TEXT$
LOOP WHILE TEXT$ <> "Bessel Function Y1"
LINE INPUT #1, NVAL$
NVAL = VAL(NVAL$)
PRINT TEXT$
PRINT "    X        Actual    BESSY1(X)"
FOR I = 1 TO NVAL
  LINE INPUT #1, DUM$
  X = VAL(MID$(DUM$, 1, 6))
  VALUE = VAL(MID$(DUM$, 7))
  PRINT USING "##.##"; X;
  PRINT USING "####.#######"; VALUE;
```

```
    PRINT USING "####.#######"; BESSY1(X)
NEXT I
CLOSE #1
END
```

Bessel function Y_n for $n > 1$:

Here is the recipe BESSY:

```
DECLARE FUNCTION BESSY! (N!, X!)
DECLARE FUNCTION BESSYO! (X!)
DECLARE FUNCTION BESSY1! (X!)

FUNCTION BESSY (N, X)
    Returns the Bessel function Y_N(X) for positive X and N≥ 2.
IF N < 2 THEN PRINT "bad argument N in BESSY": EXIT FUNCTION
TOX = 2! / X
BY = BESSY1(X)                      Starting values for the recurrence.
BYM = BESSYO(X)
FOR J = 1 TO N - 1                  Recurrence (6.4.7).
  BYP = J * TOX * BY - BYM
  BYM = BY
  BY = BYP
NEXT J
BESSY = BY
END FUNCTION
```

A sample program using BESSY is the following:

```
DECLARE FUNCTION BESSY! (N!, X!)

'PROGRAM D6R19
'Driver for routine BESSY
CLS
OPEN "FNCVAL.DAT" FOR INPUT AS #1
DO
  LINE INPUT #1, TEXT$
  TEXT$ = LEFT$(TEXT$, 18)
LOOP WHILE TEXT$ <> "Bessel Function Yn"
LINE INPUT #1, NVAL$
NVAL = VAL(NVAL$)
PRINT TEXT$
PRINT "   N      X         Actual       BESSY(N,X)"
FOR I = 1 TO NVAL
  LINE INPUT #1, DUM$
  N = VAL(MID$(DUM$, 1, 6))
  X = VAL(MID$(DUM$, 7, 6))
  VALUE = VAL(MID$(DUM$, 13))
  PRINT USING "####"; N;
  PRINT USING "#####.##"; X;
  PRINT "    ";
  PRINT USING "#.######^^^^"; VALUE;
  PRINT "    ";
  PRINT USING "#.######^^^^"; BESSY(N, X)
```

```
NEXT I
CLOSE #1
END
```

Bessel function J_n for $n > 1$:

Here is the recipe BESSJ:

```
DECLARE FUNCTION BESSJ! (N!, X!)
DECLARE FUNCTION BESSJ0! (X!)
DECLARE FUNCTION BESSJ1! (X!)

FUNCTION BESSJ (N, X)
```
        Returns the Bessel function $J_N(X)$ for any real X and $N \geq 2$.
```
IACC = 40
BIGNO = 1E+10
BIGNI = 1E-10
IF N < 2 THEN PRINT "bad argument N in BESSJ": EXIT FUNCTION
AX = ABS(X)
IF AX = 0! THEN
  BESSJ = 0!
ELSEIF AX > CSNG(N) THEN          Use upward recurrence from J₀ and J₁.
  TOX = 2! / AX
  BJM = BESSJ0(AX)
  BJ = BESSJ1(AX)
  FOR J = 1 TO N - 1
    BJP = J * TOX * BJ - BJM
    BJM = BJ
    BJ = BJP
  NEXT J
  BESSJ = BJ
ELSE                              Use downward recurrence from an
  TOX = 2! / AX                   even value M here computed. Make
  M = 2 * INT((N + INT(SQR(CSNG(IACC * N)))) / 2)
                                  IACC larger to increase accuracy.
  BESJ = 0!
  JSUM = 0                        JSUM will alternate between 0 and 1; when it is 1, we accu-
  SUM = 0!                        mulate in SUM the even terms in (5.4.6).
  BJP = 0!
  BJ = 1!
  FOR J = M TO 1 STEP -1          The downward recurrence.
    BJM = J * TOX * BJ - BJP
    BJP = BJ
    BJ = BJM
    IF ABS(BJ) > BIGNO THEN       Renormalize to prevent overflows.
      BJ = BJ * BIGNI
      BJP = BJP * BIGNI
      BESJ = BESJ * BIGNI
      SUM = SUM * BIGNI
    END IF
    IF JSUM <> 0 THEN SUM = SUM + BJ      Accumulate the sum.
    JSUM = 1 - JSUM               Change 0 to 1 or vice versa.
    IF J = N THEN BESJ = BJP      Save the unnormalized answer.
  NEXT J
  SUM = 2! * SUM - BJ            Compute (5.4.6)
```

```
    BESSJ = BESJ / SUM            and use it to normalize the answer.
END IF
END FUNCTION
```

A sample program using BESSJ is the following:

```
DECLARE FUNCTION BESSJ! (N!, X!)

'PROGRAM D6R20
'Driver for routine BESSJ
CLS
OPEN "FNCVAL.DAT" FOR INPUT AS #1
DO
  LINE INPUT #1, TEXT$
  TEXT$ = LEFT$(TEXT$, 18)
LOOP WHILE TEXT$ <> "Bessel Function Jn"
LINE INPUT #1, NVAL$
NVAL = VAL(NVAL$)
PRINT TEXT$
PRINT "    N       X         Actual        BESSJ(N,X)"
FOR I = 1 TO NVAL
  LINE INPUT #1, DUM$
  N = VAL(MID$(DUM$, 1, 6))
  X = VAL(MID$(DUM$, 7, 6))
  VALUE = VAL(MID$(DUM$, 13))
  PRINT USING "####"; N;
  PRINT USING "#####.##"; X;
  PRINT "   ";
  PRINT USING "#.######^^^^"; VALUE;
  PRINT "   ";
  PRINT USING "#.######^^^^"; BESSJ(N, X)
NEXT I
CLOSE #1
END
```

Bessel function I_0:

Here is the recipe BESSI0:

```
DECLARE FUNCTION BESSI0! (X!)
DATA 1.0D0,3.5156229D0,3.0899424D0,1.2067492D0
DATA 0.2659732D0,0.360768D-1,0.45813D-2
DATA 0.39894228D0,0.1328592D-1,0.225319D-2
DATA -0.157565D-2,0.916281D-2,-0.2057706D-1,0.2635537D-1,-0.1647633D-1
DATA 0.392377D-2

FUNCTION BESSI0 (X)
    Returns the modified Bessel function I₀(X) for any real X.
RESTORE
READ P1#, P2#, P3#, P4#, P5#, P6#, P7#       Accumulate polynomials in double precision.
READ Q1#, Q2#, Q3#, Q4#, Q5#, Q6#, Q7#, Q8#, Q9#
IF ABS(X) < 3.75 THEN
  Y# = (X / 3.75) ^ 2
  DUM# = Y# * (P4# + Y# * (P5# + Y# * (P6# + Y# * P7#)))
```

```
  BESSIO = P1# + Y# * (P2# + Y# * (P3# + DUM#))
ELSE
  AX = ABS(X)
  Y# = 3.75 / AX
  DUM# = Y# * (Q5# + Y# * (Q6# + Y# * (Q7# + Y# * (Q8# + Y# * Q9#))))
  DUM# = Q1# + Y# * (Q2# + Y# * (Q3# + Y# * (Q4# + DUM#)))
  BESSIO = (EXP(AX) / SQR(AX)) * DUM#
END IF
END FUNCTION
```

A sample program using BESSIO is the following:

```
DECLARE FUNCTION BESSI0! (X!)

'PROGRAM D6R21
'Driver for routine BESSI0
CLS
OPEN "FNCVAL.DAT" FOR INPUT AS #1
DO
  LINE INPUT #1, TEXT$
LOOP WHILE TEXT$ <> "Modified Bessel Function I0"
LINE INPUT #1, NVAL$
NVAL = VAL(NVAL$)
PRINT TEXT$
PRINT "   X        Actual        BESSI0(X)"
FOR I = 1 TO NVAL
  LINE INPUT #1, DUM$
  X = VAL(MID$(DUM$, 1, 6))
  VALUE = VAL(MID$(DUM$, 7))
  PRINT USING "##.##"; X;
  PRINT "    ";
  PRINT USING "#.#######^^^^"; VALUE;
  PRINT "    ";
  PRINT USING "#.#######^^^^"; BESSI0(X)
NEXT I
CLOSE #1
END
```

Bessel function K_0:

Here is the recipe BESSK0:

```
DECLARE FUNCTION BESSK0! (X!)
DECLARE FUNCTION BESSI0! (X!)
DATA -0.57721566D0,0.42278420D0,0.23069756D0
DATA 0.3488590D-1,0.262698D-2,0.10750D-3,0.74D-5
DATA 1.25331414D0,-0.7832358D-1,0.2189568D-1
DATA -0.1062446D-1,0.587872D-2,-0.251540D-2,0.53208D-3

FUNCTION BESSK0 (X)
```
    Returns the modified Bessel function $K_0(X)$ for positive real X.
```
RESTORE
READ P1#, P2#, P3#, P4#, P5#, P6#, P7#          Accumulate polynomials in double precision.
READ Q1#, Q2#, Q3#, Q4#, Q5#, Q6#, Q7#
```

```
IF X <= 2! THEN                          Polynomial fit.
  Y# = X * X / 4!
  DUM# = P2# + Y# * (P3# + Y# * (P4# + Y# * (P5# + Y# * (P6# + Y# * P7#))))
  BESSK0 = (-LOG(X / 2!) * BESSI0(X)) + (P1# + Y# * DUM#)
ELSE
  Y# = 2! / X
  DUM# = Q2# + Y# * (Q3# + Y# * (Q4# + Y# * (Q5# + Y# * (Q6# + Y# * Q7#))))
  BESSK0 = (EXP(-X) / SQR(X)) * (Q1# + Y# * DUM#)
END IF
END FUNCTION
```

A sample program using BESSK0 is the following:

```
DECLARE FUNCTION BESSK0! (X!)

'PROGRAM D6R22
'Driver for routine BESSK0
CLS
OPEN "FNCVAL.DAT" FOR INPUT AS #1
DO
  LINE INPUT #1, TEXT$
LOOP WHILE TEXT$ <> "Modified Bessel Function K0"
LINE INPUT #1, NVAL$
NVAL = VAL(NVAL$)
PRINT TEXT$
PRINT "   X          Actual          BESSK0(X)"
FOR I = 1 TO NVAL
  LINE INPUT #1, DUM$
  X = VAL(MID$(DUM$, 1, 6))
  VALUE = VAL(MID$(DUM$, 7))
  PRINT USING "##.##"; X;
  PRINT "   ";
  PRINT USING "#.#######^^^^"; VALUE;
  PRINT "   ";
  PRINT USING "#.#######^^^^"; BESSK0(X)
NEXT I
CLOSE #1
END
```

Bessel function I_1:

Here is the recipe BESSI1:

```
DECLARE FUNCTION BESSI1! (X!)
DATA 0.5D0,0.87890594D0,0.51498869D0,0.15084934D0
DATA 0.2658733D-1,0.301532D-2,0.32411D-3
DATA 0.39894228D0,-0.3988024D-1,-0.362018D-2
DATA 0.163801D-2,-0.1031555D-1,0.2282967D-1,-0.2895312D-1,0.1787654D-1
DATA -0.420059D-2
```

```
FUNCTION BESSI1 (X)
```
      Returns the modified Bessel function $I_1(X)$ for any real X.
```
RESTORE
READ P1#, P2#, P3#, P4#, P5#, P6#, P7#         Accumulate polynomials in double precision.
```

```
READ Q1#, Q2#, Q3#, Q4#, Q5#, Q6#, Q7#, Q8#, Q9#
IF ABS(X) < 3.75 THEN                Polynomial fit.
  Y# = (X / 3.75) ^ 2
  DUM# = P2# + Y# * (P3# + Y# * (P4# + Y# * (P5# + Y# * (P6# + Y# * P7#))))
  BESSI1 = X * (P1# + Y# * DUM#)
ELSE
  AX = ABS(X)
  Y# = 3.75 / AX
  DUM# = Q4# + Y# * (Q5# + Y# * (Q6# + Y# * (Q7# + Y# * (Q8# + Y# * Q9#))))
  DUM# = (EXP(AX) / SQR(AX)) * (Q1# + Y# * (Q2# + Y# * (Q3# + Y# * DUM#)))
  IF X < 0! THEN BESSI1 = -DUM# ELSE BESSI1 = DUM#
END IF
END FUNCTION
```

A sample program using BESSI1 is the following:

```
DECLARE FUNCTION BESSI1! (X!)

'PROGRAM D6R23
'Driver for routine BESSI1
CLS
OPEN "FNCVAL.DAT" FOR INPUT AS #1
DO
  LINE INPUT #1, TEXT$
LOOP WHILE TEXT$ <> "Modified Bessel Function I1"
LINE INPUT #1, NVAL$
NVAL = VAL(NVAL$)
PRINT TEXT$
PRINT "   X        Actual        BESSI1(X)"
FOR I = 1 TO NVAL
  LINE INPUT #1, DUM$
  X = VAL(MID$(DUM$, 1, 6))
  VALUE = VAL(MID$(DUM$, 7))
  PRINT USING "##.##"; X;
  PRINT "   ";
  PRINT USING "#.#######^^^^"; VALUE;
  PRINT "   ";
  PRINT USING "#.#######^^^^"; BESSI1(X)
NEXT I
CLOSE #1
END
```

Bessel function K_1:

Here is the recipe BESSK1:

```
DECLARE FUNCTION BESSK1! (X!)
DECLARE FUNCTION BESSI1! (X!)
DATA 1.0D0,0.15443144D0,-0.67278579D0,-0.18156897D0
DATA -0.1919402D-1,-0.110404D-2,-0.4686D-4
DATA 1.25331414D0,0.23498619D0,-0.3655620D-1
DATA 0.1504268D-1,-0.780353D-2,0.325614D-2,-0.68245D-3

FUNCTION BESSK1 (X)
```

Returns the modified Bessel function $K_1(X)$ for positive real X.

```
RESTORE
READ P1#, P2#, P3#, P4#, P5#, P6#, P7#        Accumulate polynomials in double precision.
READ Q1#, Q2#, Q3#, Q4#, Q5#, Q6#, Q7#
IF X <= 2! THEN                              Polynomial fit.
  Y# = X * X / 4!
  DUM# = P2# + Y# * (P3# + Y# * (P4# + Y# * (P5# + Y# * (P6# + Y# * P7#))))
  BESSK1 = (LOG(X / 2!) * BESSI1(X)) + (1! / X) * (P1# + Y# * DUM#)
ELSE
  Y# = 2! / X
  DUM# = Q2# + Y# * (Q3# + Y# * (Q4# + Y# * (Q5# + Y# * (Q6# + Y# * Q7#))))
  BESSK1 = (EXP(-X) / SQR(X)) * (Q1# + Y# * DUM#)
END IF
END FUNCTION
```

A sample program using BESSK1 is the following:

```
DECLARE FUNCTION BESSK1! (X!)

'PROGRAM D6R24
'Driver for routine BESSK1
CLS
OPEN "FNCVAL.DAT" FOR INPUT AS #1
DO
  LINE INPUT #1, TEXT$
LOOP WHILE TEXT$ <> "Modified Bessel Function K1"
LINE INPUT #1, NVAL$
NVAL = VAL(NVAL$)
PRINT TEXT$
PRINT "   X         Actual          BESSK1(X)"
FOR I = 1 TO NVAL
  LINE INPUT #1, DUM$
  X = VAL(MID$(DUM$, 1, 6))
  VALUE = VAL(MID$(DUM$, 7))
  PRINT USING "##.##"; X;
  PRINT "   ";
  PRINT USING "#.#######^^^^"; VALUE;
  PRINT "   ";
  PRINT USING "#.#######^^^^"; BESSK1(X)
NEXT I
CLOSE #1
END
```

Bessel function K_n for $n > 1$:

Here is the recipe BESSK:

```
DECLARE FUNCTION BESSK! (N!, X!)
DECLARE FUNCTION BESSKO! (X!)
DECLARE FUNCTION BESSK1! (X!)

FUNCTION BESSK (N, X)
```
    Returns the modified Bessel function $K_N(x)$ for positive X and $N \geq 2$.
```
IF N < 2 THEN PRINT "bad argument N in BESSK": EXIT FUNCTION
TOX = 2! / X
BKM = BESSKO(X)                   Upward recurrence for all X
BK = BESSK1(X)
FOR J = 1 TO N - 1                ...and here it is.
  BKP = BKM + J * TOX * BK
  BKM = BK
  BK = BKP
NEXT J
BESSK = BK
END FUNCTION
```

A sample program using `BESSK` is the following:

```
DECLARE FUNCTION BESSK! (N!, X!)

'PROGRAM D6R25
'Driver for routine BESSK
CLS
OPEN "FNCVAL.DAT" FOR INPUT AS #1
DO
  LINE INPUT #1, TEXT$
  TEXT$ = LEFT$(TEXT$, 27)
LOOP WHILE TEXT$ <> "Modified Bessel Function Kn"
LINE INPUT #1, NVAL$
NVAL = VAL(NVAL$)
PRINT TEXT$
PRINT "   N       X        Actual          BESSK(N,X)"
FOR I = 1 TO NVAL
  LINE INPUT #1, DUM$
  N = VAL(MID$(DUM$, 1, 6))
  X = VAL(MID$(DUM$, 7, 6))
  VALUE = VAL(MID$(DUM$, 13))
  PRINT USING "####"; N;
  PRINT USING "#####.##"; X;
  PRINT "    ";
  PRINT USING "#.#######^^^^"; VALUE;
  PRINT "    ";
  PRINT USING "#.#######^^^^"; BESSK(N, X)
NEXT I
CLOSE #1
END
```

Bessel function I_n for $n > 1$:

Here is the recipe BESSI:

```
DECLARE FUNCTION BESSI! (N!, X!)
DECLARE FUNCTION BESSI0! (X!)

FUNCTION BESSI (N, X)
        Returns the modified Bessel function I_N(X) for any real X and N >= 2.
IACC = 40
BIGNO = 1E+10
BIGNI = 1E-10
IF N < 2 THEN PRINT "bad argument N in BESSI": EXIT FUNCTION
IF X = 0! THEN
  BESSI = 0!
ELSE
  TOX = 2! / X
  BIP = 0!
  BI = 1!
  BESI = 0!
  M = 2 * (N + INT(SQR(CSNG(IACC * N))))    Downward recurrence from an even value M.
  FOR J = M TO 1 STEP -1                     Make IACC larger to increase accuracy.
    BIM = BIP + CSNG(J) * TOX * BI           The downward recurrence.
    BIP = BI
    BI = BIM
    IF ABS(BI) > BIGNO THEN                   Renormalize to prevent overflows.
      BESI = BESI * BIGNI
      BI = BI * BIGNI
      BIP = BIP * BIGNI
    END IF
    IF J = N THEN BESI = BIP
  NEXT J
  DUM = BESI * BESSI0(X) / BI                 Normalize with BESSI0.
  IF X < 0! AND N MOD 2 = 1 THEN BESSI = -DUM ELSE BESSI = DUM
END IF
END FUNCTION
```

A sample program using BESSI is the following:

```
DECLARE FUNCTION BESSI! (N!, X!)

'PROGRAM D6R26
'Driver for routine BESSI
CLS
OPEN "FNCVAL.DAT" FOR INPUT AS #1
DO
  LINE INPUT #1, TEXT$
  TEXT$ = LEFT$(TEXT$, 27)
LOOP WHILE TEXT$ <> "Modified Bessel Function In"
LINE INPUT #1, NVAL$
NVAL = VAL(NVAL$)
PRINT TEXT$
PRINT "   N      X        Actual        BESSI(N,X)"
FOR I = 1 TO NVAL
```

```
LINE INPUT #1, DUM$
N = VAL(MID$(DUM$, 1, 6))
X = VAL(MID$(DUM$, 7, 6))
VALUE = VAL(MID$(DUM$, 13))
PRINT USING "####"; N;
PRINT USING "#####.##"; X;
PRINT "   ";
PRINT USING "#.#######^^^^"; VALUE;
PRINT "   ";
PRINT USING "#.#######^^^^"; BESSI(N, X)
NEXT I
CLOSE #1
END
```

Legendre polynomials:

Here is the recipe PLGNDR:

```
DECLARE FUNCTION PLGNDR! (L!, M!, X!)

FUNCTION PLGNDR (L, M, X)
```
Computes the associated Legendre polynomial $P_l^m(x)$. Here m and l are integers satisfying $0 \leq m \leq l$, while x lies in the range $-1 \leq x \leq 1$.
```
IF M < 0 OR M > L OR ABS(X) > 1! THEN PRINT "bad arguments": EXIT FUNCTION
PMM = 1!                         Compute Pₘᵐ.
IF M > 0 THEN
  SOMX2 = SQR((1! - X) * (1! + X))
  FACT = 1!
  FOR I = 1 TO M
    PMM = -PMM * FACT * SOMX2
    FACT = FACT + 2!
  NEXT I
END IF
IF L = M THEN
  PLGNDR = PMM
ELSE                             Compute Pₘ₊₁ᵐ.
  PMMP1 = X * (2 * M + 1) * PMM
  IF L = M + 1 THEN
    PLGNDR = PMMP1
  ELSE                           Compute Pₗᵐ, l > m + 1.
    FOR LL = M + 2 TO L
      PLL = (X * (2 * LL - 1) * PMMP1 - (LL + M - 1) * PMM) / (LL - M)
      PMM = PMMP1
      PMMP1 = PLL
    NEXT LL
    PLGNDR = PLL
  END IF
END IF
END FUNCTION
```

A sample program using PLGNDR is the following:

```
DECLARE FUNCTION PLGNDR! (N!, M!, X!)

'PROGRAM D6R27
'Driver for routine PLGNDR
CLS
OPEN "FNCVAL.DAT" FOR INPUT AS #1
DO
  LINE INPUT #1, TEXT$
LOOP WHILE TEXT$ <> "Legendre Polynomials"
LINE INPUT #1, NVAL$
NVAL = VAL(NVAL$)
PRINT TEXT$
PRINT "   N   M           X              Actual          PLGNDR(N,M,X)"
FOR I = 1 TO NVAL
  LINE INPUT #1, DUM$
  N = VAL(MID$(DUM$, 1, 6))
  M = VAL(MID$(DUM$, 7, 6))
  X = VAL(MID$(DUM$, 13, 12))
  VALUE = VAL(MID$(DUM$, 25))
  FAC = 1!
  IF M > 0 THEN
    FOR J = N - M + 1 TO N + M
      FAC = FAC * J
    NEXT J
  END IF
  FAC = 2! * FAC / (2! * N + 1!)
  VALUE = VALUE * SQR(FAC)
  PRINT USING "####"; N;
  PRINT USING "####"; M;
  PRINT "   ";
  PRINT USING "#.######^^^^"; X;
  PRINT "   ";
  PRINT USING "#.######^^^^"; VALUE;
  PRINT "   ";
  PRINT USING "#.######^^^^"; PLGNDR(N, M, X)
NEXT I
CLOSE #1
END
```

There are three programs that operate in a slightly different fashion. Sample programs **D6R28** and **D6R29** for routines **EL2** and **CEL**, which calculate elliptic integrals, do not refer to tables at all. Instead, they make twenty random choices of argument and compare the output of the function evaluation routine for these arguments to the result of actually performing the integration that defines them. The routine QSIMP from Chapter 4 of *Numerical Recipes* is used for the integration.

Here is the recipe EL2:

```
DECLARE FUNCTION EL2! (X!, QQC!, AA!, BB!)

FUNCTION EL2 (X, QQC, AA, BB)
```
   Returns the general elliptic integral of the second kind, $el2(x, k_c, a, b)$ with $X = x \geq 0$, QQC $= k_c$, AA $= a$, and BB $= b$.
```
PI = 3.14159265#
CA = .0003
CB = 1E-09
```
   The desired accuracy is the square of CA, while CB should be set to 0.01 times the desired accuracy.
```
IF X = 0! THEN
  EL2 = 0!
ELSEIF QQC <> 0! THEN
  QC = QQC
  A = AA
  B = BB
  C = X ^ 2
  D = 1! + C
  P = SQR((1! + QC ^ 2 * C) / D)
  D = X / D
  C = D / (2! * P)
  Z = A - B
  EYE = A
  A = .5 * (B + A)
  Y = ABS(1! / X)
  F = 0!
  L = 0
  EM = 1!
  QC = ABS(QC)
  DO
    B = EYE * QC + B
    E = EM * QC
    G = E / P
    D = F * G + D
    F = C
    EYE = A
    P = G + P
    C = .5 * (D / P + C)
    G = EM
    EM = QC + EM
    A = .5 * (B / EM + A)
    Y = -E / Y + Y
    IF Y = 0! THEN Y = SQR(E) * CB
    IF ABS(G - QC) > CA * G THEN
      QC = SQR(E) * 2!
      L = L + L
      IF Y < 0! THEN L = L + 1
      DONE% = 0
    ELSE
      DONE% = -1
    END IF
  LOOP WHILE NOT DONE%
  IF Y < 0! THEN L = L + 1
  E = (ATN(EM / Y) + PI * L) * A / EM
```

```
   IF X < 0! THEN E = -E
   EL2 = E + C * Z
ELSE
   PRINT "failure in EL2"           Argument QQC was zero.
   EXIT FUNCTION
END IF
END FUNCTION
```

A sample program using **EL2** is the following:

```
DECLARE FUNCTION FUNC! (PHI!)
DECLARE FUNCTION RAN3! (IDUM&)
DECLARE FUNCTION EL2! (X!, AKC!, A!, B!)
DECLARE SUB QSIMP (DUM!, A!, B!, S!)

'PROGRAM D6R28
'Driver for routine EL2
CLS
PRINT "General Elliptic Integral of Second Kind"
PRINT "     x          kc          a          b          EL2      Integral"
IDUM& = -55
AGO = 0!
FOR I = 1 TO 20
  AKC = 5! * RAN3(IDUM&)
  A = 10! * RAN3(IDUM&)
  B = 10! * RAN3(IDUM&)
  X = 10! * RAN3(IDUM&)
  ASTOP = ATN(X)
  CALL QSIMP(DUM, AGO, ASTOP, S)
  PRINT USING "###.######"; X; AKC; A; B; EL2(X, AKC, A, B); S
NEXT I
END


FUNCTION FUNC (PHI)
SHARED AKC, A, B
TN = TAN(PHI)
TSQ = TN * TN
FUNC = (A + B * TSQ) / SQR((1! + TSQ) * (1! + AKC * AKC * TSQ))
END FUNCTION
```

Here is the recipe **CEL**:

```
DECLARE FUNCTION CEL! (QQC!, PP!, AA!, BB!)

FUNCTION CEL (QQC, PP, AA, BB)
```
   Returns the general complete elliptic integral $cel(k_c, p, a, b)$ with QQC $= k_c$, PP $= p$, AA $= a$, and BB $= b$.
```
CA = .0003                    The desired accuracy is the square of CA.
PIO2 = 1.5707963268#
IF QQC = 0! THEN PRINT "failure in CEL": EXIT FUNCTION
QC = ABS(QQC)
A = AA
B = BB
```

```
P = PP
E = QC
EM = 1!
IF P > 0! THEN
  P = SQR(P)
  B = B / P
ELSE
  F = QC * QC
  Q = 1! - F
  G = 1! - P
  F = F - P
  Q = Q * (B - A * P)
  P = SQR(F / G)
  A = (A - B) / G
  B = -Q / (G * G * P) + A * P
END IF
DO
  F = A
  A = A + B / P
  G = E / P
  B = B + F * G
  B = B + B
  P = G + P
  G = EM
  EM = QC + EM
  IF ABS(G - QC) > G * CA THEN
    QC = SQR(E)
    QC = QC + QC
    E = QC * EM
    DONE% = 0
  ELSE
    DONE% = -1
  END IF
LOOP WHILE NOT DONE%
CEL = PIO2 * (B + A * EM) / (EM * (EM + P))
END FUNCTION
```

A sample program using CEL is the following:

```
DECLARE FUNCTION CEL! (AKC!, P!, A!, B!)
DECLARE FUNCTION FUNC! (PHI!)
DECLARE FUNCTION RAN3! (IDUM&)
DECLARE SUB QSIMP (DUM!, AGO!, ASTOP!, S!)

'PROGRAM D6R29
'Driver for routine CEL
CLS
PIO2 = 1.5707963#
PRINT "Complete Elliptic Integral"
PRINT "    kc         p         a         b         CEL     Integral"
IDUM& = -55
AGO = 0!
ASTOP = PIO2
FOR I = 1 TO 20
  AKC = .1 + RAN3(IDUM&)
```

```
IDUM = 0
A = 10! * RAN3(IDUM&)
B = 10! * RAN3(IDUM&)
P = .1 + RAN3(IDUM&)
CALL QSIMP(DUM, AGO, ASTOP, S)
PRINT USING "###.######"; AKC; P; A; B; CEL(AKC, P, A, B); S
NEXT I
END


FUNCTION FUNC (PHI)
SHARED A, B, P, AKC
CS = COS(PHI)
CSQ = CS * CS
SSQ = 1! - CSQ
FUNC = (A * CSQ + B * SSQ) / (CSQ + P * SSQ) / SQR(CSQ + AKC * AKC * SSQ)
END FUNCTION
```

Routine SNCNDN returns Jacobian elliptic functions. The file FNCVAL.DAT contains information only about function SN. However, the values of CN and DN satisfy the relationships

$$SN^2 + CN^2 = 1, \qquad k^2SN^2 + DN^2 = 1.$$

The program D6R30 works exactly as the others in terms of testing SN, but for verifying CN and DN it lists the values RESULT1 and RESULT2 of the left sides of the two equations above. Each of them should have the value 1.0 for all choices of arguments.

Here is the recipe SNCNDN:

```
DECLARE FUNCTION COSH! (U!)
DECLARE FUNCTION TANH! (U!)

FUNCTION COSH (U)
COSH = .5 * (EXP(U) + EXP(-U))
END FUNCTION

SUB SNCNDN (UU, EMMC, SN, CN, DN)
```
   Returns the Jacobian elliptic functions $sn(u, k_c)$, $cn(u, k_c)$ and $dn(u, k_c)$. Here $UU = u$, while $EMMC = k_c^2$.
```
CA = .0003              The accuracy is the square of CA.
DIM EM(13), EN(13)
EMC = EMMC
U = UU
IF EMC <> 0! THEN
  BO& = EMC < 0!
  IF BO& THEN
    D = 1! - EMC
    EMC = -EMC / D
    D = SQR(D)
    U = D * U
  END IF
  A = 1!
  DN = 1!
  FOR I = 1 TO 13
    L = I
```

```
   EM(I) = A
   EMC = SQR(EMC)
   EN(I) = EMC
   C = .5 * (A + EMC)
   IF ABS(A - EMC) <= CA * A THEN EXIT FOR
   EMC = A * EMC
   A = C
 NEXT I
 U = C * U
 SN = SIN(U)
 CN = COS(U)
 IF SN <> 0! THEN
   A = CN / SN
   C = A * C
   FOR II = L TO 1 STEP -1
     B = EM(II)
     A = C * A
     C = DN * C
     DN = (EN(II) + A) / (B + A)
     A = C / B
   NEXT II
   A = 1! / SQR(C ^ 2 + 1!)
   IF SN < 0! THEN
     SN = -A
   ELSE
     SN = A
   END IF
   CN = C * SN
 END IF
 IF BO& THEN
   A = DN
   DN = CN
   CN = A
   SN = SN / D
 END IF
ELSE
 CN = 1! / COSH(U)
 DN = CN
 SN = TANH(U)
END IF
ERASE EN, EM
END SUB

FUNCTION TANH (U)
EPU = EXP(U)
EMU = 1! / EPU
IF ABS(U) < .3 THEN
  U2 = U * U
  DUM = 1! + U2 / 6! * (1! + U2 / 20! * (1! + U2 / 42! * (1! + U2 / 72!)))
  TANH = 2! * U * DUM / (EPU + EMU)
ELSE
  TANH = (EPU - EMU) / (EPU + EMU)
END IF
END FUNCTION
```

A sample program using SNCNDN is the following:

```
DECLARE SUB SNCNDN (UU!, EMMC!, SN!, CN!, DN!)

'PROGRAM D6R30
'Driver for routine SNCNDN
CLS
OPEN "FNCVAL.DAT" FOR INPUT AS #1
DO
    LINE INPUT #1, TEXT$
LOOP WHILE TEXT$ <> "Jacobian Elliptic Function"
LINE INPUT #1, NVAL$
NVAL = VAL(NVAL$)
PRINT TEXT$
PRINT "  Mc       U       Actual          SN          SN^2+CN^2  ";
PRINT "(Mc)*(SN^2)+DN^2"
FOR I = 1 TO NVAL
    LINE INPUT #1, DUM$
    EM = VAL(MID$(DUM$, 1, 6))
    UU = VAL(MID$(DUM$, 7, 6))
    VALUE = VAL(MID$(DUM$, 13))
    EMMC = 1! - EM
    CALL SNCNDN(UU, EMMC, SN, CN, DN)
    RESUL1 = SN * SN + CN * CN
    RESUL2 = EM * SN * SN + DN * DN
    PRINT USING "##.##"; EMMC;
    PRINT USING "#####.##"; UU;
    PRINT "     ";
    PRINT USING "#.#####^^^^"; VALUE;
    PRINT "     ";
    PRINT USING "#.#####^^^^"; SN;
    PRINT USING "#######.#####"; RESUL1;
    PRINT USING "########.#####"; RESUL2
NEXT I
CLOSE #1
END
```

Appendix

File **FNCVAL.DAT**:

```
Values of Special Functions in format x,F(x) or x,y,F(x,y)
Gamma Function
17 Values
    1.0    1.000000
    1.2    0.918169
    1.4    0.887264
    1.6    0.893515
    1.8    0.931384
    2.0    1.000000
    0.2    4.590845
    0.4    2.218160
    0.6    1.489192
    0.8    1.164230
   -0.2    5.2005665E01
   -0.4    4.617091E01
```

```
-0.6    4.0128959E01
-0.8    3.4231564E01
10.0    3.6288000E05
20.0    1.2164510E17
30.0    8.8417620E30
```
N-factorial
18 Values
```
1     1
2     2
3     6
4     24
5     120
6     720
7     5040
8     40320
9     362880
10    3628800
11    39916800
12    479001600
13    6227020800
14    87178291200
15    1.3076755E12
20    2.4329042E18
25    1.5511222E25
30    2.6525281E32
```
Binomial Coefficients
20 Values
```
1     0     1
6     1     6
6     3     20
6     5     6
15    1     15
15    3     455
15    5     3003
15    7     6435
15    9     5005
15    11    1365
15    13    105
25    1     25
25    3     2300
25    5     53130
25    7     480700
25    9     2042975
25    11    4457400
25    13    5200300
25    15    3268760
25    17    1081575
```
Beta Function
15 Values
```
1.0   1.0   1.000000
0.2   1.0   5.000000
1.0   0.2   5.000000
0.4   1.0   2.500000
1.0   0.4   2.500000
0.6   1.0   1.666667
0.8   1.0   1.250000
6.0   6.0   3.607504E-04
6.0   5.0   7.936508E-04
6.0   4.0   1.984127E-03
```

```
6.0    3.0      5.952381E-03
6.0    2.0      0.238095E-01
7.0    7.0      8.325008E-05
5.0    5.0      1.587302E-03
4.0    4.0      7.142857E-03
3.0    3.0      0.333333E-01
2.0    2.0      1.666667E-01
```
Incomplete Gamma Function
20 Values
```
0.1    3.1622777E-02    0.7420263
0.1    3.1622777E-01    0.9119753
0.1    1.5811388        0.9898955
0.5    7.0710678E-02    0.2931279
0.5    7.0710678E-01    0.7656418
0.5    3.5355339        0.9921661
1.0    0.1000000        0.0951626
1.0    1.0000000        0.6321206
1.0    5.0000000        0.9932621
1.1    1.0488088E-01    0.0757471
1.1    1.0488088        0.6076457
1.1    5.2440442        0.9933425
2.0    1.4142136E-01    0.0091054
2.0    1.4142136        0.4130643
2.0    7.0710678        0.9931450
6.0    2.4494897        0.0387318
6.0    12.247449        0.9825937
11.0   16.583124        0.9404267
26.0   25.495098        0.4863866
41.0   44.821870        0.7359709
```
Error Function
20 Values
```
0.0    0.000000
0.1    0.1124629
0.2    0.2227026
0.3    0.3286268
0.4    0.4283924
0.5    0.5204999
0.6    0.6038561
0.7    0.6778012
0.8    0.7421010
0.9    0.7969082
1.0    0.8427008
1.1    0.8802051
1.2    0.9103140
1.3    0.9340079
1.4    0.9522851
1.5    0.9661051
1.6    0.9763484
1.7    0.9837905
1.8    0.9890905
1.9    0.9927904
```
Incomplete Beta Function
20 Values
```
0.5    0.5    0.01    0.0637686
0.5    0.5    0.10    0.2048328
0.5    0.5    1.00    1.0000000
1.0    0.5    0.01    0.0050126
1.0    0.5    0.10    0.0513167
1.0    0.5    1.00    1.0000000
```

```
1.0     1.0     0.5     0.5000000
5.0     5.0     0.5     0.5000000
10.0    0.5     0.9     0.1516409
10.0    5.0     0.5     0.0897827
10.0    5.0     1.0     1.0000000
10.0    10.0    0.5     0.5000000
20.0    5.0     0.8     0.4598773
20.0    10.0    0.6     0.2146816
20.0    10.0    0.8     0.9507365
20.0    20.0    0.5     0.5000000
20.0    20.0    0.6     0.8979414
30.0    10.0    0.7     0.2241297
30.0    10.0    0.8     0.7586405
40.0    20.0    0.7     0.7001783
```
Bessel Function J0
20 Values
```
-5.0    -0.1775968
-4.0    -0.3971498
-3.0    -0.2600520
-2.0     0.2238908
-1.0     0.7651976
 0.0     1.0000000
 1.0     0.7651977
 2.0     0.2238908
 3.0    -0.2600520
 4.0    -0.3971498
 5.0    -0.1775968
 6.0     0.1506453
 7.0     0.3000793
 8.0     0.1716508
 9.0    -0.0903336
10.0    -0.2459358
11.0    -0.1711903
12.0     0.0476893
13.0     0.2069261
14.0     0.1710735
15.0    -0.0142245
```
Bessel Function Y0
15 Values
```
 0.1    -1.5342387
 1.0     0.0882570
 2.0     0.51037567
 3.0     0.37685001
 4.0    -0.0169407
 5.0    -0.3085176
 6.0    -0.2881947
 7.0    -0.0259497
 8.0     0.2235215
 9.0     0.2499367
10.0     0.0556712
11.0    -0.1688473
12.0    -0.2252373
13.0    -0.0782079
14.0     0.1271926
15.0     0.2054743
```
Bessel Function J1
20 Values
```
-5.0     0.3275791
-4.0     0.0660433
```

```
-3.0    -0.3390590
-2.0    -0.5767248
-1.0    -0.4400506
 0.0     0.0000000
 1.0     0.4400506
 2.0     0.5767248
 3.0     0.3390590
 4.0    -0.0660433
 5.0    -0.3275791
 6.0    -0.2766839
 7.0    -0.0046828
 8.0     0.2346364
 9.0     0.2453118
10.0     0.0434728
11.0    -0.1767853
12.0    -0.2234471
13.0    -0.0703181
14.0     0.1333752
15.0     0.2051040
Bessel Function Y1
15 Values
 0.1    -6.4589511
 1.0    -0.7812128
 2.0    -0.1070324
 3.0     0.3246744
 4.0     0.3979257
 5.0     0.1478631
 6.0    -0.1750103
 7.0    -0.3026672
 8.0    -0.1580605
 9.0     0.1043146
10.0     0.2490154
11.0     0.1637055
12.0    -0.0570992
13.0    -0.2100814
14.0    -0.1666448
15.0     0.0210736
Bessel Function Jn, n>=2
20 Values
 2      1.0      1.149034849E-01
 2      2.0      3.528340286E-01
 2      5.0      4.656511628E-02
 2     10.0      2.546303137E-01
 2     50.0     -5.971280079E-02
 5      1.0      2.497577302E-04
 5      2.0      7.039629756E-03
 5      5.0      2.611405461E-01
 5     10.0     -2.340615282E-01
 5     50.0     -8.140024770E-02
10      1.0      2.630615124E-10
10      2.0      2.515386283E-07
10      5.0      1.467802647E-03
10     10.0      2.074861066E-01
10     50.0     -1.138478491E-01
20      1.0      3.873503009E-25
20      2.0      3.918972805E-19
20      5.0      2.770330052E-11
20     10.0      1.151336925E-05
20     50.0     -1.167043528E-01
```

```
Bessel Function Yn, n>=2
20 Values
2      1.0      -1.650682607
2      2.0      -6.174081042E-01
2      5.0       3.676628826E-01
2     10.0      -5.868082460E-03
2     50.0       9.579316873E-02
5      1.0      -2.604058666E02
5      2.0      -9.935989128
5      5.0      -4.536948225E-01
5     10.0       1.354030477E-01
5     50.0      -7.854841391E-02
10     1.0      -1.216180143E08
10     2.0      -1.291845422E05
10     5.0      -2.512911010E01
10    10.0      -3.598141522E-01
10    50.0       5.723897182E-03
20     1.0      -4.113970315E22
20     2.0      -4.081651389E16
20     5.0      -5.933965297E08
20    10.0      -1.597483848E03
20    50.0       1.644263395E-02
Modified Bessel Function I0
20 Values
0.0    1.0000000
0.2    1.0100250
0.4    1.0404018
0.6    1.0920453
0.8    1.1665149
1.0    1.2660658
1.2    1.3937256
1.4    1.5533951
1.6    1.7499807
1.8    1.9895593
2.0    2.2795852
2.5    3.2898391
3.0    4.8807925
3.5    7.3782035
4.0    11.301922
4.5    17.481172
5.0    27.239871
6.0    67.234406
8.0    427.56411
10.0   2815.7167
Modified Bessel Function K0
20 Values
0.1    2.4270690
0.2    1.7527038
0.4    1.1145291
0.6    0.77752208
0.8    0.56534710
1.0    0.42102445
1.2    0.31850821
1.4    0.24365506
1.6    0.18795475
1.8    0.14593140
2.0    0.11389387
2.5    6.2347553E-02
3.0    3.4739500E-02
```

```
3.5    1.9598897E-02
4.0    1.1159676E-02
4.5    6.3998572E-03
5.0    3.6910983E-03
6.0    1.2439943E-03
8.0    1.4647071E-04
10.0   1.7780062E-05
```
Modified Bessel Function I1
20 Values
```
0.0    0.00000000
0.2    0.10050083
0.4    0.20402675
0.6    0.31370403
0.8    0.43286480
1.0    0.56515912
1.2    0.71467794
1.4    0.88609197
1.6    1.0848107
1.8    1.3171674
2.0    1.5906369
2.5    2.5167163
3.0    3.9533700
3.5    6.2058350
4.0    9.7594652
4.5    15.389221
5.0    24.335643
6.0    61.341937
8.0    399.87313
10.0   2670.9883
```
Modified Bessel Function K1
20 Values
```
0.1    9.8538451
0.2    4.7759725
0.4    2.1843544
0.6    1.3028349
0.8    0.86178163
1.0    0.60190724
1.2    0.43459241
1.4    0.32083589
1.6    0.24063392
1.8    0.18262309
2.0    0.13986588
2.5    7.3890816E-02
3.0    4.0156431E-02
3.5    2.2239393E-02
4.0    1.2483499E-02
4.5    7.0780949E-03
5.0    4.0446134E-03
6.0    1.3439197E-03
8.0    1.5536921E-04
10.0   1.8648773E-05
```
Modified Bessel Function Kn, n>=2
28 Values
```
2      0.2      49.512430
2      1.0      1.6248389
2      2.0      2.5375975E-01
2      2.5      1.2146021E-01
2      3.0      6.1510459E-02
2      5.0      5.3089437E-03
```

2	10.0	2.1509817E-05
2	20.0	6.3295437E-10
3	1.0	7.101262825
3	2.0	6.473853909E-01
3	5.0	8.291768415E-03
3	10.0	2.725270026E-05
3	50.0	3.72793677E-23
5	1.0	3.609605896E02
5	2.0	9.431049101
5	5.0	3.270627371E-02
5	10.0	5.754184999E-05
5	50.0	4.36718224E-23
10	1.0	1.807132899E08
10	2.0	1.624824040E05
10	5.0	9.758562829
10	10.0	1.614255300E-03
10	50.0	9.15098819E-23
20	1.0	6.294369369E22
20	2.0	5.770856853E16
20	5.0	4.827000521E08
20	10.0	1.787442782E02
20	50.0	1.70614838E-21

Modified Bessel Function In, n>=2
28 Values

2	0.2	5.0166876E-03
2	1.0	1.3574767E-01
2	2.0	6.8894844E-01
2	2.5	1.2764661
2	3.0	2.2452125
2	5.0	17.505615
2	10.0	2281.5189
2	20.0	3.9312785E07
3	1.0	2.216842492E-02
3	2.0	2.127399592E-01
3	5.0	1.033115017E01
3	10.0	1.758380717E01
3	50.0	2.67776414E20
5	1.0	2.714631560E-04
5	2.0	9.825679323E-03
5	5.0	2.157974547
5	10.0	7.771882864E02
5	50.0	2.27854831E20
10	1.0	2.752948040E-10
10	2.0	3.016963879E-07
10	5.0	4.580044419E-03
10	10.0	2.189170616E01
10	50.0	1.07159716E20
20	1.0	3.966835986E-25
20	2.0	4.310560576E-19
20	5.0	5.024239358E-11
20	10.0	1.250799736E-04
20	50.0	5.44200840E18

Legendre Polynomials
19 Values

1	0	1.0	1.224745
10	0	1.0	3.240370
20	0	1.0	4.527693
1	0	0.7071067	0.866025
10	0	0.7071067	0.373006

20	0	0.7071067	-0.874140
1	0	0.0	0.000000
10	0	0.0	-0.797435
20	0	0.0	0.797766
2	2	0.7071067	0.484123
10	2	0.7071067	-0.204789
20	2	0.7071067	0.910208
2	2	0.0	0.968246
10	2	0.0	0.804785
20	2	0.0	-0.799672
10	10	0.7071067	0.042505
20	10	0.7071067	-0.707252
10	10	0.0	1.360172
20	10	0.0	-0.853705

Jacobian Elliptic Function
20 Values

0.0	0.1	0.099833
0.0	0.2	0.19867
0.0	0.5	0.47943
0.0	1.0	0.84147
0.0	2.0	0.90930
0.5	0.1	0.099751
0.5	0.2	0.19802
0.5	0.5	0.47075
0.5	1.0	0.80300
0.5	2.0	0.99466
1.0	0.1	0.099668
1.0	0.2	0.19738
1.0	0.5	0.46212
1.0	1.0	0.76159
1.0	2.0	0.96403
1.0	4.0	0.99933
1.0	-0.2	-0.19738
1.0	-0.5	-0.46212
1.0	-1.0	-0.76159
1.0	-2.0	-0.96403

Chapter 7: Random Numbers

Chapter 7 of *Numerical Recipes* deals with the generation of random numbers drawn from various distributions. The first four subroutines produce uniform deviates with a range of 0.0 to 1.0. RAN0 is a subroutine for improving the randomness of a system-supplied random number generator by shuffling the output. RAN1 is a portable random number generator based on three linear congruential generators and a shuffler. RAN2 contains a single congruential generator and a shuffler, and has the advantage of being somewhat faster, if less plentiful in possible output values. RAN3 is another portable generator, based on a subtractive rather than a congruential method.

The transformation method is used to generate some nonuniform distributions. Resulting from this are routines EXPDEV, which gives exponentially distributed deviates, and GASDEV for Gaussian deviates. The rejection method of producing nonuniform deviates is also discussed, and is used in GAMDEV (deviate with a gamma function distribution), POIDEV (deviate with a Poisson distribution), and BNLDEV (deviate with a binomial distribution). For generating random sequences of zeros and ones, there are two subroutines, IRBIT1 and IRBIT2, both based on the 14 lowest significant bits in the seed ISEED, but each using a different recurrence to proceed from step to step. The national Data Encryption Standard (DES) is discussed as the basis for a random number generator which we call RAN4. DES itself is carried out by routines DES, KS, and CYFUN.

$\star$ $\star$ $\star$ $\star$

Sample programs D7R1 to D7R4 are all really the same program, except that each calls a different random number generator (RAN0 to RAN3, respectively). They first draw four consecutive random numbers $X_1, \ldots, X_4$ from the generator in question. Then they treat the numbers as coordinates of a point. For example, they take (X_1, X_2) as a point in two dimensions, (X_1, X_2, X_3) as a point in three dimensions, etc. These points are inside boxes of unit dimension in their respective n-space. They may, however, be either inside or outside of the unit sphere in that space. For $n = 2, 3, 4$ we seek the probability that a point is inside the unit n-sphere. This number is easily calculated theoretically. For $n = 2$ it is $\pi/4$, for $n = 3$ it is $\pi/6$, and for $n = 4$ it is $\pi^2/32$. If the random number generator is not faulty, the points will fall within the unit n-sphere this fraction of the time, and the result should become increasingly accurate as the number of points increases. In programs D7R1 to D7R4 we have taken out a factor of 2^n for convenience, and used the random number generators as a statistical means of determining the value of π, $4\pi/3$, and $\pi^2/2$.

Here is the recipe RAN0:

```
DECLARE FUNCTION RANO! (IDUM&)
```

```
FUNCTION RANO (IDUM&) STATIC
```
Returns a uniform random deviate between 0.0 and 1.0 using a system-supplied routine RND(ISEED&). Set IDUM& to any negative value to initialize or reinitialize the sequence.

```
DIM V(97)                          The exact number 97 is unimportant.
IF IDUM& < 0 OR IFF = 0 THEN        As a precaution against misuse, we will always initialize on
  IFF = 1                          the first call, even if IDUM& is not set negative.
  ISEED& = ABS(IDUM&)
  IDUM& = 1
  FOR J = 1 TO 97                  Exercise the system routine, especially important if the sys-
    DUM = RND(ISEED&)              tem's multiplier is small.
  NEXT J
  FOR J = 1 TO 97                  Then save 97 values
    V(J) = RND(ISEED&)
  NEXT J
  Y = RND(ISEED&)                  and a 98th.
END IF
```
This is where we start if not initializing. Use the previously saved random number Y to get an index J between 1 and 97. Then use the corresponding V(J) for both the next Y and as the output number.

```
J = 1 + INT(97! * Y)
IF J > 97 OR J < 1 THEN PRINT "Abnormal exit": EXIT FUNCTION
Y = V(J)
RANO = Y
V(J) = RND(ISEED&)                 Finally, refill the table entry with the next random number
END FUNCTION                       from RND.
```

A sample program using RANO is the following:

```
DECLARE FUNCTION RANO! (IDUM&)

'PROGRAM D7R4
'Driver for routine RANO
'Calculates pi statistically using volume of unit n-sphere
CLS
PI = 3.1415926#
DIM IY(3), YPROB(3)
DEF FNC (X1, X2, X3, X4) = SQR(X1 ^ 2 + X2 ^ 2 + X3 ^ 2 + X4 ^ 2)
IDUM& = -1
FOR I = 1 TO 3
  IY(I) = 0
NEXT I
PRINT "          Volume of unit n-sphere, n=2,3,4"
PRINT
PRINT "# points       pi        (4/3)*pi    (1/2)*pi^2"
FOR J = 1 TO 15
  FOR K = 2 ^ (J - 1) TO 2 ^ J
    X1 = RANO (IDUM&)
    X2 = RANO (IDUM&)
    X3 = RANO (IDUM&)
    X4 = RANO (IDUM&)
    IF FNC(X1, X2, 0!, 0!) < 1! THEN IY(1) = IY(1) + 1
    IF FNC(X1, X2, X3, 0!) < 1! THEN IY(2) = IY(2) + 1
```

```
    IF FNC(X1, X2, X3, X4) < 1! THEN IY(3) = IY(3) + 1
 NEXT K
 FOR I = 1 TO 3
    YPROB(I) = 1! * 2 ^ (I + 1) * IY(I) / 2 ^ J
 NEXT I
 PRINT USING "#######"; 2 ^ J;
 FOR I = 1 TO 3
    PRINT USING "#####.######"; YPROB(I);
 NEXT I
 PRINT
 NEXT J
 PRINT
 PRINT " actual"; USING "#####.######"; PI; 4! * PI / 3!; .5 * PI ^ 2
 END
```

Here is the recipe RAN1:

```
DECLARE FUNCTION RAN1! (IDUM&)

FUNCTION RAN1 (IDUM&) STATIC
```
     Returns a uniform random deviate between 0.0 and 1.0. Set IDUM& to any negative value to
     initialize or reinitialize the sequence.
```
DIM R(97)
M1& = 259200
IA1& = 7141
IC1& = 54773
RM1 = .0000038580247#
M2& = 134456
IA2& = 8121
IC2& = 28411
RM2 = .0000074373773#
M3& = 243000
IA3& = 4561
IC3& = 51349
IF IDUM& < 0 OR IFF = 0 THEN        As above, initialize on first call even if IDUM& is not negative.
   IFF = 1
   IX1& = (IC1& - IDUM&) MOD M1&    Seed the first routine,
   IX1& = (IA1& * IX1& + IC1&) MOD M1&
   IX2& = IX1& MOD M2&             and use it to seed the second
   IX1& = (IA1& * IX1& + IC1&) MOD M1&
   IX3& = IX1& MOD M3&             and third routines.
   FOR J = 1 TO 97                 Fill the table with sequential uni-
      IX1& = (IA1& * IX1& + IC1&) MOD M1&    form deviates generated by the first
      IX2& = (IA2& * IX2& + IC2&) MOD M2&    two routines.
      R(J) = (CSNG(IX1&) + CSNG(IX2&) * RM2) * RM1    Low- and high-order pieces com-
   NEXT J                         bined here.
   IDUM& = 1
END IF
IX1& = (IA1& * IX1& + IC1&) MOD M1&    Except when initializing, this is where we start. Gen-
IX2& = (IA2& * IX2& + IC2&) MOD M2&    erate the next number for each sequence.
IX3& = (IA3& * IX3& + IC3&) MOD M3&
J = 1 + INT((97 * IX3&) / M3&)        Use the third sequence to get an integer between 1 and 97.
IF J > 97 OR J < 1 THEN PRINT "Abnormal exit": EXIT FUNCTION
RAN1 = R(J)                         Return that table entry,
```

```
R(J) = (CSNG(IX1&) + CSNG(IX2&) * RM2) * RM1        and refill it.
END FUNCTION
```

A sample program using RAN1 is the following:

```
DECLARE FUNCTION RAN1! (IDUM&)

'PROGRAM D7R4
'Driver for routine RAN1
'Calculates pi statistically using volume of unit n-sphere
CLS
PI = 3.1415926#
DIM IY(3), YPROB(3)
DEF FNC (X1, X2, X3, X4) = SQR(X1 ^ 2 + X2 ^ 2 + X3 ^ 2 + X4 ^ 2)
IDUM& = -1
FOR I = 1 TO 3
  IY(I) = 0
NEXT I
PRINT "           Volume of unit n-sphere, n=2,3,4"
PRINT
PRINT "# points      pi        (4/3)*pi    (1/2)*pi^2"
FOR J = 1 TO 15
  FOR K = 2 ^ (J - 1) TO 2 ^ J
    X1 = RAN1(IDUM&)
    X2 = RAN1(IDUM&)
    X3 = RAN1(IDUM&)
    X4 = RAN1(IDUM&)
    IF FNC(X1, X2, 0!, 0!) < 1! THEN IY(1) = IY(1) + 1
    IF FNC(X1, X2, X3, 0!) < 1! THEN IY(2) = IY(2) + 1
    IF FNC(X1, X2, X3, X4) < 1! THEN IY(3) = IY(3) + 1
  NEXT K
  FOR I = 1 TO 3
    YPROB(I) = 1! * 2 ^ (I + 1) * IY(I) / 2 ^ J
  NEXT I
  PRINT USING "#######"; 2 ^ J;
  FOR I = 1 TO 3
    PRINT USING "#####.######"; YPROB(I);
  NEXT I
  PRINT
NEXT J
PRINT
PRINT " actual"; USING "#####.######"; PI; 4! * PI / 3!; .5 * PI ^ 2
END
```

Here is the recipe RAN2:

```
DECLARE FUNCTION RAN2! (IDUM&)

FUNCTION RAN2 (IDUM&) STATIC
```

Returns a uniform random deviate between 0.0 and 1.0. Set IDUM& to any negative value to initialize or reinitialize the sequence.

```
M& = 714025
IA& = 1366
IC& = 150889
```

```
RM = .0000014005112#
DIM IR&(97)
IF IDUM& < 0 OR IFF = 0 THEN        As above.
  IFF = 1
  IDUM& = (IC& - IDUM&) MOD M&
  FOR J = 1 TO 97                   Initialize the shuffle table.
    IDUM& = (IA& * IDUM& + IC&) MOD M&
    IR&(J) = IDUM&
  NEXT J
  IDUM& = (IA& * IDUM& + IC&) MOD M&
  IY& = IDUM&                       Compare to RAN0, above.
END IF
J = 1 + INT((97 * IY&) / M&)        Here is where we start except on initialization.
IF J > 97 OR J < 1 THEN PRINT "Abnormal exit": EXIT FUNCTION
IY& = IR&(J)
RAN2 = IY& * RM
IDUM& = (IA& * IDUM& + IC&) MOD M&
IR&(J) = IDUM&
END FUNCTION
```

A sample program using `RAN2` is the following:

```
DECLARE FUNCTION RAN2! (IDUM&)

'PROGRAM D7R4
'Driver for routine RAN2
'Calculates pi statistically using volume of unit n-sphere
CLS
PI = 3.1415926#
DIM IY(3), YPROB(3)
DEF FNC (X1, X2, X3, X4) = SQR(X1 ^ 2 + X2 ^ 2 + X3 ^ 2 + X4 ^ 2)
IDUM& = -1
FOR I = 1 TO 3
  IY(I) = 0
NEXT I
PRINT "          Volume of unit n-sphere, n=2,3,4"
PRINT
PRINT "# points      pi        (4/3)*pi    (1/2)*pi^2"
FOR J = 1 TO 15
  FOR K = 2 ^ (J - 1) TO 2 ^ J
    X1 = RAN2(IDUM&)
    X2 = RAN2(IDUM&)
    X3 = RAN2(IDUM&)
    X4 = RAN2(IDUM&)
    IF FNC(X1, X2, 0!, 0!) < 1! THEN IY(1) = IY(1) + 1
    IF FNC(X1, X2, X3, 0!) < 1! THEN IY(2) = IY(2) + 1
    IF FNC(X1, X2, X3, X4) < 1! THEN IY(3) = IY(3) + 1
  NEXT K
  FOR I = 1 TO 3
    YPROB(I) = 1! * 2 ^ (I + 1) * IY(I) / 2 ^ J
  NEXT I
  PRINT USING "#######"; 2 ^ J;
  FOR I = 1 TO 3
    PRINT USING "#####.######"; YPROB(I);
  NEXT I
```

```
PRINT
NEXT J
PRINT
PRINT " actual"; USING "#####.######"; PI; 4! * PI / 3!; .5 * PI ^ 2
END
```

Here is the recipe RAN3:

```
DECLARE FUNCTION RAN3! (IDUM&)

FUNCTION RAN3 (IDUM&) STATIC
```
    Returns a uniform random deviate between 0.0 and 1.0. Set IDUM& to any negative value to
    initialize or reinitialize the sequence.
```
MBIG& = 1000000000
MSEED& = 161803398
```
    According to Knuth, any large MBIG, and any smaller (but still large) MSEED can be substituted
    for the above values.
```
MZ& = 0
FAC = 1E-09
DIM MA&(55)                              This value is special and should not be modified; see Knuth.
IF IDUM& < 0 OR IFF = 0 THEN             Initialization.
  IFF = 1
  MJ& = MSEED& - ABS(IDUM&)             Initialize MA(55) using the seed IDUM and
  MJ& = MJ& - MBIG& * INT(MJ& / MBIG&)  the large number MSEED.
  MA&(55) = MJ&
  MK& = 1
  FOR I = 1 TO 54                        Now initialize the rest of the table,
    II = 21 * I - 55 * INT((21 * I) / 55)   in a slightly random order,
    MA&(II) = MK&                        with numbers that are not especially random.
    MK& = MJ& - MK&
    IF MK& < MZ& THEN MK& = MK& + MBIG&
    MJ& = MA&(II)
  NEXT I
  FOR K = 1 TO 4                         We randomize them by "warming up the generator."
    FOR I = 1 TO 55
      MA&(I) = MA&(I) - MA&(1 + I + 30 - 55 * INT((I + 30) / 55))
      IF MA&(I) < MZ& THEN MA&(I) = MA&(I) + MBIG&
    NEXT I
  NEXT K
  INEXT = 0                              Prepare indices for our first generated number.
  INEXTP = 31                            The constant 31 is special; see Knuth.
  IDUM& = 1
END IF
INEXT = INEXT + 1                        Here is where we start, except on initialization. Increment
IF INEXT = 56 THEN INEXT = 1            INEXT, wrapping around 56 to 1.
INEXTP = INEXTP + 1                      Ditto for INEXTP.
IF INEXTP = 56 THEN INEXTP = 1
MJ& = MA&(INEXT) - MA&(INEXTP)          Now generate a new random number subtractively.
IF MJ& < MZ& THEN MJ& = MJ& + MBIG&            Be sure that it is in range.
MA&(INEXT) = MJ&                         Store it,
RAN3 = MJ& * FAC                         and output the derived uniform deviate.
END FUNCTION
```

A sample program using RAN3 is the following:

```
DECLARE FUNCTION RAN3! (IDUM&)

'PROGRAM D7R4
'Driver for routine RAN3
'Calculates pi statistically using volume of unit n-sphere
CLS
PI = 3.1415926#
DIM IY(3), YPROB(3)
DEF FNC (X1, X2, X3, X4) = SQR(X1 ^ 2 + X2 ^ 2 + X3 ^ 2 + X4 ^ 2)
IDUM& = -1
FOR I = 1 TO 3
 IY(I) = 0
NEXT I
PRINT "           Volume of unit n-sphere, n=2,3,4"
PRINT
PRINT "# points      pi        (4/3)*pi    (1/2)*pi^2"
FOR J = 1 TO 15
  FOR K = 2 ^ (J - 1) TO 2 ^ J
    X1 = RAN3(IDUM&)
    X2 = RAN3(IDUM&)
    X3 = RAN3(IDUM&)
    X4 = RAN3(IDUM&)
    IF FNC(X1, X2, 0!, 0!) < 1! THEN IY(1) = IY(1) + 1
    IF FNC(X1, X2, X3, 0!) < 1! THEN IY(2) = IY(2) + 1
    IF FNC(X1, X2, X3, X4) < 1! THEN IY(3) = IY(3) + 1
  NEXT K
  FOR I = 1 TO 3
    YPROB(I) = 1! * 2 ^ (I + 1) * IY(I) / 2 ^ J
  NEXT I
  PRINT USING "#######"; 2 ^ J;
  FOR I = 1 TO 3
    PRINT USING "#####.######"; YPROB(I);
  NEXT I
  PRINT
NEXT J
PRINT
PRINT " actual"; USING "#####.######"; PI; 4! * PI / 3!; .5 * PI ^ 2
END
```

Routine EXPDEV generates random numbers drawn from an exponential deviate. Sample program D7R5 makes ten thousand calls to EXPDEV and bins the results into 21 bins, the contents of which are tallied in array X(I). Then the sum TOTAL of all bins is taken, since some of the numbers will be too large to have fallen in any of the bins. The X(I) are scaled to TOTAL, and then compared to a similarly normalized exponential which is called EXPECT.

Here is the recipe EXPDEV:

```
DECLARE FUNCTION EXPDEV! (IDUM&)
DECLARE FUNCTION RAN1! (IDUM&)

FUNCTION EXPDEV (IDUM&)
```
   Returns an exponentially distributed, positive, random deviate of unit mean, using
   RAN1(IDUM&) as the source of uniform deviates.
```
DUM = 0!
WHILE DUM = 0!
  DUM = RAN1(IDUM&)
WEND
EXPDEV = -LOG(DUM)
END FUNCTION
```

A sample program using EXPDEV is the following:

```
DECLARE FUNCTION EXPDEV! (IDUM&)

'PROGRAM D7R5
'Driver for routine EXPDEV
CLS
NPTS = 10000
EE = 2.718281828#
DIM TRIG(21), X(21)
FOR I = 1 TO 21
  TRIG(I) = (I - 1) / 20!
  X(I) = 0!
NEXT I
IDUM& = -1
FOR I = 1 TO NPTS
  Y = EXPDEV(IDUM&)
  FOR J = 2 TO 21
    IF Y < TRIG(J) AND Y > TRIG(J - 1) THEN
      X(J) = X(J) + 1!
    END IF
  NEXT J
NEXT I
TOTAL = 0!
FOR I = 2 TO 21
  TOTAL = TOTAL + X(I)
NEXT I
PRINT "Exponential distribution with"; NPTS; "points:"
PRINT "    interval      observed     expected"
FOR I = 2 TO 21
  X(I) = X(I) / TOTAL
  EXPECT = EXP(-(TRIG(I - 1) + TRIG(I)) / 2!)
  EXPECT = EXPECT * .05 * EE / (EE - 1)
  PRINT USING "###.##"; TRIG(I - 1); TRIG(I);
  PRINT USING "#######.####"; X(I); EXPECT
NEXT I
END
```

GASDEV generates random numbers from a Gaussian deviate. Example D7R6 takes ten thousand of these and puts them into 21 bins. For the purpose of binning, the center of the Gaussian is shifted over by NOVER2=10 bins, to put it in the middle bin. The remainder of the program simply plots the contents of the bins, to illustrate that they have the characteristic Gaussian bell shape. This allows a quick, though superficial, check of the integrity of the routine.

Here is the recipe GASDEV:

```
DECLARE FUNCTION GASDEV! (IDUM&)
DECLARE FUNCTION RAN1! (IDUM&)

FUNCTION GASDEV (IDUM&)
```
Returns a normally distributed deviate with zero mean and unit variance, using RAN1(IDUM&) as the source of uniform deviates.
```
STATIC ISET, GSET
IF ISET = 0 THEN            We don't have an extra deviate handy, so
  DO                       pick two uniform numbers in the square extending from -1 to
    V1 = 2! * RAN1(IDUM&) - 1!    +1 in each direction,
    V2 = 2! * RAN1(IDUM&) - 1!
    R = V1 ^ 2 + V2 ^ 2            see if they are in the unit circle,
  LOOP WHILE R >= 1! OR R = 0!  and if they are not, try again.
  FAC = SQR(-2! * LOG(R) / R)    Now make the Box-Muller transformation
  GSET = V1 * FAC              to get two normal deviates. Return one and save the other
  GASDEV = V2 * FAC            for next time.
  ISET = 1                    Set flag.
ELSE                          We have an extra deviate handy,
  GASDEV = GSET               so return it,
  ISET = 0                    and unset the flag.
END IF
END FUNCTION
```

A sample program using GASDEV is the following:

```
DECLARE FUNCTION GASDEV! (IDUM&)

'PROGRAM D7R6
'Driver for routine GASDEV
CLS
N = 20
NP1 = N + 1
NOVER2 = N / 2
NPTS = 10000
ISCAL = 400
LLEN = 50
DIM DIST(NP1), TEXT$(50)
IDUM& = -13
FOR J = 1 TO NP1
  DIST(J) = 0!
NEXT J
FOR I = 1 TO NPTS
  J = CINT(.25 * N * GASDEV(IDUM&)) + NOVER2 + 1
  IF J >= 1 AND J <= NP1 THEN DIST(J) = DIST(J) + 1
NEXT I
PRINT "Normally distributed deviate of"; NPTS; "points"
```

```
PRINT "    x         p(x)       graph:"
FOR J = 1 TO NP1
  DIST(J) = DIST(J) / NPTS
  FOR K = 1 TO 50
    TEXT$(K) = " "
  NEXT K
  KLIM = INT(ISCAL * DIST(J))
  IF KLIM > LLEN THEN KLIM = LLEN
  FOR K = 1 TO KLIM
    TEXT$(K) = "*"
  NEXT K
  X = CSNG(J) / (.25 * N)
  PRINT USING "####.##"; X;
  PRINT USING "#####.####"; DIST(J);
  PRINT "    ";
  FOR K = 1 TO 50
    PRINT TEXT$(K);
  NEXT K
  PRINT
NEXT J
END
```

The next three sample programs **D7R7** to **D7R9** are identical to the previous one, but each drives a different random number generator and produces a different graph. **D7R7** drives **GAMDEV** and displays a gamma distribution of order **IA** specified by the user. **D7R8** drives **POIDEV** and produces a Poisson distribution with mean **XM** specified by the user. **D7R9** drives **BNLDEV** and produces a binomial distribution, also with specified **XM**.

Here is the recipe **GAMDEV**:

```
DECLARE FUNCTION GAMDEV! (IA!, IDUM&)
DECLARE FUNCTION RAN1! (IDUM&)

FUNCTION GAMDEV (IA, IDUM&)
        Returns a deviate distributed as a gamma distribution of integer order IA, i.e. a waiting time
        to the IAth event in a Poisson process of unit mean, using RAN1(IDUM&) as the source of
        uniform deviates.
IF IA < 1 THEN PRINT "Abnormal exit": EXIT FUNCTION
IF IA < 6 THEN                        Use direct method, adding waiting times.
  X = 1!
  FOR J = 1 TO IA
    X = X * RAN1(IDUM&)
  NEXT J
  X = -LOG(X)
ELSE                                  Use rejection method.
  DO
    DO
      DO
        V1 = 2! * RAN1(IDUM&) - 1!    These four lines generate the tangent of a random
        V2 = 2! * RAN1(IDUM&) - 1!    angle, i.e. are equivalent to Y = TAN(PI * RAN1(IDU
      LOOP WHILE V1 ^ 2 + V2 ^ 2 > 1!
      Y = V2 / V1
      AM = IA - 1
      S = SQR(2! * AM + 1!)
```

```
    X = S * Y + AM              We decide whether to reject X:
    LOOP WHILE X <= 0!          Reject in region of zero probability.
    E = (1! + Y ^ 2) * EXP(AM * LOG(X / AM) - S * Y)    Ratio to comparison fn.
    LOOP WHILE RAN1(IDUM&) > E  Reject on basis of a second uniform deviate.
  END IF
  GAMDEV = X
END FUNCTION
```

A sample program using **GAMDEV** is the following:

```
DECLARE FUNCTION GAMDEV! (IA!, IDUM&)

'PROGRAM D7R7
'Driver for routine GAMDEV
CLS
N = 20
NPTS = 10000
ISCAL = 200
LLEN = 50
DIM DIST(21), TEXT$(50)
IDUM& = -13
DO
  DO
    FOR J = 1 TO 21
      DIST(J) = 0!
    NEXT J
    PRINT "Order of Gamma distribution (n=1..20); -1 to end."
    INPUT IA
    IF IA <= 0 THEN END
  LOOP WHILE IA > 20
  FOR I = 1 TO NPTS
    J = INT(GAMDEV(IA, IDUM&)) + 1
    IF J >= 1 AND J <= 21 THEN DIST(J) = DIST(J) + 1
  NEXT I
  PRINT "Gamma-distribution deviate; order"; IA; "of"; NPTS; "points"
  PRINT "     x          p(x)        graph:"
  FOR J = 1 TO 20
    DIST(J) = DIST(J) / NPTS
    FOR K = 1 TO 50
      TEXT$(K) = " "
    NEXT K
    KLIM = INT(ISCAL * DIST(J))
    IF KLIM > LLEN THEN KLIM = LLEN
    FOR K = 1 TO KLIM
      TEXT$(K) = "*"
    NEXT K
    PRINT USING "####.##"; CSNG(J);
    PRINT USING "#####.####"; DIST(J);
    PRINT "    ";
    FOR K = 1 TO 50
      PRINT TEXT$(K);
    NEXT K
    PRINT
  NEXT J
LOOP
```

END

Here is the recipe POIDEV:

```
DECLARE FUNCTION POIDEV! (XM!, IDUM&)
DECLARE FUNCTION GAMMLN! (X!)
DECLARE FUNCTION RAN1! (IDUM&)
```

```
FUNCTION POIDEV (XM, IDUM&) STATIC
```
      Returns as a floating-point number an integer value that is a random deviate drawn from a Poisson distribution of mean XM, using RAN1(IDUM&) as a source of uniform random deviates.

`PI = 3.141592654#`	
`IF XM < 12! THEN`	Use direct method.
`  IF XM <> OLDM THEN`	Flag for whether XM has changed since last call.
`    OLDM = XM`	
`    G = EXP(-XM)`	If XM is new, compute the exponential.
`  END IF`	
`  EM = -1`	
`  T = 1!`	
`  DO`	
`    EM = EM + 1!`	Instead of adding exponential deviates it is equivalent to mul-
`    T = T * RAN1(IDUM&)`	tiply uniform deviates. Then we never actually have to take
`  LOOP WHILE T > G`	the log, merely compare to the precomputed exponential.
`ELSE`	Use rejection method.
`  IF XM <> OLDM THEN`	If XM has changed since the last call, then precompute some
`    OLDM = XM`	functions which occur below.
`    SQ = SQR(2! * XM)`	
`    ALXM = LOG(XM)`	
`    G = XM * ALXM - GAMMLN(XM + 1!)`	The function GAMMLN is the natural log of the
`  END IF`	gamma function, as given in §6.2.
`  DO`	
`    DO`	
`      Y = TAN(PI * RAN1(IDUM&))`	Y is a deviate from a Lorentzian comparison function.
`      EM = SQ * Y + XM`	EM is Y, shifted and scaled.
`    LOOP WHILE EM < 0!`	Reject if in regime of zero probability.
`    EM = INT(EM)`	The trick for integer-valued distributions.
`    DUM = EM * ALXM - GAMMLN(EM + 1!) - G`	
`    T = .9 * (1! + Y ^ 2) * EXP(DUM)`	The ratio of the desired distribution to the com-
`  LOOP WHILE RAN1(IDUM&) > T`	parison function; we accept or reject by compar-
`END IF`	ing it to another uniform deviate. The factor 0.9
`POIDEV = EM`	is chosen so that T never exceeds 1.
`END FUNCTION`	

A sample program using POIDEV is the following:

```
DECLARE FUNCTION POIDEV! (XM!, IDUM&)

'PROGRAM D7R8
'Driver for routine POIDEV
CLS
N = 20
NPTS = 10000
ISCAL = 200
LLEN = 50
```

```
DIM DIST(21), TEXT$(50)
IDUM& = -13
DO
  DO
    FOR J = 1 TO 21
      DIST(J) = 0!
    NEXT J
    PRINT "Mean of Poisson distrib. (x=0 to 20); neg. to end"
    INPUT XM
    IF XM < 0! THEN END
  LOOP WHILE XM > 20!
  FOR I = 1 TO NPTS
    J = INT(POIDEV(XM, IDUM&)) + 1
    IF J >= 1 AND J <= 21 THEN DIST(J) = DIST(J) + 1
  NEXT I
  PRINT "Poisson-distributed deviate; mean"; XM; "of"; NPTS; "points"
  PRINT "    x        p(x)      graph:"
  FOR J = 1 TO 20
    DIST(J) = DIST(J) / NPTS
    FOR K = 1 TO 50
      TEXT$(K) = " "
    NEXT K
    KLIM = INT(ISCAL * DIST(J))
    IF KLIM > LLEN THEN KLIM = LLEN
    FOR K = 1 TO KLIM
      TEXT$(K) = "*"
    NEXT K
    PRINT USING "####.##"; CSNG(J);
    PRINT USING "#####.####"; DIST(J);
    PRINT "    ";
    FOR K = 1 TO 50
      PRINT TEXT$(K);
    NEXT K
    PRINT
  NEXT J
LOOP
END
```

Here is the recipe BNLDEV:

```
DECLARE FUNCTION BNLDEV! (PP!, N!, IDUM&)
DECLARE FUNCTION GAMMLN! (X!)
DECLARE FUNCTION RAN1! (IDUM&)

FUNCTION BNLDEV (PP, N, IDUM&) STATIC
```
Returns as a floating-point number an integer value that is a random deviate drawn from a binomial distribution of N trials each of probability PP, using RAN1(IDUM&) as a source of uniform random deviates.
```
PI = 3.141592654#
IF PP <= .5 THEN
  P = PP
ELSE
  P = 1! - PP
END IF
```
The binomial distribution is invariant under changing PP to 1-PP, if we also change the answer to N minus itself; we'll remember to do this below.

```
AM = N * P                                  This is the mean of the deviate to be produced.
IF N < 25 THEN                              Use the direct method while N is not too large. This can
  BNL = 0!                                  require up to 25 calls to RAN1.
  FOR J = 1 TO N
    IF RAN1(IDUM&) < P THEN BNL = BNL + 1!
  NEXT J
ELSEIF AM < 1! THEN                         If fewer than one event is expected out of 25 or more trials,
  G = EXP(-AM)                              then the distribution is quite accurately Poisson. Use direct
  T = 1!                                    Poisson method.
  FOR J = 0 TO N
    T = T * RAN1(IDUM&)
    IF T < G THEN EXIT FOR
  NEXT J
  IF T >= G THEN J = N
  BNL = J
ELSE                                        Use the rejection method.
  IF N <> NOLD THEN                         If N has changed, then compute useful quantities.
    EN = N
    OLDG = GAMMLN(EN + 1!)
    NOLD = N
  END IF
  IF P <> POLD THEN                         If P has changed, then compute useful quantities.
    PC = 1! - P
    PLOG = LOG(P)
    PCLOG = LOG(PC)
    POLD = P
  END IF
  SQ = SQR(2! * AM * PC)                    The following code should by now seem familiar: rejection
  DO                                        method with a Lorentzian comparison function.
    DO
      Y = TAN(PI * RAN1(IDUM&))
      EM = SQ * Y + AM
    LOOP WHILE EM < 0! OR EM >= EN + 1!       Reject.
    EM = INT(EM)                            Trick for integer-valued distribution.
    T = EN - EM
    T = EXP(OLDG - GAMMLN(EM + 1!) - GAMMLN(T + 1!) + EM * PLOG + T * PCLOG)
    T = 1.2 * SQ * (1! + Y ^ 2) * T
  LOOP WHILE RAN1(IDUM&) > T                Reject. This happens about 1.5 times per deviate, on aver-
  BNL = EM                                  age.
END IF
IF P <> PP THEN BNL = N - BNL              Remember to undo the symmetry transformation.
BNLDEV = BNL
END FUNCTION
```

A sample program using BNLDEV is the following:

```
DECLARE FUNCTION BNLDEV! (PP!, N!, IDUM&)

'PROGRAM D7R9
'Driver for routine BNLDEV
CLS
N = 20
NPTS = 1000
ISCAL = 200
NN = 100
```

```
DIM DIST(21), TEXT$(50)
IDUM& = -133
LLEN = 50
DO
  FOR J = 1 TO 21
    DIST(J) = 0!
  NEXT J
  PRINT "Mean of binomial distribution (0 to 20) (neg to end)"
  INPUT XM
  IF XM < 0 THEN EXIT DO
  PP = XM / NN
  FOR I = 1 TO NPTS
    J = INT(BNLDEV(PP, NN, IDUM&))
    IF J >= 0 AND J <= 20 THEN DIST(J + 1) = DIST(J + 1) + 1
  NEXT I
  PRINT "   x      p(x)      graph:"
  FOR J = 1 TO 20
    DIST(J) = DIST(J) / NPTS
    FOR K = 1 TO 50
      TEXT$(K) = " "
    NEXT K
    TEXT$(1) = "*"
    KLIM = INT(ISCAL * DIST(J))
    IF KLIM > LLEN THEN KLIM = LLEN
    FOR K = 1 TO KLIM
      TEXT$(K) = "*"
    NEXT K
    PRINT USING "###.#"; CSNG(J - 1);
    PRINT USING "###.####"; DIST(J);
    PRINT "   ";
    FOR K = 1 TO 50
      PRINT TEXT$(K);
    NEXT K
    PRINT
  NEXT J
LOOP
END
```

Subroutines `IRBIT1` and `IRBIT2` both generate random series of ones and zeros. The sample programs `D7R10` and `D7R11` for the two are the same, and they check that the series have correct statistical properties (or more exactly, that they have at least one correct property). They look for a 1 in the series and count how many zeros follow it before the next 1 appears. The result is stored as a distribution. There should be, for example, a 50% chance of no zeros, a 25% chance of exactly one zero, and so on.

Here is the recipe `IRBIT1`:

```
DECLARE FUNCTION IRBIT1! (ISEED!)

FUNCTION IRBIT1 (ISEED) STATIC
        Returns as an integer a random bit, based on the 14 low-significance bits in ISEED (which is
        modified for the next call).
IB1 = 1                         Powers of 2.
IB3 = 4
```

```
IB5 = 16
IB14 = 8192
NEWBIT% = (ISEED AND IB14) <> 0      Get bit I4.
IF (ISEED AND IB5) <> 0 THEN NEWBIT% = NOT NEWBIT%      XOR with bit 5.
IF (ISEED AND IB3) <> 0 THEN NEWBIT% = NOT NEWBIT%      XOR with bit 3.
IF (ISEED AND IB1) <> 0 THEN NEWBIT% = NOT NEWBIT%      XOR with bit 1.
IF ISEED > 2 ^ 14 THEN ISEED = ISEED - 2 ^ 14      Leftshift the seed.
ISEED = 2 * ISEED
IF NEWBIT% THEN                                    Put the result of the XOR
   IRBIT1 = 1                                      calculation in bit 1.
   ISEED = ISEED OR IB1
ELSE
   IRBIT1 = 0
   ISEED = ISEED AND (NOT IB1)
END IF
END FUNCTION
```

A sample program using IRBIT1 is the following:

```
DECLARE FUNCTION IRBIT1! (ISEED!)

'PROGRAM D7R10
'Driver for routine IRBIT1
'Calculate distribution of runs of zeros
CLS
NBIN = 15
NTRIES = 10000
DIM DELAQ(NBIN)
ISEED = 12345
FOR I = 1 TO NBIN
  DELAQ(I) = 0!
NEXT I
IPTS = 0
FOR I = 1 TO NTRIES
  IF IRBIT1(ISEED) = 1 THEN
    IPTS = IPTS + 1
    IFLG = 0
    FOR J = 1 TO NBIN
      IF IRBIT1(ISEED) = 1 AND IFLG = 0 THEN
        IFLG = 1
        DELAQ(J) = DELAQ(J) + 1!
      END IF
    NEXT J
  END IF
NEXT I
PRINT "Distribution of runs of N zeros"
PRINT "     N          Probability          Expected"
FOR N = 1 TO NBIN
  PRINT USING "######"; N - 1;
  PRINT USING "###########.######"; DELAQ(N) / IPTS; 1 / 2! ^ N
NEXT N
END
```

Here is the recipe `IRBIT2`:

```
DECLARE FUNCTION IRBIT2! (ISEED!)

FUNCTION IRBIT2 (ISEED) STATIC
```
   Returns as an integer a random bit, based on the 14 low-significance bits in `ISEED` (which is modified for the next call).
```
IB1 = 1
IB3 = 4
IB5 = 16
IB14 = 8192
MASK = 21
IF (ISEED AND IB14) <> 0 THEN        Change all masked bits, shift, and put 1 into bit 1.
  ISEED = ISEED XOR MASK
  IF ISEED > 2 ^ 14 THEN ISEED = ISEED - 2 ^ 14
  ISEED = 2 * ISEED OR IB1
  IRBIT2 = 1
ELSE                                 Shift and put 0 into bit 1.
  IF ISEED > 2 ^ 14 THEN ISEED = ISEED - 2 ^ 14
  ISEED = 2 * ISEED AND (NOT IB1)
  IRBIT2 = 0
END IF
END FUNCTION
```

A sample program using `IRBIT2` is the following:

```
DECLARE FUNCTION IRBIT2! (ISEED!)

'PROGRAM D7R11
'Driver for routine IRBIT2
'Calculate distribution of runs of zeros
CLS
NBIN = 15
NTRIES = 10000
DIM DELAQ(NBIN)
ISEED = 111
FOR I = 1 TO NBIN
  DELAQ(I) = 0!
NEXT I
IPTS = 0
FOR I = 1 TO NTRIES
  IF IRBIT2(ISEED) = 1 THEN
    IPTS = IPTS + 1
    IFLG = 0
    FOR J = 1 TO NBIN
      IF IRBIT2(ISEED) = 1 AND IFLG = 0 THEN
        IFLG = 1
        DELAQ(J) = DELAQ(J) + 1!
      END IF
    NEXT J
  END IF
NEXT I
PRINT "Distribution of runs of N zeros"
PRINT "     N         Probability         Expected"
FOR N = 1 TO NBIN
```

```
PRINT USING "######"; N - 1;
PRINT USING "##########.######"; DELAQ(N) / IPTS; 1 / 2! ^ N
NEXT N
END
```

The next routine, RAN4, is a random number generator with a uniform deviate, based on the data encryption standard DES. When applied to RAN4, the routine we used to demonstrate RAN0 to RAN3 is outrageously time consuming. RAN4 is random but very slow. To try out RAN4 we simply list the first ten random numbers for a given seed IDUM=-123. Compare your results to these; they should be exactly the same. We also generate 50 more random numbers and find their average and variance with AVEVAR.

Here is the recipe RAN4:

```
DECLARE FUNCTION RAN4! (IDUM&)
DECLARE SUB DES (INPUQ&(), KEQ&(), NEWKEY!, ISW!, JOT&())

FUNCTION RAN4 (IDUM&) STATIC
      Returns a uniform random deviate between 0.0 and 1.0 using DES. Set IDUM& negative to
      initialize. There are IM possible initializations. This routine is extremely slow and should be
      used for demonstration purposes only.
IM& = 11979
IA& = 430
IC& = 2531
NACC = 24
      The first three parameters are used in initializing, cf. §7.1. NACC is the number of bits of
      floating point precision desired on the random deviate.
DIM INQ&(64), JOT&(64), KEQ&(64), POW(65)
IF IDUM& < 0 OR IFF = 0 THEN       Initialize:
  IFF = 1
  IDUM& = IDUM& MOD IM&
  IF IDUM& < 0 THEN IDUM& = IDUM& + IM&
  POW(1) = .5
  FOR J = 1 TO 64                            Set both the 64 bits of key and also the
    IDUM& = (IDUM& * IA& + IC&) MOD IM&      starting configuration of the 64-bit input ar-
    KEQ&(J) = (2 * IDUM&) \ IM&     Highest order bit. ray.
    INQ&(J) = ((4 * IDUM&) \ IM&) MOD 2      Next highest order bit.
    POW(J + 1) = .5 * POW(J)        Inverse powers of 2 in floating point.
  NEXT J
  NEWKEY = 1                        Set this flag.
END IF
ISAV& = INQ&(64)                    Start here except on initialization.
IF ISAV& <> 0 THEN                  Generate the next input bit configuration; cf. §7.4.
  INQ&(4) = 1 - INQ&(4)             I.e., change 1 to 0 and 0 to 1.
  INQ&(3) = 1 - INQ&(3)
  INQ&(1) = 1 - INQ&(1)
END IF
FOR J = 64 TO 2 STEP -1
  INQ&(J) = INQ&(J - 1)
NEXT J
INQ&(1) = ISAV&                     Input bit configuration now ready.
CALL DES(INQ&(), KEQ&(), NEWKEY, 0, JOT&())    Here is the real business.
DUM = 0!                            It remains only to make a floating number out of random
FOR J = 1 TO NACC                   bits.
```

```
IF JOT&(J) <> 0 THEN DUM = DUM + POW(J)
NEXT J
RAN4 = DUM
END FUNCTION
```

A sample program using RAN4 is the following:

```
DECLARE FUNCTION RAN4! (IDUM&)
DECLARE SUB AVEVAR (Y!(), N!, AVE!, VAR!)

'PROGRAM D7R12
'Driver for routine RAN4
CLS
NPT = 50
DIM Y(NPT)
IDUM& = -123
AVE = 0!
PRINT "First 10 random numbers with IDUM = "; IDUM&
PRINT
PRINT "   #        RAN4"
FOR J = 1 TO 10
  PRINT USING "####"; J;
  PRINT USING "#####.######"; RAN4(IDUM&)
NEXT J
PRINT
PRINT "Average and Variance of next"; NPT
PRINT
FOR J = 1 TO NPT
  Y(J) = RAN4(IDUM&)
NEXT J
CALL AVEVAR(Y(), NPT, AVE, VAR)
PRINT "Average: "; USING "#####.####"; AVE
PRINT "Variance:"; USING "#####.####"; VAR
PRINT
PRINT
PRINT "Expected Result for an Infinite Sample:"
PRINT
PRINT "Average:    0.5000"
PRINT "Variance:   0.0833"
END
```

```
First 10 random numbers with IDUM = -123

  #        RAN4
  1      0.076597
  2      0.533635
  3      0.919756
  4      0.317618
  5      0.187471
  6      0.629516
  7      0.588766
  8      0.953446
  9      0.366207
 10      0.915449
Average and Variance of next 50
```

```
Average:     0.4542
Variance:    0.0786
```

DES is a software implementation of the national data encryption standard. The complete formal test for this standard, though long and detailed, is included in program D7R13. This test consists of feeding in a long series of input codes and checking the output for agreement with expected output codes. The input-output pairs comprising the test are contained in file DESTST.DAT, listed in the Appendix to this chapter. To save you the job of comparing the many 16-character strings for accuracy, we have ended each line with the phrase "o.k." or "wrong", depending on the outcome.

Here is the recipe DES:

```
DECLARE SUB CYFUN (TITMP&(), K&(), IOUT&())
DECLARE SUB KS (KEQ&(), N, KN&())
DATA 58,50,42,34,26,18,10,2,60,52,44,36,28,20,12,4,62,54,46,38,30,22,14,6,64
DATA 56,48,40,32,24,16,8,57,49,41,33,25,17,9,1,59,51,43,35,27,19,11,3,61,53,
DATA 37,29,21,13,5,63,55,47,39,31,23,15,7
DATA 40,8,48,16,56,24,64,32,39,7,47,15,55,23,63,31,38,6,46,14,54,22,62,30,37
DATA 5,45,13,53,21,61,29,36,4,44,12,52,20,60,28,35,3,43,11,51,19,59,27,34,2,
DATA 10,50,18,58,26,33,1,41,9,49,17,57,25

SUB DES (INPUQ&(), KEQ&(), NEWKEY, ISW, JOTPUT&()) STATIC
          Data Encryption Standard. Encrypts 64 bits, stored one bit per word, in array INPUQ& into
          JOTPUT& using KEQ& as the key. Set NEWKEY=1 when the key is new. Set ISW=0 for en-
          cryption, =1 for decryption.
DIM ITMP&(64), TITMP&(32), IP(64), IPM(64), ICF&(32), TKNS&(48), KNS&(48, 16
RESTORE
FOR I = 1 TO 64
  READ IP(I)
NEXT I
FOR I = 1 TO 64
  READ IPM(I)
NEXT I
IF NEWKEY <> 0 THEN              Get the 16 sub-master keys from the master key.
  NEWKEY = 0
  FOR I = 1 TO 16
    CALL KS(KEQ&(), I, TKNS&())
    FOR J = 1 TO 48
      KNS&(J, I) = TKNS&(J)
    NEXT J
  NEXT I
END IF
FOR J = 1 TO 64                  The initial permutation.
  ITMP&(J) = INPUQ&(IP(J))
NEXT J
FOR I = 1 TO 16                  The 16 stages of encryption.
  II = I
  IF ISW = 1 THEN II = 17 - I    Use the sub-master keys in reverse order for decryption.
  FOR J = 1 TO 48
    TKNS&(J) = KNS&(J, II)
  NEXT J
  FOR J = 1 TO 32
    TITMP&(J) = ITMP&(32 + J)
```

```
NEXT J
CALL CYFUN(TITMP&(), TKNS&(), ICF&())        Get cipher function.
FOR J = 1 TO 32                              Pass one half-word through unchanged, while encrypting the
  IC& = ICF&(J) + ITMP&(J)                   other half-word and exchanging the two half-words output.
  ITMP&(J) = ITMP&(J + 32)
  ITMP&(J + 32) = ABS(IC& MOD 2)
NEXT J
NEXT I                                       Done with the 16 stages.
FOR J = 1 TO 32                              A final exchange of the two half-words is required.
  IC& = ITMP&(J)
  ITMP&(J) = ITMP&(J + 32)
  ITMP&(J + 32) = IC&
NEXT J
FOR J = 1 TO 64                              Final output permutation.
  JOTPUT&(J) = ITMP&(IPM(J))
NEXT J
END SUB
```

A sample program using DES is the following:

```
DECLARE SUB DES (INPUQ&(), KEQ&(), NEWKEY!, ISW!, JOTPUT&())

'PROGRAM D7R13
'Driver for routine DES
CLS
DIM IN&(64), KEQ&(64), IOUT&(64), ICMP&(64)
DIM HIN$(17), HKEY$(17), HOUT$(17), HCMP$(17)
OPEN "DESTST.DAT" FOR INPUT AS #1
LINE INPUT #1, TEXT$
PRINT TEXT$
PRINT
DO
  LINE INPUT #1, TEXT$
  PRINT TEXT$
  PRINT
  IF EOF(1) THEN EXIT DO
  LINE INPUT #1, TEXT$
  NCIPHR = VAL(TEXT$)
  LINE INPUT #1, TEXT2$
  IF TEXT2$ = "encode" THEN IDIREC = 0
  IF TEXT2$ = "decode" THEN IDIREC = 1
  DO
    PRINT "      Key          Plaintext      Expected Cipher    Actual Cipher"
    IF NCIPHR < 16 THEN MM = NCIPHR ELSE MM = 16
    NCIPHR = NCIPHR - 16
    FOR M = 1 TO MM
      LINE INPUT #1, DUM$
      FOR K = 1 TO 17
        HKEY$(K) = MID$(DUM$, K, 1)
        HIN$(K) = MID$(DUM$, K + 17, 1)
        HCMP$(K) = MID$(DUM$, K + 34, 1)
      NEXT K
      FOR I = 1 TO 16
        J = I + 1
        IDUM& = VAL("&H" + HIN$(J))
```

```
        JDUM& = VAL("&H" + HKEY$(J))
        FOR K = 1 TO 4
          L = 4 * I + 1 - K
          IN&(L) = IDUM& MOD 2
          IDUM& = INT(IDUM& / 2)
          KEQ&(L) = JDUM& MOD 2
          JDUM& = INT(JDUM& / 2)
        NEXT K
      NEXT I
      NEWKEY = 1
      CALL DES(IN&(), KEQ&(), NEWKEY, IDIREC, IOUT&())
      HOUT$(1) = " "
      FOR I = 1 TO 16
        JDUM& = 0
        FOR J = 1 TO 4
          JDUM& = JDUM& + (2 ^ (4 - J)) * IOUT&(4 * (I - 1) + J)
        NEXT J
        HOUT$(I + 1) = HEX$(JDUM&)
      NEXT I
      VERDCT$ = "  o.k."
      FOR I = 1 TO 17
        IF HCMP$(I) <> HOUT$(I) THEN VERDCT$ = "  wrong"
      NEXT I
      FOR K = 1 TO 17
        PRINT HKEY$(K);
      NEXT K
      FOR K = 1 TO 17
        PRINT HIN$(K);
      NEXT K
      FOR K = 1 TO 17
        PRINT HCMP$(K);
      NEXT K
      FOR K = 1 TO 17
        PRINT HOUT$(K);
      NEXT K
      PRINT VERDCT$
    NEXT M
    PRINT
    PRINT "press RETURN to continue..."
    LINE INPUT DUM$
  LOOP WHILE NCIPHR > 0
LOOP
CLOSE #1
END
```

DESKS contains two auxiliary routines for DES, namely CYFUN and KS. Technically, these were fully tested by the previous formal procedure. Sample program D7R14 is simply an additional routine to help you track down problems. It first feeds a single KEY to KS and generates 16 subkeys. These subkeys are strings of zeros and ones which we print out as strings of "-" and "*". In this form the results are easier to compare with printed text. Next an input vector is fed to the cipher function CYFUN along with each of these subkeys. The results are again printed in "-" and "*" characters. If all is well, you will observe these patterns:

```
Legend:
            -=0    *=1
Master key:
        *-*-*-*-*-*-*-*-*-*-*-*-*-*-*-*-*-*-*-*-*-*-*-*-*-*-*-*-*-*-
Sub-master keys:
     1  -*--****-**-**-*--**-*-*---*-*-*-*****-**-*-*-**
     2  -*--*******--*-*---*-**-*---*-***--****-**-*-*-**
     3  **--*-***----*-**-*-*****-*-***--*-**--*-***--**
     4  *****--**---*-*-*-*-****---*****--*-**-***--*-
     5  *-**---**-***-*-*-*-*-**-*-***---*-***-*-**---
     6  *-**------****-**-*-**-**-**--**--*-**-*-**--
     7  -***-*---*****--*-*-*---*-**--****-**-*-*-**--
     8  -*---**-****-*-*-***-*----***----****--*-*-**-*
     9  **---**-***--*-*-***-*-*-*-***-*--*****--*-**-*-*
    10  **--******---***--**--***-*-*-**-**-*---**-**-**
    11  ***--*****--*--**-*-***-*--**-**-*-**---*--**
    12  *-***-***--*--*-**--*-*****--***------**-*-*-**-
    13  --***--*-*-**-*-**-**-*-**-*-*-**----****--***-
    14  --**-*----****--***-***---*-*-*--*-**-**-**--**-*
    15  ---*-**--**-**-*-*-*-*-****-*-*-**-*--***-**-*
    16  -*-****--**-**-*-*-*-**-*-*-****-*--*-*-****
Legend:
            -=0    *=1
Input to cipher function:
        *--*-*--*--*--*--*--*--*--*--*-
Ciphered output:
     1   -*----**-*-------****-----***-*
     2   --*--***-**--*-*-****-*-*---*-*
     3   --*****--*-*--*-**----*---*--*
     4   -**-*--*---*-**---***-****---*-*
     5   *-*---*---****--*-*-*-*--*-**--
     6   --*--**-*--*---**-*-**----*----*
     7   *-*-****-*-**-*------*-*----****
     8   -***--***----**-*-----*---*--**
     9   -**-*-****-***-****-----*--
    10   *--***--**-*****-*--**-*-***-*-
    11   -*****-*----**--******-*-**-*---
    12   ***-*-----***---**----*-*-*-*-
    13   -**-*-----*-****-*----------**-*
    14   ----*---**-***----*-*---**-----
    15   *-**--*-*-**--**-*-*--***-*-****
    16   *---*--*--*---*-********-***-*-***-
```

Here is the recipe DESKS:

```
1 'RESTORE 1 begins reading here
DATA 57,49,41,33,25,17,9,1,58,50,42,34,26,18,10,2,59,51,43,35,27,19,11,3,60
DATA 52,44,36,63,55,47,39,31,23,15,7,62,54,46,38,30,22,14,6,61,53,45,37,29,21
DATA 13,5,28,20,12,4
DATA 14,17,11,24,1,5,3,28,15,6,21,10,23,19,12,4,26,8,16,7,27,20,13,2,41,52,31
DATA 37,47,55,30,40,51,45,33,48,44,49,39,56,34,53,46,42,50,36,29,32
2 'RESTORE 2 begins reading here
DATA 32,1,2,3,4,5,4,5,6,7,8,9,8,9,10,11,12,13,12,13,14,15,16,17,16,17,18,19
DATA 20,21,20,21,22,23,24,25,24,25,26,27,28,29,28,29,30,31,32,1
DATA 16,7,20,21,29,12,28,17,1,15,23,26,5,18,31,10,2,8,24,14,32,27,3,9,19,13,30
DATA 6,22,11,4,25
DATA 14,4,13,1,2,15,11,8,3,10,6,12,5,9,0,7,0,15,7,4,14,2,13,1,10,6,12,11,9,5,3
```

```
DATA 8,4,1,14,8,13,6,2,11,15,12,9,7,3,10,5,0,15,12,8,2,4,9,1,7,5,11,3,14,10,
DATA 6,13,15,1,8,14,6,11,3,4,9,7,2,13,12,0,5,10,3,13,4,7,15,2,8,14,12,0,1,10
DATA 9,11,5,0,14,7,11,10,4,13,1,5,8,12,6,9,3,2,15,13,8,10,1,3,15,4,2,11,6,7,
DATA 0,5,14,9
DATA 10,0,9,14,6,3,15,5,1,13,12,7,11,4,2,8,13,7,0,9,3,4,6,10,2,8,5,14,12,11,
DATA 1,13,6,4,9,8,15,3,0,11,1,2,12,5,10,14,7,1,10,13,0,6,9,8,7,4,15,14,3,11,
DATA 12,7,13,14,3,0,6,9,10,1,2,8,5,11,12,4,15,13,8,11,5,6,15,0,3,4,7,2,12,1,
DATA 14,9,10,6,9,0,12,11,7,13,15,1,3,14,5,2,8,4,3,15,0,6,10,1,13,8,9,4,5,11,
DATA 7,2,14
DATA 2,12,4,1,7,10,11,6,8,5,3,15,13,0,14,9,14,11,2,12,4,7,13,1,5,0,15,10,3,9
DATA 6,4,2,1,11,10,13,7,8,15,9,12,5,6,3,0,14,11,8,12,7,1,14,2,13,6,15,0,9,10
DATA 5,3,12,1,10,15,9,2,6,8,0,13,3,4,14,7,5,11,10,15,4,2,7,12,9,5,6,1,13,14,
DATA 11,3,8,9,14,15,5,2,8,12,3,7,0,4,10,1,13,11,6,4,3,2,12,9,5,15,10,11,14,1
DATA 6,0,8,13
DATA 4,11,2,14,15,0,8,13,3,12,9,7,5,10,6,1,13,0,11,7,4,9,1,10,14,3,5,12,2,15
DATA 6,1,4,11,13,12,3,7,14,10,15,6,8,0,5,9,2,6,11,13,8,1,4,10,7,9,5,0,15,14,
DATA 12,13,2,8,4,6,15,11,1,10,9,3,14,5,0,12,7,1,15,13,8,10,3,7,4,12,5,6,11,0
DATA 9,2,7,11,4,1,9,12,14,2,0,6,10,13,15,3,5,8,2,1,14,7,4,10,8,13,15,12,9,0,
DATA 6,11
DATA 0,0,0,0,0,0,0,1,0,0,1,0,0,0,1,1,0,1,0,0,0,1,0,1,0,1,1,0,0,1,1,1,1,0,0,0
DATA 0,0,1,1,0,1,0,1,0,1,1,1,1,0,0,1,1,0,1,1,1,1,0,1,1,1,1
```

```basic
SUB CYFUN (IR&(), K&(), IOUT&()) STATIC
        Returns the cipher function of IR& and K& in IOUT&.
DIM IE(48), IET(48), IP(32), ITMP(32), IQ(16, 4, 8), IBIN(4, 16)
RESTORE 2
FOR I = 1 TO 48
  READ IET(I)
NEXT I
FOR I = 1 TO 32
  READ IP(I)
NEXT I
        Here we input the S-Box, in full glory. Alternatively, you might read IQ from a data file.
FOR K = 1 TO 8
  FOR J = 1 TO 4
    FOR I = 1 TO 16
      READ IQ(I, J, K)
    NEXT I
  NEXT J
NEXT K
        Next we get the table of bits in the integers 0 to 15:
FOR J = 1 TO 16
  FOR I = 1 TO 4
    READ IBIN(I, J)
  NEXT I
NEXT J
FOR J = 1 TO 48                     Expand IR to 48 bits and combine it with K.
  IE(J) = IR&(IET(J)) + K&(J) AND 1
NEXT J
FOR JJ = 1 TO 8                     Loop over 8 groups of 6 bits.
  J = 6 * JJ - 5
  IROW = 2 * IE(J) + IE(J + 5)      Find place in the S-box table.
  ICOL = 8 * IE(J + 1) + 4 * IE(J + 2) + 2 * IE(J + 3) + IE(J + 4)
  ISS = IQ(ICOL + 1, IROW + 1, JJ)  Look up the number in the S-box table
```

```
KK = 4 * (JJ - 1)
FOR KI = 1 TO 4                    and plug its bits into the output.
   ITMP(KK + KI) = IBIN(KI, ISS + 1)
NEXT KI
NEXT JJ
FOR J = 1 TO 32                    Final permutation.
   IOUT&(J) = ITMP(IP(J))
NEXT J
END SUB


SUB KS (KEQ&(), N, KN&()) STATIC
        Key schedule calculation, returns KN& given KEQ& and N=1,2,...,16; must be called with N in
        that order.
DIM ICD&(56), IPC1(56), IPC2(48)
RESTORE 1
FOR I = 1 TO 56
   READ IPC1(I)
NEXT I
FOR I = 1 TO 48
   READ IPC2(I)
NEXT I
IF N = 1 THEN                      Initial selection and permutation.
   FOR J = 1 TO 56
      ICD&(J) = KEQ&(IPC1(J))
   NEXT J
END IF
IT = 2                            For most values of N perform two shifts,
IF N = 1 OR N = 2 OR N = 9 OR N = 16 THEN IT = 1      but for these perform only one.
FOR I = 1 TO IT                   Circular left-shifts of the two halves of the array ICD&.
   IC& = ICD&(1)
   ID& = ICD&(29)
   FOR J = 1 TO 27
      ICD&(J) = ICD&(J + 1)
      ICD&(J + 28) = ICD&(J + 29)
   NEXT J
   ICD&(28) = IC&
   ICD&(56) = ID&
NEXT I                            Done with the shifts.
FOR J = 1 TO 48                   The sub-master key is a selection of bits from the shifted
   KN&(J) = ICD&(IPC2(J))         ICD&.
NEXT
END SUB
```

A sample program using DESKS is the following:

```
DECLARE SUB KS (KEQ&(), N!, KN&())
DECLARE SUB CYFUN (IR&(), K&(), IOUT&())


'PROGRAM D7R14
'Driver for routines KS and CYFUN in file DESKS.FOR
CLS
DIM KEQ&(64), KN&(48), IR&(32), IOUT&(32), TEXT$(64)
'First test routine KS
FOR I = 1 TO 64
```

```
   KEQ&(I) = I - 2 * INT(I / 2)
   IF KEQ&(I) = 0 THEN TEXT$(I) = "-"
   IF KEQ&(I) = 1 THEN TEXT$(I) = "*"
NEXT I
PRINT "Legend:"
PRINT "          -=0    *=1"
PRINT "Master key:"
PRINT "              ";
FOR I = 1 TO 56
  PRINT TEXT$(I);
NEXT I
PRINT
PRINT "Sub-master keys:"
FOR I = 1 TO 16
  CALL KS(KEQ&(), I, KN&())
  FOR K = 1 TO 48
    IF KN&(K) = 0 THEN TEXT$(K) = "-"
    IF KN&(K) = 1 THEN TEXT$(K) = "*"
  NEXT K
  PRINT USING "######"; I;
  PRINT "   ";
  FOR J = 1 TO 48
    PRINT TEXT$(J);
  NEXT J
  PRINT
NEXT I
PRINT "press RETURN to continue..."
LINE INPUT DUM$
'Now test routine CYFUN
FOR I = 1 TO 32
  IR&(I) = I - 3 * INT(I / 3)
  IR&(I) = IR&(I) - 2 * INT(IR&(I) / 2)
  IF IR&(I) = 0 THEN TEXT$(I) = "-"
  IF IR&(I) = 1 THEN TEXT$(I) = "*"
NEXT I
PRINT "Legend:"
PRINT "          -=0    *=1 "
PRINT "Input to cipher function:"
PRINT "              ";
FOR I = 1 TO 32
  PRINT TEXT$(I);
NEXT I
PRINT
PRINT "Ciphered output:"
FOR I = 1 TO 16
  CALL KS(KEQ&(), I, KN&())
  CALL CYFUN(IR&(), KN&(), IOUT&())
  FOR K = 1 TO 32
    IF IOUT&(K) = 0 THEN TEXT$(K) = "-"
    IF IOUT&(K) = 1 THEN TEXT$(K) = "*"
  NEXT K
  PRINT USING "######"; I;
  PRINT "   ";
  FOR J = 1 TO 32
    PRINT TEXT$(J);
```

```
NEXT J
PRINT
NEXT I
END
```

Appendix

File DESTST.DAT:

DES Validation, as per NBS publication 500-20
*** Initial Permutation and Expansion test: ***

```
    Key             Plaintext        Expected Cipher
0101010101010101  95F8A5E5DD31D900  8000000000000000
0101010101010101  DD7F121CA5015619  4000000000000000
0101010101010101  2E8653104F3834EA  2000000000000000
0101010101010101  4BD388FF6CD81D4F  1000000000000000
0101010101010101  20B9E767B2FB1456  0800000000000000
0101010101010101  55579380D77138EF  0400000000000000
0101010101010101  6CC5DEFAAF04512F  0200000000000000
0101010101010101  0D9F279BA5D87260  0100000000000000
0101010101010101  D9031B0271BD5A0A  0080000000000000
0101010101010101  424250B37C3DD951  0040000000000000
0101010101010101  B8061B7ECD9A21E5  0020000000000000
0101010101010101  F15D0F286B65BD28  0010000000000000
0101010101010101  ADD0CC8D6E5DEBA1  0008000000000000
0101010101010101  E6D5F82752AD63D1  0004000000000000
0101010101010101  ECBFE3BD3F591A5E  0002000000000000
0101010101010101  F356834379D165CD  0001000000000000
0101010101010101  2B9F982F20037FA9  0000800000000000
0101010101010101  889DE068A16F0BE6  0000400000000000
0101010101010101  E19E275D846A1298  0000200000000000
0101010101010101  329A8ED523D71AEC  0000100000000000
0101010101010101  E7FCE22557D23C97  0000080000000000
0101010101010101  12A9F5817FF2D65D  0000040000000000
0101010101010101  A484C3AD38DC9C19  0000020000000000
0101010101010101  FBE00A8A1EF8AD72  0000010000000000
0101010101010101  750D079407521363  0000008000000000
0101010101010101  64FEED9C724C2FAF  0000004000000000
0101010101010101  F02B263B328E2B60  0000002000000000
0101010101010101  9D64555A9A10B852  0000001000000000
0101010101010101  D106FF0BED5255D7  0000000800000000
0101010101010101  E1652C6B138C64A5  0000000400000000
0101010101010101  E428581186EC8F46  0000000200000000
0101010101010101  AEB5F5EDE22D1A36  0000000100000000
0101010101010101  E943D7568AEC0C5C  0000000080000000
0101010101010101  DF98C8276F54B04B  0000000040000000
0101010101010101  B160E4680F6C696F  0000000020000000
0101010101010101  FA0752B07D9C4AB8  0000000010000000
0101010101010101  CA3A2B036DBC8502  0000000008000000
0101010101010101  5E0905517BB59BCF  0000000004000000
0101010101010101  814EEB3B91D90726  0000000002000000
0101010101010101  4D49DB1532919C9F  0000000001000000
0101010101010101  25EB5FC3F8CF0621  0000000000800000
0101010101010101  AB6A20C0620D1C6F  0000000000400000
0101010101010101  79E90DBC98F92CCA  0000000000200000
0101010101010101  866ECEDD8072BB0E  0000000000100000
0101010101010101  8B54536F2F3E64A8  0000000000080000
0101010101010101  EA51D3975595B86B  0000000000040000
0101010101010101  CAFFC6AC4542DE31  0000000000020000
```

```
0101010101010101 8DD45A2DDF90796C 0000000000010000
0101010101010101 1029D55E880EC2D0 0000000000008000
0101010101010101 5D86CB23639DBEA9 0000000000004000
0101010101010101 1D1CA853AE7C0C5F 0000000000002000
0101010101010101 CE332329248F3228 0000000000001000
0101010101010101 8405D1ABE24FB942 0000000000000800
0101010101010101 E643D78090CA4207 0000000000000400
0101010101010101 48221B9937748A23 0000000000000200
0101010101010101 DD7C0BBD61FAFD54 0000000000000100
0101010101010101 2FBC291A570DB5C4 0000000000000080
0101010101010101 E07C30D7E4E26E12 0000000000000040
0101010101010101 0953E2258E8E90A1 0000000000000020
0101010101010101 5B711BC4CEEBF2EE 0000000000000010
0101010101010101 CC083F1E6D9E85F6 0000000000000008
0101010101010101 D2FD8867D50D2DFE 0000000000000004
0101010101010101 06E7EA22CE92708F 0000000000000002
0101010101010101 166B40B44ABA4BD6 0000000000000001
*** Inverse Permutation and Expansion test ***
       Key            Plaintext       Expected Cipher
0101010101010101 8000000000000000 95F8A5E5DD31D900
0101010101010101 4000000000000000 DD7F121CA5015619
0101010101010101 2000000000000000 2E8653104F3834EA
0101010101010101 1000000000000000 4BD388FF6CD81D4F
0101010101010101 0800000000000000 20B9E767B2FB1456
0101010101010101 0400000000000000 55579380D77138EF
0101010101010101 0200000000000000 6CC5DEFAAF04512F
0101010101010101 0100000000000000 0D9F279BA5D87260
0101010101010101 0080000000000000 D9031B0271BD5A0A
0101010101010101 0040000000000000 424250B37C3DD951
0101010101010101 0020000000000000 B8061B7ECD9A21E5
0101010101010101 0010000000000000 F15D0F286B65BD28
0101010101010101 0008000000000000 ADD0CC8D6E5DEBA1
0101010101010101 0004000000000000 E6D5F82752AD63D1
0101010101010101 0002000000000000 ECBFE3BD3F591A5E
0101010101010101 0001000000000000 F356834379D165CD
0101010101010101 0000800000000000 2B9F982F20037FA9
0101010101010101 0000400000000000 889DE068A16F0BE6
0101010101010101 0000200000000000 E19E275D846A1298
0101010101010101 0000100000000000 329A8ED523D71AEC
0101010101010101 0000080000000000 E7FCE22557D23C97
0101010101010101 0000040000000000 12A9F5817FF2D65D
0101010101010101 0000020000000000 A484C3AD38DC9C19
0101010101010101 0000010000000000 FBE00A8A1EF8AD72
0101010101010101 0000008000000000 750D079407521363
0101010101010101 0000004000000000 64FEED9C724C2FAF
0101010101010101 0000002000000000 F02B263B328E2B60
0101010101010101 0000001000000000 9D64555A9A10B852
0101010101010101 0000000800000000 D106FF0BED5255D7
0101010101010101 0000000400000000 E1652C6B138C64A5
0101010101010101 0000000200000000 E428581186EC8F46
0101010101010101 0000000100000000 AEB5F5EDE22D1A36
0101010101010101 0000000080000000 E943D7568AEC0C5C
0101010101010101 0000000040000000 DF98C8276F54B04B
0101010101010101 0000000020000000 B160E4680F6C696F
0101010101010101 0000000010000000 FA0752B07D9C4AB8
0101010101010101 0000000008000000 CA3A2B036DBC8502
0101010101010101 0000000004000000 5E0905517BB59BCF
0101010101010101 0000000002000000 814EEB3B91D90726
0101010101010101 0000000001000000 4D49DB1532919C9F
```

```
0101010101010101 0000000000800000 25EB5FC3F8CF0621
0101010101010101 0000000000400000 AB6A20C0620D1C6F
0101010101010101 0000000000200000 79E90DBC98F92CCA
0101010101010101 0000000000100000 866ECEDD8072BB0E
0101010101010101 0000000000080000 8B54536F2F3E64A8
0101010101010101 0000000000040000 EA51D3975595B86B
0101010101010101 0000000000020000 CAFFC6AC4542DE31
0101010101010101 0000000000010000 8DD45A2DDF90796C
0101010101010101 0000000000008000 1029D55E880EC2D0
0101010101010101 0000000000004000 5D86CB23639DBEA9
0101010101010101 0000000000002000 1D1CA853AE7C0C5F
0101010101010101 0000000000001000 CE332329248F3228
0101010101010101 0000000000000800 8405D1ABE24FB942
0101010101010101 0000000000000400 E643D78090CA4207
0101010101010101 0000000000000200 48221B9937748A23
0101010101010101 0000000000000100 DD7C0BBD61FAFD54
0101010101010101 0000000000000080 2FBC291A570DB5C4
0101010101010101 0000000000000040 E07C30D7E4E26E12
0101010101010101 0000000000000020 0953E2258E8E90A1
0101010101010101 0000000000000010 5B711BC4CEEBF2EE
0101010101010101 0000000000000008 CC083F1E6D9E85F6
0101010101010101 0000000000000004 D2FD8867D50D2DFE
0101010101010101 0000000000000002 06E7EA22CE92708F
0101010101010101 0000000000000001 166B40B44ABA4BD6
*** Key Permutation tests: ***
     Key          Plaintext       Expected Cipher
8001010101010101 0000000000000000 95A8D72813DAA94D
4001010101010101 0000000000000000 0EEC1487DD8C26D5
2001010101010101 0000000000000000 7AD16FFB79C45926
1001010101010101 0000000000000000 D3746294CA6A6CF3
0801010101010101 0000000000000000 809F5F873C1FD761
0401010101010101 0000000000000000 C02FAFFEC989D1FC
0201010101010101 0000000000000000 4615AA1D33E72F10
0180010101010101 0000000000000000 2055123350C00858
0140010101010101 0000000000000000 DF3B99D6577397C8
0120010101010101 0000000000000000 31FE17369B5288C9
0110010101010101 0000000000000000 DFDD3CC64DAE1642
0108010101010101 0000000000000000 178C83CE2B399D94
0104010101010101 0000000000000000 50F636324A9B7F80
0102010101010101 0000000000000000 A8468EE3BC18F06D
0101800101010101 0000000000000000 A2DC9E92FD3CDE92
0101400101010101 0000000000000000 CAC09F797D031287
0101200101010101 0000000000000000 90BA680B22AEB525
0101100101010101 0000000000000000 CE7A24F350E280B6
0101080101010101 0000000000000000 882BFF0AA01A0B87
0101040101010101 0000000000000000 25610288924511C2
0101020101010101 0000000000000000 C71516C29C75D170
0101018001010101 0000000000000000 5199C29A52C9F059
0101014001010101 0000000000000000 C22F0A294A71F29F
0101012001010101 0000000000000000 EE371483714C02EA
0101011001010101 0000000000000000 A81FBD448F9E522F
0101010801010101 0000000000000000 4F644C92E192DFED
0101010401010101 0000000000000000 1AFA9A66A6DF92AE
0101010201010101 0000000000000000 B3C1CC715CB879D8
0101010180010101 0000000000000000 19D032E64AB0BD8B
0101010140010101 0000000000000000 3CFAA7A7DC8720DC
0101010120010101 0000000000000000 B7265F7F447AC6F3
0101010110010101 0000000000000000 9DB73B3C0D163F54
0101010108010101 0000000000000000 8181B65BABF4A975
```

```
0101010104010101 0000000000000000 93C9B64042EAA240
0101010102010101 0000000000000000 5570530829705592
0101010101800101 0000000000000000 8638809E878787A0
0101010101400101 0000000000000000 41B9A79AF79AC208
0101010101200101 0000000000000000 7A9BE42F2009A892
0101010101100101 0000000000000000 29038D56BA6D2745
0101010101080101 0000000000000000 5495C6ABF1E5DF51
0101010101040101 0000000000000000 AE13DBD561488933
0101010101020101 0000000000000000 024D1FFA8904E389
0101010101018001 0000000000000000 D1399712F99BF02E
0101010101014001 0000000000000000 14C1D7C1CFFEC79E
0101010101012001 0000000000000000 1DE5279DAE3BED6F
0101010101011001 0000000000000000 E941A33F85501303
0101010101010801 0000000000000000 DA99DBBC9A03F379
0101010101010401 0000000000000000 B7FC92F91D8E92E9
0101010101010201 0000000000000000 AE8E5CAA3CA04E85
0101010101010180 0000000000000000 9CC62DF43B6EED74
0101010101010140 0000000000000000 D863DBB5C59A91A0
0101010101010120 0000000000000000 A1AB2190545B91D7
0101010101010110 0000000000000000 0875041E64C570F7
0101010101010108 0000000000000000 5A594528BEBEF1CC
0101010101010104 0000000000000000 FCDB3291DE21F0C0
0101010101010102 0000000000000000 869EFD7F9F265A09
*** Test of right-shifts in Decryption ***
     Key            Plaintext        Expected Cipher
8001010101010101 95A8D72813DAA94D 0000000000000000
4001010101010101 0EEC1487DD8C26D5 0000000000000000
2001010101010101 7AD16FFB79C45926 0000000000000000
1001010101010101 D3746294CA6A6CF3 0000000000000000
0801010101010101 809F5F873C1FD761 0000000000000000
0401010101010101 C02FAFFEC989D1FC 0000000000000000
0201010101010101 4615AA1D33E72F10 0000000000000000
0180010101010101 2055123350C00858 0000000000000000
0140010101010101 DF3B99D6577397C8 0000000000000000
0120010101010101 31FE17369B5288C9 0000000000000000
0110010101010101 DFDD3CC64DAE1642 0000000000000000
0108010101010101 178C83CE2B399D94 0000000000000000
0104010101010101 50F636324A9B7F80 0000000000000000
0102010101010101 A8468EE3BC18F06D 0000000000000000
0101800101010101 A2DC9E92FD3CDE92 0000000000000000
0101400101010101 CAC09F797D031287 0000000000000000
0101200101010101 90BA680B22AEB525 0000000000000000
0101100101010101 CE7A24F350E280B6 0000000000000000
0101080101010101 882BFF0AA01A0B87 0000000000000000
0101040101010101 25610288924511C2 0000000000000000
0101020101010101 C71516C29C75D170 0000000000000000
0101018001010101 5199C29A52C9F059 0000000000000000
0101014001010101 C22F0A294A71F29F 0000000000000000
0101012001010101 EE371483714C02EA 0000000000000000
0101011001010101 A81FBD448F9E522F 0000000000000000
0101010801010101 4F644C92E192DFED 0000000000000000
0101010401010101 1AFA9A66A6DF92AE 0000000000000000
0101010201010101 B3C1CC715CB879D8 0000000000000000
0101010180010101 19D032E64AB0BD8B 0000000000000000
0101010140010101 3CFAA7A7DC8720DC 0000000000000000
0101010120010101 B7265F7F447AC6F3 0000000000000000
0101010110010101 9DB73B3C0D163F54 0000000000000000
0101010108010101 8181B65BABF4A975 0000000000000000
0101010104010101 93C9B64042EAA240 0000000000000000
```

```
0101010102010101 5570530829705592 0000000000000000
0101010101800101 8638809E878787A0 0000000000000000
0101010101400101 41B9A79AF79AC208 0000000000000000
0101010101200101 7A9BE42F2009A892 0000000000000000
0101010101100101 29038D56BA6D2745 0000000000000000
0101010101080101 5495C6ABF1E5DF51 0000000000000000
0101010101040101 AE13DBD561488933 0000000000000000
0101010101020101 024D1FFA8904E389 0000000000000000
0101010101018001 D1399712F99BF02E 0000000000000000
0101010101014001 14C1D7C1CFFEC79E 0000000000000000
0101010101012001 1DE5279DAE3BED6F 0000000000000000
0101010101011001 E941A33F85501303 0000000000000000
0101010101010801 DA99DBBC9A03F379 0000000000000000
0101010101010401 B7FC92F91D8E92E9 0000000000000000
0101010101010201 AE8E5CAA3CA04E85 0000000000000000
0101010101010180 9CC62DF43B6EED74 0000000000000000
0101010101010140 D863DBB5C59A91A0 0000000000000000
0101010101010120 A1AB2190545B91D7 0000000000000000
0101010101010110 0875041E64C570F7 0000000000000000
0101010101010108 5A594528BEBEF1CC 0000000000000000
0101010101010104 FCDB3291DE21F0C0 0000000000000000
0101010101010102 869EFD7F9F265A09 0000000000000000
```

*** Data permutation test: ***

```
      Key            Plaintext       Expected Cipher
1046913489980131 0000000000000000 88D55E54F54C97B4
1007103489988020 0000000000000000 0C0CC00C83EA48FD
10071034C8980120 0000000000000000 83BC8EF3A6570183
1046103489988020 0000000000000000 DF725DCAD94EA2E9
1086911519190101 0000000000000000 E652B53B550BE8B0
1086911519580101 0000000000000000 AF527120C485CBB0
5107B01519580101 0000000000000000 0F04CE393DB926D5
1007B01519190101 0000000000000000 C9F00FFC74079067
3107915498080101 0000000000000000 7CFD82A593252B4E
3107919498080101 0000000000000000 CB49A2F9E91363E3
10079115B9080140 0000000000000000 00B588BE70D23F56
3107911598080140 0000000000000000 406A9A6AB43399AE
1007D01589980101 0000000000000000 6CB773611DCA9ADA
9107911589980101 0000000000000000 67FD21C17DBB5D70
9107D01589190101 0000000000000000 9592CB4110430787
1007D01598980120 0000000000000000 A6B7FF68A318DDD3
1007940498190101 0000000000000000 4D102196C914CA16
0107910491190401 0000000000000000 2DFA9F4573594965
0107910491190101 0000000000000000 B46604816C0E0774
0107940491190401 0000000000000000 6E7E6221A4F34E87
19079210981A0101 0000000000000000 AA85E74643233199
1007911998190801 0000000000000000 2E5A19DB4D1962D6
10079119981A0801 0000000000000000 23A866A809D30894
1007921098190101 0000000000000000 D812D961F017D320
100791159819010B 0000000000000000 055605816E58608F
1004801598190101 0000000000000000 ABD88E8B1B7716F1
1004801598190102 0000000000000000 537AC95BE69DA1E1
1004801598190108 0000000000000000 AED0F6AE3C25CDD8
1002911498100104 0000000000000000 B3E35A5EE53E7B8D
1002911598190104 0000000000000000 61C79C71921A2EF8
1002911598100201 0000000000000000 E2F5728F0995013C
1002911698100101 0000000000000000 1AEAC39A61F0A464
```

*** S-Box test: ***

```
      Key            Plaintext       Expected Cipher
7CA110454A1A6E57 01A1D6D039776742 690F5B0D9A26939B
```

```
0131D9619DC1376E  5CD54CA83DEF57DA  7A389D10354BD271
07A1133E4A0B2686  0248D43806F67172  868EBB51CAB4599A
3849674C2602319E  51454B582DDF440A  7178876E01F19B2A
04B915BA43FEB5B6  42FD443059577FA2  AF37FB421F8C4095
0113B970FD34F2CE  059B5E0851CF143A  86A560F10EC6D85B
0170F175468FB5E6  0756D8E0774761D2  0CD3DA020021DC09
43297FAD38E373FE  762514B829BF486A  EA676B2CB7DB2B7A
07A7137045DA2A16  3BDD119049372802  DFD64A815CAF1A0F
04689104C2FD3B2F  26955F6835AF609A  5C513C9C4886C088
37D06BB516CB7546  164D5E404F275232  0A2AEEAE3FF4AB77
1F08260D1AC2465E  6B056E18759F5CCA  EF1BF03E5DFA575A
584023641ABA6176  004BD6EF09176062  88BF0DB6D70DEE56
025816164629B007  480D39006EE762F2  A1F9915541020B56
49793EBC79B3258F  437540C8698F3CFA  6FBF1CAFCFFD0556
4FB05E1515AB73A7  072D43A077075292  2F22E49BAB7CA1AC
49E95D6D4CA229BF  02FE55778117F12A  5A6B612CC26CCE4A
018310DC409B26D6  1D9D5C5018F728C2  5F4C038ED12B2E41
1C587F1C13924FEF  305532286D6F295A  63FAC0D034D9F793
*** End of Test ***
```

Chapter 8: Sorting

Chapter 8 of *Numerical Recipes* covers a variety of sorting tasks including sorting arrays into numerical order, preparing an index table for the order of an array, and preparing a rank table showing the rank order of each element in the array. PIKSRT sorts a single array by the straight insertion method. PIKSR2 sorts by the same method but makes the corresponding rearrangement of a second array as well. SHELQ carries out a Shell sort. SORT and SORT2 both do a Heapsort, and they are related in the same way as PIKSRT and PIKSR2. That is, SORT sorts a single array; SORT2 sorts an array while correspondingly rearranging a second array. QCKSRT sorts an array by the Quicksort algorithm, which is fast (on average) but requires a small amount of auxiliary storage.

INDEXX indexes an array. That is, it produces a second array that contains pointers to the elements of the original array in the order of their size. SORT3 uses INDEXX and illustrates its value by sorting one array while making corresponding rearrangements in two others. RANK produces the rank table for an array of data. The rank table is a second array whose elements list the rank order of the corresponding elements of the original array.

Finally, the routines ECLASS and ECLAZZ deal with equivalence classes. ECLASS gives the equivalence class of each element in an array based on a list of equivalent pairs which it is given as input. ECLAZZ gives the same output but bases it on a procedure named EQUIV(J,K) which tells whether two array elements J and K are in the same equivalence class.

$$\star \quad \star \quad \star \quad \star$$

Routine PIKSRT sorts an array by straight insertion. Sample program D8R1 provides it with a 100-element array from file TARRAY.DAT which is listed in the Appendix to this chapter. The program prints both the original and the sorted array for comparison.

Here is the recipe PIKSRT:

```
SUB PIKSRT (N, ARR())
        Sorts an array ARR of length N into ascending numerical order, by straight insertion. N is input;
        ARR is replaced on output by its sorted rearrangement.
FOR J = 2 TO N                 Pick out each element in turn.
  A = ARR(J)
  FOR I = J - 1 TO 1 STEP -1    Look for the place to insert it.
    IF ARR(I) <= A THEN EXIT FOR
    ARR(I + 1) = ARR(I)
  NEXT I
  IF ARR(I) > A THEN I = 0
  ARR(I + 1) = A                Insert it.
NEXT J
END SUB
```

A sample program using `PIKSRT` is the following:

```
DECLARE SUB PIKSRT (N!, ARR!())

'PROGRAM D8R1
'Driver for routine PIKSRT
CLS
DIM A(100)
OPEN "TARRAY.DAT" FOR INPUT AS #1
FOR I = 1 TO 10
  LINE INPUT #1, DUM$
  FOR J = 1 TO 10
    A(10 * (I - 1) + J) = VAL(MID$(DUM$, 6 * J - 5, 6))
  NEXT J
NEXT I
CLOSE #1
'Print original array
PRINT "Original array:"
FOR I = 1 TO 10
  FOR J = 1 TO 10
    PRINT USING "###.##"; A(10 * (I - 1) + J);
  NEXT J
  PRINT
NEXT I
'Sort array
CALL PIKSRT(100, A())
'Print sorted array
PRINT "Sorted array:"
FOR I = 1 TO 10
  FOR J = 1 TO 10
    PRINT USING "###.##"; A(10 * (I - 1) + J);
  NEXT J
  PRINT
NEXT I
END
```

`PIKSR2` sorts an array, and simultaneously rearranges a second array (of the same size) correspondingly. In program `D8R2`, the first array `A(I)` is again taken from `TARRAY.DAT`. The second is defined by `B(I)=I`. In other words, `B` is originally sorted and `A` is not. After a call to `PIKSR2`, the situation should be reversed. With a second call, this time with `B` as the first argument and `A` as the second, the two arrays should be returned to their original form.

Here is the recipe `PIKSR2`:

```
SUB PIKSR2 (N, ARR(), BRR())
    Sorts an array ARR of length N into ascending numerical order, by straight insertion, while
    making the corresponding rearrangement of the array BRR.
FOR J = 2 TO N                       Pick out each element in turn.
  A = ARR(J)
  B = BRR(J)
  FOR I = J - 1 TO 1 STEP -1         Look for the place to insert it.
    IF ARR(I) <= A THEN EXIT FOR
    ARR(I + 1) = ARR(I)
```

```
    BRR(I + 1) = BRR(I)
   NEXT I
   IF ARR(I) > A THEN I = 0
   ARR(I + 1) = A                    Insert it.
   BRR(I + 1) = B
 NEXT J
END SUB
```

A sample program using PIKSR2 is the following:

```
DECLARE SUB PIKSR2 (N!, ARR!(), BRR!())

'PROGRAM D8R2
'Driver for routine PIKSR2
CLS
DIM A(100), B(100)
OPEN "TARRAY.DAT" FOR INPUT AS #1
FOR I = 1 TO 10
  LINE INPUT #1, DUM$
  FOR J = 1 TO 10
    A(10 * (I - 1) + J) = VAL(MID$(DUM$, 6 * J - 5, 6))
  NEXT J
NEXT I
CLOSE #1
'Generate B-array
FOR I = 1 TO 100
  B(I) = I
NEXT I
'Sort A and mix B
CALL PIKSR2(100, A(), B())
PRINT "After sorting A and mixing B, array A is:"
FOR I = 1 TO 10
  FOR J = 1 TO 10
    PRINT USING "###.##"; A(10 * (I - 1) + J);
  NEXT J
  PRINT
NEXT I
PRINT "...and array B is:"
FOR I = 1 TO 10
  FOR J = 1 TO 10
    PRINT USING "###.##"; B(10 * (I - 1) + J);
  NEXT J
  PRINT
NEXT I
PRINT "press RETURN to continue..."
LINE INPUT DUM$
'Sort B and mix A
CALL PIKSR2(100, B(), A())
PRINT "After sorting B and mixing A, array A is:"
FOR I = 1 TO 10
  FOR J = 1 TO 10
    PRINT USING "###.##"; A(10 * (I - 1) + J);
  NEXT J
  PRINT
NEXT I
```

```
PRINT "...and array B is:"
FOR I = 1 TO 10
  FOR J = 1 TO 10
    PRINT USING "###.##"; B(10 * (I - 1) + J);
  NEXT J
  PRINT
NEXT I
END
```

Subroutine SHELQ does a Shell sort of a data array. The calling format is identical to that of PIKSRT, and so we use the same sample program, now called D8R9.

Here is the recipe SHELQ:

```
SUB SHELQ (N, ARR())
```
       Sorts an array ARR of length N into ascending numerical order, by the Shell-Mezgar algorithm (diminishing increment sort). N is input; ARR is replaced on output by its sorted rearrangement. The routine has been renamed to SHELQ since SHELL is a reserved word in BASIC.
```
ALN2I = 1.442695#
TINY = .00001
M = N
LOGNB2 = INT(LOG(CSNG(N)) * ALN2I + TINY)
FOR NN = 1 TO LOGNB2              Loop over the partial sorts.
  M = INT(M / 2)
  K = N - M
  FOR J = 1 TO K                  Outer loop of straight insertion.
    I = J
    DO                           Inner loop of straight insertion.
      DONE% = -1
      L = I + M
      IF ARR(L) < ARR(I) THEN
        T = ARR(I)
        ARR(I) = ARR(L)
        ARR(L) = T
        I = I - M
        IF I >= 1 THEN DONE% = 0
      END IF
    LOOP WHILE NOT DONE%
  NEXT J
NEXT NN
END SUB
```

A sample program using SHELQ is the following:

```
DECLARE SUB SHELQ (N!, ARR!())

'PROGRAM D8R9
'Driver for routine SHELQ
CLS
DIM A(100)
OPEN "TARRAY.DAT" FOR INPUT AS #1
FOR I = 1 TO 10
  LINE INPUT #1, DUM$
  FOR J = 1 TO 10
```

```
    A(10 * (I - 1) + J) = VAL(MID$(DUM$, 6 * J - 5, 6))
  NEXT J
NEXT I
CLOSE #1
'Print original array
PRINT "Original array:"
FOR I = 1 TO 10
  FOR J = 1 TO 10
    PRINT USING "###.##"; A(10 * (I - 1) + J);
  NEXT J
  PRINT
NEXT I
'Sort array
CALL SHELQ(100, A())
'Print sorted array
PRINT "Sorted array:"
FOR I = 1 TO 10
  FOR J = 1 TO 10
    PRINT USING "###.##"; A(10 * (I - 1) + J);
  NEXT J
  PRINT
NEXT I
END
```

By the same token, routines SORT and SORT2 employ the same programs as routines PIKSRT and PIKSR2, respectively. (Here they are called D8R3 and D8R4.) Both routines use the Heapsort algorithm. SORT, however, works on a single array. SORT2 sorts one array while making corresponding rearrangements to a second.

Here is the recipe SORT:

```
SUB SORT (N, RA())
```
    Sorts an array RA of length N into ascending numerical order using the Heapsort algorithm. N
    is input; RA is replaced on output by its sorted rearrangement.
```
L = INT(N / 2) + 1
IR = N
```
    The index L will be decremented from its initial value down to 1 during the "hiring" (heap
    creation) phase. Once it reaches 1, the index IR will be decremented from its initial value
    down to 1 during the "retirement-and-promotion" (heap selection) phase.
```
DO
  IF L > 1 THEN                    Still in hiring phase.
    L = L - 1
    RRA = RA(L)
  ELSE                            In retirement-and-promotion phase.
    RRA = RA(IR)                  Clear a space at end of array.
    RA(IR) = RA(1)                Retire the top of the heap into it.
    IR = IR - 1                   Decrease the size of the corporation.
    IF IR = 1 THEN                Done with the last promotion.
      RA(1) = RRA                 The least competent worker of all!
      EXIT SUB
    END IF
  END IF
  I = L                           Whether we are in the hiring phase or promotion phase, we
  J = L + L                       here set up to sift down element RRA to its proper level.
```

```
  WHILE J <= IR                              "Do while J less than or equal to IR:"
    IF J < IR THEN IF RA(J) < RA(J + 1) THEN J = J + 1   Compare to the better un-
    IF RRA < RA(J) THEN                      Demote RRA.              derling.
      RA(I) = RA(J)
      I = J
      J = J + J
    ELSE                                     This is RRA's level. Set J to terminate the sift-down.
      J = IR + 1
    END IF
  WEND
  RA(I) = RRA                                Put RRA into its slot.
LOOP
END SUB
```

A sample program using SORT is the following:

```
DECLARE SUB SORT (N!, RA!())

'PROGRAM D8R3
'Driver for routine SORT
CLS
DIM A(100)
OPEN "TARRAY.DAT" FOR INPUT AS #1
FOR I = 1 TO 10
  LINE INPUT #1, DUM$
  FOR J = 1 TO 10
    A(10 * (I - 1) + J) = VAL(MID$(DUM$, 6 * J - 5, 6))
  NEXT J
NEXT I
CLOSE #1
'Print original array
PRINT "Original array:"
FOR I = 1 TO 10
  FOR J = 1 TO 10
    PRINT USING "###.##"; A(10 * (I - 1) + J);
  NEXT J
  PRINT
NEXT I
'Sort array
CALL SORT(100, A())
'Print sorted array
PRINT "Sorted array:"
FOR I = 1 TO 10
  FOR J = 1 TO 10
    PRINT USING "###.##"; A(10 * (I - 1) + J);
  NEXT J
  PRINT
NEXT I
END
```

Here is the recipe SORT2:

```
SUB SORT2 (N, RA(), RB())
```
Sorts an array RA of length N into ascending numerical order using the Heapsort algorithm, while making the corresponding rearrangement of the array RB.
```
L = INT(N / 2) + 1
IR = N
DO
  IF L > 1 THEN
    L = L - 1
    RRA = RA(L)
    RRB = RB(L)
  ELSE
    RRA = RA(IR)
    RRB = RB(IR)
    RA(IR) = RA(1)
    RB(IR) = RB(1)
    IR = IR - 1
    IF IR = 1 THEN
      RA(1) = RRA
      RB(1) = RRB
      EXIT SUB
    END IF
  END IF
  I = L
  J = L + L
  WHILE J <= IR
    IF J < IR THEN
      IF RA(J) < RA(J + 1) THEN J = J + 1
    END IF
    IF RRA < RA(J) THEN
      RA(I) = RA(J)
      RB(I) = RB(J)
      I = J
      J = J + J
    ELSE
      J = IR + 1
    END IF
  WEND
  RA(I) = RRA
  RB(I) = RRB
LOOP
END SUB
```

A sample program using SORT2 is the following:

```
DECLARE SUB SORT2 (N!, RA!(), RB!())

'PROGRAM D8R4
'Driver for routine SORT2
CLS
DIM A(100), B(100)
OPEN "TARRAY.DAT" FOR INPUT AS #1
FOR I = 1 TO 10
  LINE INPUT #1, DUM$
```

```
    FOR J = 1 TO 10
      A(10 * (I - 1) + J) = VAL(MID$(DUM$, 6 * J - 5, 6))
    NEXT J
  NEXT I
  CLOSE #1
  'Generate B-array
  FOR I = 1 TO 100
    B(I) = I
  NEXT I
  'Sort A and mix B
  CALL SORT2(100, A(), B())
  PRINT "After sorting A and mixing B, array A is:"
  FOR I = 1 TO 10
    FOR J = 1 TO 10
      PRINT USING "###.##"; A(10 * (I - 1) + J);
    NEXT J
    PRINT
  NEXT I
  PRINT "...and array B is:"
  FOR I = 1 TO 10
    FOR J = 1 TO 10
      PRINT USING "###.##"; B(10 * (I - 1) + J);
    NEXT J
    PRINT
  NEXT I
  PRINT "press RETURN to continue..."
  LINE INPUT DUM$
  'Sort B and mix A
  CALL SORT2(100, B(), A())
  PRINT "After sorting B and mixing A, array A is:"
  FOR I = 1 TO 10
    FOR J = 1 TO 10
      PRINT USING "###.##"; A(10 * (I - 1) + J);
    NEXT J
    PRINT
  NEXT I
  PRINT "...and array B is:"
  FOR I = 1 TO 10
    FOR J = 1 TO 10
      PRINT USING "###.##"; B(10 * (I - 1) + J);
    NEXT J
    PRINT
  NEXT I
  END
```

The subroutine INDEXX generates the index array for a given input array. The index array INDX(J) gives, for each J, the index of the element of the input array which will assume position J if the array is sorted. That is, for an input array A, the sorted version of A will be A(INDX(J)). To demonstrate this, sample program D8R5 produces an index for the array in TARRAY.DAT. It then prints the array in the order A(INDX(J)), J=1,..,100 for inspection.

Here is the recipe INDEXX:

```
SUB INDEXX (N, ARRIN(), INDX())
    Indexes an array ARRIN of length N, i.e. outputs the array INDX such that ARRIN(INDX(J))
    is in ascending order for J= 1, 2, ..., N. The input quantities N and ARRIN are not changed.
FOR J = 1 TO N                          Initialize the index array with consecutive integers.
  INDX(J) = J
NEXT J
IF N = 1 THEN EXIT SUB
L = INT(N / 2 + 1)                      From here on, we just have Heapsort, but with indirect in-
IR = N                                  dexing through INDX in all references to ARRIN.
DO
  IF L > 1 THEN
    L = L - 1
    INDXT = INDX(L)
    Q = ARRIN(INDXT)
  ELSE
    INDXT = INDX(IR)
    Q = ARRIN(INDXT)
    INDX(IR) = INDX(1)
    IR = IR - 1
    IF IR = 1 THEN
      INDX(1) = INDXT
      EXIT SUB
    END IF
  END IF
  I = L
  J = L + L
  WHILE J <= IR
    IF J < IR THEN
      IF ARRIN(INDX(J)) < ARRIN(INDX(J + 1)) THEN J = J + 1
    END IF
    IF Q < ARRIN(INDX(J)) THEN
      INDX(I) = INDX(J)
      I = J
      J = J + J
    ELSE
      J = IR + 1
    END IF
  WEND
  INDX(I) = INDXT
LOOP
END SUB
```

A sample program using INDEXX is the following:

```
DECLARE SUB INDEXX (N!, ARRIN!(), INDX!())

'PROGRAM D8R5
'Driver for routine INDEXX
CLS
DIM A(100), INDX(100)
OPEN "TARRAY.DAT" FOR INPUT AS #1
FOR I = 1 TO 10
  LINE INPUT #1, DUM$
```

```
FOR J = 1 TO 10
  A(10 * (I - 1) + J) = VAL(MID$(DUM$, 6 * J - 5, 6))
NEXT J
NEXT I
CLOSE #1
'Generate index for sorted array
CALL INDEXX(100, A(), INDX())
'Print original array
PRINT "Original array:"
FOR I = 1 TO 10
  FOR J = 1 TO 10
    PRINT USING "###.##"; A(10 * (I - 1) + J);
  NEXT J
  PRINT
NEXT I
'Print sorted array
PRINT "Sorted array:"
FOR I = 1 TO 10
  FOR J = 1 TO 10
    PRINT USING "###.##"; A(INDX(10 * (I - 1) + J));
  NEXT J
  PRINT
NEXT I
END
```

One use for INDEXX is the management of more than two arrays. SORT3, for example, sorts one array while making corresponding reorderings of two other arrays. In sample program D8R6, the first array is taken as the first 64 elements of TARRAY.DAT (see Appendix). The second and third arrays are taken to be the numbers 1 to 64 in forward order and reverse order, respectively. When the first array is ordered, the second and third are scrambled, but scrambled in exactly the same way. To prove this, a text message is assigned to a character array. Then the letters are scrambled according to the order of numbers found in the rearranged second array. They are subsequently unscrambled according to the order of numbers found in the rearranged third array. If SORT3 works properly, this ought to leave the message reading in the reverse order.

Here is the recipe SORT3:

```
DECLARE SUB INDEXX (N!, ARRIN!(), INDX!())

SUB SORT3 (N, RA(), RB(), RC(), WKSP(), IWKSP())
      Sorts an array RA of length N into ascending numerical order while making the corresponding
      rearrangements of the arrays RB and RC. An index table is constructed via the routine INDEXX.
CALL INDEXX(N, RA(), IWKSP())        Make the index table.
FOR J = 1 TO N                       Save the array RA.
  WKSP(J) = RA(J)
NEXT J
FOR J = 1 TO N                       Copy it back in the rearranged order.
  RA(J) = WKSP(IWKSP(J))
NEXT J
FOR J = 1 TO N                       Ditto RB.
  WKSP(J) = RB(J)
NEXT J
FOR J = 1 TO N
```

```
  RB(J) = WKSP(IWKSP(J))
NEXT J
FOR J = 1 TO N                    Ditto RC.
  WKSP(J) = RC(J)
NEXT J
FOR J = 1 TO N
  RC(J) = WKSP(IWKSP(J))
NEXT J
END SUB
```

A sample program using SORT3 is the following:

```
DECLARE SUB SORT3 (N!, RA!(), RB!(), RC!(), WKSP!(), IWKSP!())

'PROGRAM D8R6
'Driver for routine SORT3
CLS
NLEN = 64
DIM A(NLEN), B(NLEN), C(NLEN), WKSP(NLEN), INDX(NLEN)
DIM AMSG$(NLEN), BMSG$(NLEN), CMSG$(NLEN)
MSG1$ = "I'd rather have a bottle in front"
MSG2$ = " of me than a frontal lobotomy."
PRINT "Original message:"
FOR J = 1 TO 64
  AMSG$(J) = MID$(MSG1$ + MSG2$, J, 1)
  PRINT AMSG$(J);
NEXT J
PRINT
'Read array of random numbers
OPEN "TARRAY.DAT" FOR INPUT AS #1
FOR I = 1 TO INT(NLEN / 10)
  LINE INPUT #1, DUM$
  FOR J = 1 TO 10
    A(10 * (I - 1) + J) = VAL(MID$(DUM$, 6 * J - 5, 6))
  NEXT J
NEXT I
LINE INPUT #1, DUM$
FOR J = 1 TO NLEN - 10 * INT(NLEN / 10)
  A(10 * INT(NLEN / 10) + J) = VAL(MID$(DUM$, 6 * J - 5, 6))
NEXT J
CLOSE #1
'Create array B and array C
FOR I = 1 TO NLEN
  B(I) = I
  C(I) = NLEN + 1 - I
NEXT I
'Sort array A while mixing B and C
CALL SORT3(NLEN, A(), B(), C(), WKSP(), INDX())
'Scramble message according to array B
FOR I = 1 TO NLEN
  J = B(I)
  BMSG$(I) = AMSG$(J)
NEXT I
PRINT
PRINT "Scrambled message:"
```

```
FOR J = 1 TO 64
  PRINT BMSG$(J);
NEXT J
PRINT
'Unscramble according to array C
FOR I = 1 TO NLEN
  J = C(I)
  CMSG$(J) = BMSG$(I)
NEXT I
PRINT
PRINT "Mirrored message:"
FOR J = 1 TO 64
  PRINT CMSG$(J);
NEXT J
PRINT
END
```

RANK is a subroutine that is similar to INDEXX. Instead of producing an indexing array, though, it produces a rank table. For an array A(J) and rank table IRANK(J), entry J in IRANK will tell what index A(J) will have if A is sorted. IRANK actually takes its input information not from the array itself, but from the index array produced by INDEXX. Sample program D8R7 begins with the array from TARRAY, and feeds it to INDEXX and RANK. The table of ranks produced is listed. To check it, the array A is copied into an array B in the rank order suggested by IRANK. B should then be in proper order.

Here is the recipe RANK:

```
SUB RANK (N, INDX(), IRANK())
     Given INDX of length N as output from the routine INDEXX, this routine returns an array
     IRANK, the corresponding table of ranks.
FOR J = 1 TO N
  IRANK(INDX(J)) = J
NEXT J
END SUB
```

A sample program using RANK is the following:

```
DECLARE SUB RANK (N!, INDX!(), IRANK!())
DECLARE SUB INDEXX (N!, ARRIN!(), INDX!())

'PROGRAM D8R7
'Driver for routine RANK
CLS
DIM A(100), B(10), INDX(100), IRANK(100)
OPEN "TARRAY.DAT" FOR INPUT AS #1
FOR I = 1 TO 10
  LINE INPUT #1, DUM$
  FOR J = 1 TO 10
    A(10 * (I - 1) + J) = VAL(MID$(DUM$, 6 * J - 5, 6))
  NEXT J
NEXT I
CLOSE #1
CALL INDEXX(100, A(), INDX())
```

```
CALL RANK(100, INDX(), IRANK())
PRINT "Original array is:"
FOR I = 1 TO 10
  FOR J = 1 TO 10
    PRINT USING "###.##"; A(10 * (I - 1) + J);
  NEXT J
  PRINT
NEXT I
PRINT "Table of ranks is:"
FOR I = 1 TO 10
  FOR J = 1 TO 10
    PRINT USING "######"; IRANK(10 * (I - 1) + J);
  NEXT J
  PRINT
NEXT I
PRINT "press RETURN to continue..."
LINE INPUT DUM$
PRINT "Array sorted according to rank table:"
FOR I = 1 TO 10
  FOR J = 1 TO 10
    K = 10 * (I - 1) + J
    FOR L = 1 TO 100
      IF IRANK(L) = K THEN B(J) = A(L)
    NEXT L
  NEXT J
  FOR J = 1 TO 10
    PRINT USING "###.##"; B(J);
  NEXT J
  PRINT
NEXT I
END
```

QCKSRT sorts an array by the Quicksort algorithm. Its calling sequence is exactly like that of PIKSRT and SORT, so we again rely on the same sample program, now called D8R8.

Here is the recipe QCKSRT:

```
SUB QCKSRT (N, ARR())
```
> Sorts an array ARR of length N into ascending numerical order using the Quicksort algorithm. N is input; ARR is replaced on output by its sorted rearrangement.

```
M = 7
NSTACK = 50
FM = 7875!
FA = 211!
FC = 1663!
FMI = .00012698413#
```
> Here M is the size of subarrays sorted by straight insertion, NSTACK is the required auxiliary storage, and the remaining constants are used by the random number generating statements.

```
DIM ISTACK(NSTACK)
JSTACK = 0
L = 1
IR = N
FX = 0!
DO
```

```
IF IR - L < M THEN                      Sort by straight insertion:
  FOR J = L + 1 TO IR
    A = ARR(J)
    FOR I = J - 1 TO 1 STEP -1
      IF ARR(I) <= A THEN EXIT FOR
      ARR(I + 1) = ARR(I)
    NEXT I
    IF ARR(I) > A THEN I = 0
    ARR(I + 1) = A
  NEXT J
  IF JSTACK = 0 THEN ERASE ISTACK: EXIT SUB
  IR = ISTACK(JSTACK)
  L = ISTACK(JSTACK - 1)
  JSTACK = JSTACK - 2
ELSE
  I = L
  J = IR
  FX = FX * FA + FC - FM * INT((FX * FA + FC) / FM)    Generate a random inte-
  IQ = L + (IR - L + 1) * (FX * FMI)                   ger IQ between L and IR,
  A = ARR(IQ)                                          inclusive.
  ARR(IQ) = ARR(L)
  DO
    DO
      IF J > 0 THEN
        IF A < ARR(J) THEN
          J = J - 1
          DONE% = 0
        ELSE
          DONE% = -1
        END IF
      END IF
    LOOP WHILE NOT DONE%
    IF J <= I THEN
      ARR(I) = A
      EXIT DO
    END IF
    ARR(I) = ARR(J)
    I = I + 1
    DO
      IF I <= N THEN
        IF A > ARR(I) THEN
          I = I + 1
          DONE% = 0
        ELSE
          DONE% = -1
        END IF
      END IF
    LOOP WHILE NOT DONE%
    IF J <= I THEN
      ARR(J) = A
      I = J
      EXIT DO
    END IF
    ARR(J) = ARR(I)
    J = J - 1
```

```
    LOOP
    JSTACK = JSTACK + 2
    IF JSTACK > NSTACK THEN PRINT "NSTACK must be made larger.": EXIT SUB
    IF IR - I >= I - L THEN
      ISTACK(JSTACK) = IR
      ISTACK(JSTACK - 1) = I + 1
      IR = I - 1
    ELSE
      ISTACK(JSTACK) = I - 1
      ISTACK(JSTACK - 1) = L
      L = I + 1
    END IF
  END IF
LOOP
END SUB
```

A sample program using QCKSRT is the following:

```
DECLARE SUB QCKSRT (N!, ARR!())

'PROGRAM D8R8
'Driver for routine QCKSRT
CLS
DIM A(100)
OPEN "TARRAY.DAT" FOR INPUT AS #1
FOR I = 1 TO 10
  LINE INPUT #1, DUM$
  FOR J = 1 TO 10
    A(10 * (I - 1) + J) = VAL(MID$(DUM$, 6 * J - 5, 6))
  NEXT J
NEXT I
CLOSE #1
'Print original array
PRINT "Original array:"
FOR I = 1 TO 10
  FOR J = 1 TO 10
    PRINT USING "###.##"; A(10 * (I - 1) + J);
  NEXT J
  PRINT
NEXT I
'Sort array
CALL QCKSRT(100, A())
'Print sorted array
PRINT "Sorted array:"
FOR I = 1 TO 10
  FOR J = 1 TO 10
    PRINT USING "###.##"; A(10 * (I - 1) + J);
  NEXT J
  PRINT
NEXT I
END
```

Subroutine ECLASS generates a list of equivalence classes for the elements of an input array, based on the arrays LISTA(J) and LISTB(J) which list equivalent pairs for each J. In sample program D8R10, these lists are

$$LISTA: \quad 1,1,5,2,6,2,7,11,3,4,12$$

$$LISTB: \quad 5,9,13,6,10,14,3,7,15,8,4$$

According to these lists, 1 is equivalent to 5, 1 is equivalent to 9, etc. If you work it out, you will find the following classes:

$$class1: \quad 1,5,9,13$$

$$class2: \quad 2,6,10,14$$

$$class3: \quad 3,7,11,15$$

$$class4: \quad 4,8,12$$

The sample program prints out the classes and ought to agree with this list.

Here is the recipe ECLASS:

```
SUB ECLASS (NF(), N, LISTA(), LISTB(), M)
        Given M equivalences between pairs of N individual elements in the form of the input arrays
        LISTA and LISTB, this routine returns in NF the number of the equivalence class of each of
        the N elements, integers between 1 and N (not all such integers used).
FOR K = 1 TO N                    Initialize each element its own class.
  NF(K) = K
NEXT K
FOR L = 1 TO M                    For each piece of input information...
  J = LISTA(L)
  WHILE NF(J) <> J                Track first element up to its ancestor.
    J = NF(J)
  WEND
  K = LISTB(L)
  WHILE NF(K) <> K                Track second element up to its ancestor.
    K = NF(K)
  WEND
  IF J <> K THEN NF(J) = K        If they are not already related, make them so.
NEXT L
FOR J = 1 TO N                    Final sweep up to highest ancestors.
  WHILE NF(J) <> NF(NF(J))
    NF(J) = NF(NF(J))
  WEND
NEXT J
END SUB
```

A sample program using ECLASS is the following:

```
DECLARE SUB ECLASS (NF!(), N!, LISTA!(), LISTB!(), M!)

'PROGRAM D8R10
'Driver for routine ECLASS
CLS
N = 15
M = 11
DIM LISTA(M), LISTB(M), NF(N), NFLAG(N), NSAV(N)
FOR I = 1 TO M
  READ LISTA(I)
NEXT I
DATA 1,1,5,2,6,2,7,11,3,4,12
FOR I = 1 TO M
  READ LISTB(I)
NEXT I
DATA 5,9,13,6,10,14,3,7,15,8,4
CALL ECLASS(NF(), N, LISTA(), LISTB(), M)
FOR I = 1 TO N
  NFLAG(I) = 1
NEXT I
PRINT "Numbers from 1-15 divided according to"
PRINT "their value modulo 4:"
PRINT
LCLAS = 0
FOR I = 1 TO N
  NCLASS = NF(I)
  IF NFLAG(NCLASS) <> 0 THEN
    NFLAG(NCLASS) = 0
    LCLAS = LCLAS + 1
    K = 0
    FOR J = I TO N
      IF NF(J) = NF(I) THEN
        K = K + 1
        NSAV(K) = J
      END IF
    NEXT J
    PRINT "Class"; LCLAS; ":   ";
    FOR J = 1 TO K
      PRINT USING "###"; NSAV(J);
    NEXT J
    PRINT
  END IF
NEXT I
END
```

ECLAZZ performs the same analysis but figures the equivalences from a logical function **EQUIV(I,J)** that tells whether I and J are in the same equivalence class. In **D8R11**, **EQUIV%** is defined as −1 if I MOD 4 and J MOD 4 are the same. It is otherwise 0.

Here is the recipe **ECLAZZ**:

```
DECLARE FUNCTION EQUIV% (J!, K!)

SUB ECLAZZ (NF(), N, DUM)
```
Given a user-supplied logical function EQUIV% which tells whether a pair of elements, each in the range 1...N, are related, return in NF equivalence class numbers for each element.
```
NF(1) = 1
FOR JJ = 2 TO N                      Loop over first element of all pairs.
  NF(JJ) = JJ
  FOR KK = 1 TO JJ - 1               Loop over second element of all pairs.
    NF(KK) = NF(NF(KK))              Sweep it up this much.
    IF EQUIV%(JJ, KK) THEN NF(NF(NF(KK))) = JJ    Good exercise for the reader to fig-
  NEXT KK                                         ure out why this much ancestry is
NEXT JJ                                           necessary!
FOR JJ = 1 TO N                      Only this much sweeping is needed finally.
  NF(JJ) = NF(NF(JJ))
NEXT JJ
END SUB
```

A sample program using ECLAZZ is the following:

```
DECLARE FUNCTION EQUIV% (I!, J!)
DECLARE SUB ECLAZZ (NF!(), N!, DUM!)

'PROGRAM D8R11
'Driver for routine ECLAZZ
CLS
N = 15
DIM NF(N), NFLAG(N), NSAV(N)
CALL ECLAZZ(NF(), N, DUM)
FOR I = 1 TO N
  NFLAG(I) = 1
NEXT I
PRINT "Numbers from 1-15 divided according to"
PRINT "their value modulo 4:"
PRINT
LCLAS = 0
FOR I = 1 TO N
  NCLASS = NF(I)
  IF NFLAG(NCLASS) <> 0 THEN
    NFLAG(NCLASS) = 0
    LCLAS = LCLAS + 1
    K = 0
    FOR J = I TO N
      IF NF(J) = NF(I) THEN
        K = K + 1
        NSAV(K) = J
      END IF
    NEXT J
    PRINT "Class"; LCLAS; ":   ";
    FOR J = 1 TO K
      PRINT USING "###"; NSAV(J);
    NEXT J
    PRINT
  END IF
```

```
NEXT I
END

FUNCTION EQUIV% (I, J)
EQUIV% = 0
IF (I MOD 4) = (J MOD 4) THEN EQUIV% = -1
END FUNCTION
```

Appendix

File **TARRAY.DAT**:

```
29.82 71.51  3.30 87.44 53.42 63.16 89.10 25.75 93.16 27.72
71.58 48.34 53.11 18.34 27.13 60.31 83.34 22.81 66.84 52.91
53.42 15.22  8.01 53.39 76.12 79.09 67.61 38.39 24.81 73.21
13.42 52.10 34.86 99.83 38.46 81.59 61.75 79.62 93.39  3.21
99.34 92.22 94.29  7.03  6.67 89.35 83.14  9.01 12.68 62.22
 2.95 85.02 95.82 73.96 49.29 77.72 36.65  3.48 48.98 71.83
 1.41  9.48 32.37 89.95 28.39 79.36 54.05 46.08 11.67 37.78
77.17 74.33 10.13  4.62 49.95 68.40 19.40 34.06  4.11 98.40
42.44 64.14 89.41 52.99 71.79  3.94 19.73 44.91 71.44 59.10
27.54 15.67 67.95 55.61 26.05 25.01 82.09 89.67 57.08 38.27
```

Chapter 9: Root Finding and Sets of Equations

Chapter 9 of *Numerical Recipes* deals primarily with the problem of finding roots to equations, and treats the problem in greatest detail in one dimension. We begin with a general-purpose routine called SCRSHO that produces a crude graph of a given function on a specified interval. It is used for low-resolution plotting to investigate the properties of the function. With this in hand, we add bracketing routines ZBRAC and ZBRAK. The first of these takes a function and an interval, and expands the interval geometrically until it brackets a root. The second breaks the interval into N subintervals of equal size. It then reports any intervals that contain at least one root. Once bracketed, roots can be found by a number of other routines. RTBIS finds such roots by bisection. RTFLSP and RTSEC use the method of false position and the secant method, respectively. ZBRENT uses a combination of methods to give assured and relatively efficient convergence. RTNEWT implements the Newton-Raphson root-finding method, while RTSAFE combines it with bisection to correct for its risky global convergence properties.

For finding the roots of polynomials, LAGUER is handy, and when combined with its driver ZROOTS it can find all roots of a polynomial having complex coefficients. When you have some tentative complex roots of a real polynomial, they can be polished by QROOT, which employs Bairstow's method.

In multiple dimensions, root finding requires some foresight. However, if you can identify the neighborhood of a root of a system of nonlinear equations, then MNEWT will help you to zero in using Newton-Raphson.

$$\star \quad \star \quad \star \quad \star$$

SCRSHO is a primitive graphing routine that will print graphs on virtually any terminal or printer. Sample program D9R1 demonstrates it by graphing the zero-order Bessel function J_0.

Here is the recipe SCRSHO:

```
DECLARE FUNCTION FUNC! (X!)

SUB SCRSHO (DUM)
        For interactive CRT terminal use. Produce a crude graph of the user-supplied function FUNC
        over the interval X1,X2. Query for another plot interval until the user signals satisfaction.
    ISCR = 60                       Number of horizontal and vertical positions in display.
    JSCR = 21
    DIM Y(ISCR), SCR$(ISCR, JSCR)
    BLANK$ = " "
    ZERO$ = "-"
    YY$ = "l"
    XX$ = "-"
    FF$ = "x"
    DO
```

```
PRINT " Enter X1,X2 (= to stop)"          Query for another plot, quit if X1=X2
INPUT X1, X2
IF X1 = X2 THEN ERASE SCR$, Y: EXIT SUB
FOR J = 1 TO JSCR                    Fill vertical sides with character '1'.
  SCR$(1, J) = YY$
  SCR$(ISCR, J) = YY$
NEXT J
FOR I = 2 TO ISCR - 1
  SCR$(I, 1) = XX$                   Fill top, bottom with character '-'.
  SCR$(I, JSCR) = XX$
  FOR J = 2 TO JSCR - 1              Fill interior with blanks.
    SCR$(I, J) = BLANK$
  NEXT J
NEXT I
DX = (X2 - X1) / CSNG(ISCR - 1)
X = X1
YBIG = 0!                            Limits will include 0.
YSML = YBIG
FOR I = 1 TO ISCR                    Evaluate the function at equal intervals.  Find the largest and
  Y(I) = FUNC(X)                     smallest values.
  IF Y(I) < YSML THEN YSML = Y(I)
  IF Y(I) > YBIG THEN YBIG = Y(I)
  X = X + DX
NEXT I
IF YBIG = YSML THEN YBIG = YSML + 1!     Be sure to separate top and bottom.
DYJ = CSNG(JSCR - 1) / (YBIG - YSML)
JZ = 1 - YSML * DYJ                  Note which row corresponds to 0.
FOR I = 1 TO ISCR                    Place an indicator at function height and 0.
  SCR$(I, JZ) = ZERO$
  J = 1 + (Y(I) - YSML) * DYJ
  SCR$(I, J) = FF$
NEXT I
PRINT USING "##.###^^^^"; YBIG;
PRINT " ";
FOR I = 1 TO ISCR
  PRINT SCR$(I, JSCR);
NEXT I
PRINT
FOR J = JSCR - 1 TO 2 STEP -1          Display.
  PRINT "          ";
  FOR I = 1 TO ISCR
    PRINT SCR$(I, J);
  NEXT I
  PRINT
NEXT J
PRINT USING "##.###^^^^"; YSML;
PRINT " ";
FOR I = 1 TO ISCR
  PRINT SCR$(I, 1);
NEXT I
PRINT
PRINT "            ";
PRINT USING "##.###^^^^"; X1;
PRINT SPACE$(40);
PRINT USING "##.###^^^^"; X2
```

```
LOOP
END SUB
```

A sample program using SCRSHO is the following:

```
DECLARE FUNCTION FUNC! (X!)
DECLARE SUB SCRSHO (DUM!)
DECLARE FUNCTION BESSJO! (X!)

'PROGRAM D9R1
'Driver for routine SCRSHO
CLS
PRINT "Graph of the Bessel Function J0:"
CALL SCRSHO(DUM)
END

FUNCTION FUNC (X)
FUNC = BESSJO(X)
END FUNCTION
```

ZBRAC is a root-bracketing routine that works by expanding the range of an interval geometrically until it brackets a root. Sample program D9R2 applies it to the Bessel function J_0. It starts with the ten intervals (1.0, 2.0), (2.0, 3.0), etc., and expands each until it contains a root. Then it prints the interval limits, and the function J_0 evaluated at these limits. The two values of J_0 should have opposite signs.

Here is the recipe ZBRAC:

```
DECLARE FUNCTION FUNC! (X!)

SUB ZBRAC (DUM, X1, X2, SUCCES%)
      Given a user-supplied function FUNC and an initial guessed range X1 to X2, the routine expands
      the range geometrically until a root is bracketed by the returned values X1 and X2 (in which
      case SUCCES% returns as -1) or until the range becomes unacceptably large (in which case
      SUCCES% returns as 0).
FACTOR = 1.6
NTRY = 50
IF X1 = X2 THEN PRINT "You have to guess an initial range": EXIT SUB
F1 = FUNC(X1)
F2 = FUNC(X2)
SUCCES% = -1
FOR J = 1 TO NTRY
  IF F1 * F2 < 0! THEN EXIT SUB
  IF ABS(F1) < ABS(F2) THEN
    X1 = X1 + FACTOR * (X1 - X2)
    F1 = FUNC(X1)
  ELSE
    X2 = X2 + FACTOR * (X2 - X1)
    F2 = FUNC(X2)
  END IF
NEXT J
SUCCES% = 0
END SUB
```

A sample program using **ZBRAC** is the following:

```
DECLARE FUNCTION FUNC! (X!)
DECLARE SUB ZBRAC (DUM!, X1!, X2!, SUCCES%)
DECLARE FUNCTION BESSJ0! (X!)

'PROGRAM D9R2
'Driver for routine ZBRAC
CLS
PRINT "  Bracketing values:         Function values:"
PRINT
PRINT "    X1         X2          BESSJ0(X1)  BESSJ0(X2)"
PRINT
FOR I = 1 TO 10
  X1 = I
  X2 = X1 + 1!
  CALL ZBRAC(DUM, X1, X2, SUCCES%)
  IF SUCCES% THEN
    PRINT USING "####.##"; X1;
    PRINT "   ";
    PRINT USING "####.##"; X2;
    PRINT "         ";
    PRINT USING "#.######"; BESSJ0(X1);
    PRINT "    ";
    PRINT USING "#.######"; BESSJ0(X2)
  END IF
NEXT I
END

FUNCTION FUNC (X)
FUNC = BESSJ0(X)
END FUNCTION
```

ZBRAK is much like **ZBRAC** except that it takes an interval and subdivides it into N equal parts. It then identifies any of the subintervals that contain roots. Sample program **D9R3** looks for roots of $J_0(x)$ between $X1 = 1.0$ and $X2 = 50.0$ by allowing **ZBRAK** to divide the interval into $N = 100$ parts. If there are no roots spaced closer than $\Delta x = 0.49$, then it will find brackets for all roots in this region. The limits of bracketing intervals, as well as function values at these limits, are printed, and again the function values at the end of each interval ought to be of opposite sign. There are 16 roots of J_0 between 1 and 50.

Here is the recipe **ZBRAK**:

```
DECLARE FUNCTION FUNC! (X!)

SUB ZBRAK (DUM, X1, X2, N, XB1(), XB2(), NB)
```
     Given a user-supplied function **FUNC** defined on the interval from **X1**–**X2** subdivide the interval into **N** equally spaced segments, and search for zero crossings of the function. **NB** is input as the maximum number of roots sought, and is reset to the number of bracketing pairs **XB1**, **XB2** that are found.

```
NBB = NB
NB = 0
X = X1
DX = (X2 - X1) / N        Determine the spacing appropriate to the mesh.
```

```
FP = FUNC(X)
FOR I = 1 TO N                          Loop over all intervals
  X = X + DX
  FC = FUNC(X)
  IF FC * FP < 0! THEN                  If a sign change occurs then record values for the bounds.
    NB = NB + 1
    XB1(NB) = X - DX
    XB2(NB) = X
  END IF
  FP = FC
  IF NBB = NB THEN EXIT SUB
NEXT I
END SUB
```

A sample program using **ZBRAK** is the following:

```
DECLARE FUNCTION FUNC! (X!)
DECLARE SUB ZBRAK (DUM!, X1!, X2!, N!, XB1!(), XB2!(), NB!)
DECLARE FUNCTION BESSJ0! (X!)

'PROGRAM D9R3
'Driver for routine ZBRAK
CLS
N = 100
NBMAX = 20
X1 = 1!
X2 = 50!
DIM XB1(NBMAX), XB2(NBMAX)
NB = NBMAX
CALL ZBRAK(DUM, X1, X2, N, XB1(), XB2(), NB)
PRINT "Brackets for roots of BESSJ0:"
PRINT
PRINT
PRINT "                 lower      upper          F(lower)   F(upper)"
PRINT
FOR I = 1 TO NB
  PRINT "Root";
  PRINT USING "###"; I;
  PRINT USING "#########.####"; XB1(I);
  PRINT USING "#####.####"; XB2(I);
  PRINT USING "###########.####"; BESSJ0(XB1(I));
  PRINT USING "#####.####"; BESSJ0(XB2(I))
NEXT I
END

FUNCTION FUNC (X)
FUNC = BESSJ0(X)
END FUNCTION
```

Routine RTBIS begins with the brackets for a root and finds the root itself by bisection, The accuracy with which the root is found is determined by parameter XACC. Sample program D9R4 finds all the roots of Bessel function $J_0(x)$ between X1 = 1.0 and X2 = 50.0. In this case XACC is specified to be about 10^{-6} of the value of the root itself (actually, 10^{-6} of the center of the interval being bisected). The roots ROOT are listed, as well as $J_0(\text{ROOT})$ to verify their accuracy.

Here is the recipe RTBIS:

```
DECLARE FUNCTION RTBIS! (DUM!, X1!, X2!, XACC!)
DECLARE FUNCTION FUNC! (X!)

FUNCTION RTBIS (DUM, X1, X2, XACC)
     Using bisection, find the root of a user-supplied function FUNC known to lie between X1 and
     X2. The root, returned as RTBIS, will be refined until its accuracy is ±XACC.
JMAX = 40                          Maximum allowed number of bisections.
FMID = FUNC(X2)
F = FUNC(X1)
IF F * FMID >= 0! THEN
  PRINT "Root must be bracketed for bisection."
  EXIT FUNCTION
END IF
IF F < 0! THEN                     Orient the search so that F>0 lies at X+DX.
  TRTBIS = X1
  DX = X2 - X1
ELSE
  TRTBIS = X2
  DX = X1 - X2
END IF
FOR J = 1 TO JMAX                  Bisection loop
  DX = DX * .5
  XMID = TRTBIS + DX
  FMID = FUNC(XMID)
  IF FMID <= 0! THEN TRTBIS = XMID
  RTBIS = TRTBIS
  IF ABS(DX) < XACC OR FMID = 0 THEN EXIT FUNCTION
NEXT J
PRINT "too many bisections"
END FUNCTION
```

A sample program using RTBIS is the following:

```
DECLARE FUNCTION FUNC! (X!)
DECLARE SUB ZBRAK (DUM!, X1!, X2!, N!, XB1!(), XB2!(), NB!)
DECLARE FUNCTION BESSJ0! (X!)
DECLARE FUNCTION RTBIS! (DUM!, X1!, X2!, XA!)

'PROGRAM D9R4
'Driver for routine RTBIS
CLS
N = 100
NBMAX = 20
X1 = 1!
X2 = 50!
DIM XB1(NBMAX), XB2(NBMAX)
```

```
NB = NBMAX
CALL ZBRAK(DUM, X1, X2, N, XB1(), XB2(), NB)
PRINT "Roots of BESSJO:"
PRINT
PRINT "                x              F(x)"
PRINT
FOR I = 1 TO NB
  XACC = .000001 * (XB1(I) + XB2(I)) / 2!
  ROOT = RTBIS(DUM, XB1(I), XB2(I), XACC)
  PRINT "Root";
  PRINT USING "###"; I;
  PRINT USING "#######.######"; ROOT;
  PRINT "     ";
  PRINT USING "#.####^^^^"; BESSJO(ROOT)
NEXT I
END


FUNCTION FUNC (X)
FUNC = BESSJO(X)
END FUNCTION
```

The next five sample programs, D9R5–D9R9, are essentially identical to the one just discussed, except for the root-finder they employ. D9R5 calls RTFLSP, finding the root by "false position". D9R6 calls RTSEC and uses the secant method. D9R7 uses ZBRENT to give reliable and efficient convergence. The Newton-Raphson method implemented in RTNEWT is demonstrated by D9R8, and D9R9 calls RTSAFE, which improves upon RTNEWT by combining it with bisection to achieve better global convergence. The latter two programs include a subroutine FUNCD that returns the value of the function and its derivative at a given x. In the case of test function $J_0(x)$ the derivative is $-J_1(x)$, and is conveniently in our collection of special functions.

Here is the recipe RTFLSP:

```
DECLARE FUNCTION RTFLSP! (DUM!, X1!, X2!, XACC!)
DECLARE FUNCTION FUNC! (X!)

FUNCTION RTFLSP (DUM, X1, X2, XACC)
        Using the false position method, find the root of a user-supplied function FUNC known to lie
        between X1 and X2. The root, returned as RTFLSP, is refined until its accuracy is ±XACC.
MAXIT = 30                     Set MAXIT to the maximum allowed number of iterations.
FL = FUNC(X1)
FH = FUNC(X2)                  Be sure the interval brackets a root.
IF FL * FH > 0! THEN
  PRINT "Root must be bracketed for false position."
  EXIT FUNCTION
END IF
IF FL < 0! THEN                Identify the limits so that XL corresponds to the low side.
  XL = X1
  XH = X2
ELSE
  XL = X2
  XH = X1
  SWAP FL, FH
END IF
```

```
DX = XH - XL
FOR J = 1 TO MAXIT                    False position loop.
  TRTFLSP = XL + DX * FL / (FL - FH)  Increment with respect to latest value.
  F = FUNC(TRTFLSP)
  IF F < 0! THEN                      Replace appropriate limit.
    DEL = XL - TRTFLSP
    XL = TRTFLSP
    FL = F
  ELSE
    DEL = XH - TRTFLSP
    XH = TRTFLSP
    FH = F
  END IF
  DX = XH - XL                        Convergence.
  RTFLSP = TRTFLSP
  IF ABS(DEL) < XACC OR F = 0! THEN EXIT FUNCTION
NEXT J
PRINT "RTFLSP exceed maximum iterations"
END FUNCTION
```

A sample program using RTFLSP is the following:

```
DECLARE FUNCTION FUNC! (X!)
DECLARE SUB ZBRAK (DUM!, X1!, X2!, N!, XB1!(), XB2!(), NB!)
DECLARE FUNCTION BESSJ0! (X!)
DECLARE FUNCTION RTFLSP! (DUM!, X1!, X2!, XA!)

'PROGRAM D9R5
'Driver for routine RTFLSP
CLS
N = 100
NBMAX = 20
X1 = 1!
X2 = 50!
DIM XB1(NBMAX), XB2(NBMAX)
NB = NBMAX
CALL ZBRAK(DUM, X1, X2, N, XB1(), XB2(), NB)
PRINT "Roots of BESSJ0:"
PRINT
PRINT "                x              F(x)"
PRINT
FOR I = 1 TO NB
  XACC = .000001 * (XB1(I) + XB2(I)) / 2!
  ROOT = RTFLSP(DUM, XB1(I), XB2(I), XACC)
  PRINT "Root";
  PRINT USING "###"; I;
  PRINT USING "#######.######"; ROOT;
  PRINT "     ";
  PRINT USING "#.####^^^^"; BESSJ0(ROOT)
NEXT I
END

FUNCTION FUNC (X)
FUNC = BESSJ0(X)
END FUNCTION
```

Here is the recipe RTSEC:

```
DECLARE FUNCTION RTSEC! (DUM!, X1!, X2!, XACC!)
DECLARE FUNCTION FUNC! (X!)

FUNCTION RTSEC (DUM, X1, X2, XACC)
```
> Using the secant method, find the root of a user-supplied function FUNC thought to lie between X1 and X2. The root, returned as RTSEC, is refined until its accuracy is ±XACC.
```
MAXIT = 30                        Maximum allowed number of iterations.
FL = FUNC(X1)
F = FUNC(X2)
IF ABS(FL) < ABS(F) THEN          Pick the bound with the smaller function value as the most
  TRTSEC = X1                     recent guess.
  XL = X2
  SWAP FL, F
ELSE
  XL = X1
  TRTSEC = X2
END IF
FOR J = 1 TO MAXIT                Secant loop.
  DX = (XL - TRTSEC) * F / (F - FL)    Increment with respect to latest value.
  XL = TRTSEC
  FL = F
  TRTSEC = TRTSEC + DX
  F = FUNC(TRTSEC)               Convergence.
  RTSEC = TRTSEC
  IF ABS(DX) < XACC OR F = 0! THEN EXIT FUNCTION
NEXT J
PRINT "RTSEC exceed maximum iterations"
END FUNCTION
```

A sample program using RTSEC is the following:

```
DECLARE FUNCTION FUNC! (X!)
DECLARE SUB ZBRAK (DUM!, X1!, X2!, N!, XB1!(), XB2!(), NB!)
DECLARE FUNCTION BESSJ0! (X!)
DECLARE FUNCTION RTSEC! (DUM!, X1!, X2!, XA!)

'PROGRAM D9R6
'Driver for routine RTSEC
CLS
N = 100
NBMAX = 20
X1 = 1!
X2 = 50!
DIM XB1(NBMAX), XB2(NBMAX)
NB = NBMAX
CALL ZBRAK(DUM, X1, X2, N, XB1(), XB2(), NB)
PRINT "Roots of BESSJ0:"
PRINT
PRINT "              x            F(x)"
PRINT
FOR I = 1 TO NB
```

```
XACC = .000001 * (XB1(I) + XB2(I)) / 2!
ROOT = RTSEC(DUM, XB1(I), XB2(I), XACC)
PRINT "Root";
PRINT USING "###"; I;
PRINT USING "######.######"; ROOT;
PRINT "     ";
PRINT USING "#.####^^^^"; BESSJ0(ROOT)
NEXT I
END


FUNCTION FUNC (X)
FUNC = BESSJ0(X)
END FUNCTION
```

Here is the recipe ZBRENT:

```
DECLARE FUNCTION ZBRENT! (DUM!, X1!, X2!, TOL!)
DECLARE FUNCTION FUNC! (X!)

FUNCTION ZBRENT (DUM, X1, X2, TOL)
```
Using Brent's method, find the root of a user-supplied function FUNC known to lie between X1 and X2. The root, returned as ZBRENT, will be refined until its accuracy is TOL.

```
ITMAX = 100                          Maximum allowed number of iterations, and machine floating
EPS = 3E-08                          point precision.
A = X1
B = X2
FA = FUNC(A)
FB = FUNC(B)
IF FB * FA > 0! THEN
  PRINT "Root must be bracketed for ZBRENT."
  EXIT FUNCTION
END IF
FC = FB
FOR ITER = 1 TO ITMAX
  IF FB * FC > 0! THEN
    C = A                            Rename A,B,C and adjust bounding interval D.
    FC = FA
    D = B - A
    E = D
  END IF
  IF ABS(FC) < ABS(FB) THEN
    A = B
    B = C
    C = A
    FA = FB
    FB = FC
    FC = FA
  END IF
  TOL1 = 2! * EPS * ABS(B) + .5 * TOL    Convergence check.
  XM = .5 * (C - B)
  IF ABS(XM) <= TOL1 OR FB = 0! THEN
    ZBRENT = B
    EXIT FUNCTION
  END IF
```

```
IF ABS(E) >= TOL1 AND ABS(FA) > ABS(FB) THEN
   S = FB / FA                       Attempt inverse quadratic interpolation.
   IF A = C THEN
      P = 2! * XM * S
      Q = 1! - S
   ELSE
      Q = FA / FC
      R = FB / FC
      P = S * (2! * XM * Q * (Q - R) - (B - A) * (R - 1!))
      Q = (Q - 1!) * (R - 1!) * (S - 1!)
   END IF
   IF P > 0! THEN Q = -Q             Check whether in bounds.
   P = ABS(P)
   DUM = 3! * XM * Q - ABS(TOL1 * Q)
   IF DUM > ABS(E * Q) THEN DUM = E * Q
   IF 2! * P < DUM THEN
      E = D                          Accept interpolation.
      D = P / Q
   ELSE
      D = XM                         Interpolation failed, use bisection.
      E = D
   END IF
ELSE                                 Bounds decreasing too slowly, use bisection.
   D = XM
   E = D
END IF
A = B                                Move last best guess to A.
FA = FB
IF ABS(D) > TOL1 THEN                Evaluate new trial root.
   B = B + D
ELSE
   B = B + ABS(TOL1) * SGN(XM)
END IF
FB = FUNC(B)
NEXT ITER
PRINT "ZBRENT exceeding maximum iterations."
ZBRENT = B
END FUNCTION
```

A sample program using ZBRENT is the following:

```
DECLARE FUNCTION FUNC! (X!)
DECLARE SUB ZBRAK (DUM!, X1!, X2!, N!, XB1!(), XB2!(), NB!)
DECLARE FUNCTION BESSJ0! (X!)
DECLARE FUNCTION ZBRENT! (DUM!, X1!, X2!, TOL!)

'PROGRAM D9R7
'Driver for routine ZBRENT
CLS
N = 100
NBMAX = 20
X1 = 1!
X2 = 50!
DIM XB1(NBMAX), XB2(NBMAX)
NB = NBMAX
```

```
CALL ZBRAK(DUM, X1, X2, N, XB1(), XB2(), NB)
PRINT "Roots of BESSJ0:"
PRINT
PRINT "              x              F(x)"
PRINT
FOR I = 1 TO NB
  TOL = .000001 * (XB1(I) + XB2(I)) / 2!
  ROOT = ZBRENT(DUM, XB1(I), XB2(I), TOL)
  PRINT "Root";
  PRINT USING "###"; I;
  PRINT USING "######.######"; ROOT;
  PRINT "        ";
  PRINT USING "#.####^^^^"; BESSJ0(ROOT)
NEXT I
END


FUNCTION FUNC (X)
FUNC = BESSJ0(X)
END FUNCTION
```

Here is the recipe RTNEWT:


```
DECLARE SUB FUNCD (X!, FQ!, DF!)
DECLARE FUNCTION RTNEWT! (DUM!, X1!, X2!, XACC!)

FUNCTION RTNEWT (DUM, X1, X2, XACC)
```
    Using the Newton-Raphson method, find the root of a function known to lie in the interval
    X1–X2. The root RTNEWT will be refined until its accuracy is known within ±XACC. The user
    must supply a subroutine FUNCD that returns both the function value and the first derivative
    of the function at the point X.
```
JMAX = 20                           Set to maximum number of iterations.
TRTNEWT = .5 * (X1 + X2)            Initial guess.
FOR J = 1 TO JMAX
  CALL FUNCD(TRTNEWT, F, DF)
  DX = F / DF
  TRTNEWT = TRTNEWT - DX
  IF (X1 - TRTNEWT) * (TRTNEWT - X2) < 0! THEN
    PRINT "jumped out of brackets"
    EXIT FUNCTION
  END IF
  RTNEWT = TRTNEWT
  IF ABS(DX) < XACC THEN EXIT FUNCTION       Convergence.
NEXT J
PRINT "RTNEWT exceeding maximum iterations"
END FUNCTION
```

A sample program using RTNEWT is the following:

```
DECLARE FUNCTION FUNC! (X!)
DECLARE SUB ZBRAK (DUM!, X1!, X2!, N!, XB1!(), XB2!(), NB!)
DECLARE FUNCTION BESSJ0! (X!)
DECLARE FUNCTION BESSJ1! (X!)
DECLARE FUNCTION RTNEWT! (DUM!, X1!, X2!, XA!)

'PROGRAM D9R8
'Driver for routine RTNEWT
CLS
N = 100
NBMAX = 20
X1 = 1!
X2 = 50!
DIM XB1(NBMAX), XB2(NBMAX)
NB = NBMAX
CALL ZBRAK(DUM, X1, X2, N, XB1(), XB2(), NB)
PRINT "Roots of BESSJ0:"
PRINT
PRINT "              x              F(x)"
PRINT
FOR I = 1 TO NB
  XACC = .000001 * (XB1(I) + XB2(I)) / 2!
  ROOT = RTNEWT(DUM, XB1(I), XB2(I), XACC)
  PRINT "Root";
  PRINT USING "###"; I;
  PRINT USING "#######.######"; ROOT;
  PRINT "      ";
  PRINT USING "#.####^^^^"; BESSJ0(ROOT)
NEXT I
END


FUNCTION FUNC (X)
FUNC = BESSJ0(X)
END FUNCTION


SUB FUNCD (X, FQ, DF)
FQ = BESSJ0(X)
DF = -BESSJ1(X)
END SUB
```

Here is the recipe RTSAFE:

```
DECLARE SUB FUNCD (X!, FQ!, DF!)
DECLARE FUNCTION RTSAFE! (FUNCD!, X1!, X2!, XACC!)

FUNCTION RTSAFE (DUM, X1, X2, XACC)
```
Using a combination of Newton-Raphson and bisection, find the root of a function bracketed between X1 and X2. The root, returned as the function value RTSAFE, will be refined until its accuracy is known within ±XACC. The user must supply a subroutine FUNCD that returns both the function value and the first derivative of the function.

```
MAXIT = 100                        Maximum allowed number of iterations.
CALL FUNCD(X1, FL, DF)
CALL FUNCD(X2, FH, DF)
IF FL * FH >= 0! THEN PRINT "root must be bracketed": EXIT FUNCTION
IF FL < 0! THEN                    Orient the search so that f(XL) < 0.
```

```
   XL = X1
   XH = X2
ELSE
   XH = X1
   XL = X2
END IF
TRTSAFE = .5 * (X1 + X2)          Initialize the guess for root,
DXOLD = ABS(X2 - X1)              the "step-size before last,"
DX = DXOLD                        and the last step.
CALL FUNCD(TRTSAFE, F, DF)
FOR J = 1 TO MAXIT                Loop over allowed iterations.
   DUM = ((TRTSAFE - XH) * DF - F) * ((TRTSAFE - XL) * DF - F)
   IF DUM >= 0! OR ABS(2! * F) > ABS(DXOLD * DF) THEN   Bisect if Newton out of range,
      DXOLD = DX                                        or not decreasing fast enough.
      DX = .5 * (XH - XL)
      TRTSAFE = XL + DX
      RTSAFE = TRTSAFE
      IF XL = TRTSAFE THEN EXIT FUNCTION     Change in root is negligible.
   ELSE                               Newton step acceptable. Take it.
      DXOLD = DX
      DX = F / DF
      TEMP = TRTSAFE
      TRTSAFE = TRTSAFE - DX
      RTSAFE = TRTSAFE
      IF TEMP = TRTSAFE THEN EXIT FUNCTION
   END IF
   RTSAFE = TRTSAFE
   IF ABS(DX) < XACC THEN EXIT FUNCTION    Convergence criterion.
   CALL FUNCD(TRTSAFE, F, DF)           The one new function evaluation per iteration.
   IF F < 0! THEN                       Maintain the bracket on the root.
      XL = TRTSAFE
   ELSE
      XH = TRTSAFE
   END IF
NEXT J
PRINT "RTSAFE exceeding maximum iterations"
END FUNCTION
```

A sample program using RTSAFE is the following:

```
DECLARE FUNCTION FUNC! (X!)
DECLARE SUB ZBRAK (DUM!, X1!, X2!, N!, XB1!(), XB2!(), NB!)
DECLARE FUNCTION BESSJ0! (X!)
DECLARE FUNCTION BESSJ1! (X!)
DECLARE FUNCTION RTSAFE! (DUM!, X1!, X2!, XA!)

'PROGRAM D9R9
'Driver for routine RTSAFE
CLS
N = 100
NBMAX = 20
X1 = 1!
X2 = 50!
DIM XB1(NBMAX), XB2(NBMAX)
NB = NBMAX
```

```
CALL ZBRAK(DUM, X1, X2, N, XB1(), XB2(), NB)
PRINT "Roots of BESSJO:"
PRINT
PRINT "                    x                  F(x)"
PRINT
FOR I = 1 TO NB
  XACC = .000001 * (XB1(I) + XB2(I)) / 2!
  ROOT = RTSAFE(DUM, XB1(I), XB2(I), XACC)
  PRINT "Root";
  PRINT USING "###"; I;
  PRINT USING "######.######"; ROOT;
  PRINT "       ";
  PRINT USING "#.####^^^^"; BESSJO(ROOT)
NEXT I
END


FUNCTION FUNC (X)
FUNC = BESSJO(X)
END FUNCTION


SUB FUNCD (X, FQ, DF)
FQ = BESSJO(X)
DF = -BESSJ1(X)
END SUB
```

Routine **LAGUER** finds the roots of a polynomial with complex coefficients. The polynomial of degree M is specified by $M + 1$ coefficients which, in sample program **D9R10**, are specified in a **DATA** statement and kept in array **A**. The polynomial in this case is

$$F(x) = x^4 - (1 + 2i)x^2 + 2i$$

The four roots of this polynomial are $x = 1.0$, $x = -1.0$, $x = 1 + i$, and $x = -(1 + i)$. **LAGUER** proceeds on the basis of a trial root, and attempts to converge to true roots. The root to which it converges depends on the trial value. The program tries a series of complex trial values along the line in the imaginary plane from $-1.0 - i$ to $1.0 + i$. The actual roots to which it converges are compared to all previously found values, and if different, are printed.

Here is the recipe **LAGUER**:

```
DECLARE FUNCTION CABS! (A1!, A2!)
DECLARE FUNCTION CDIV1! (A1!, A2!, B1!, B2!)
DECLARE FUNCTION CDIV2! (A1!, A2!, B1!, B2!)
DECLARE FUNCTION CSQR1! (X!, Y!)
DECLARE FUNCTION CSQR2! (X!, Y!)

FUNCTION CABS (A1, A2)              Magnitude of complex number.
X = ABS(A1)
Y = ABS(A2)
IF X = 0! THEN
  CABS = Y
ELSEIF Y = 0! THEN
  CABS = X
ELSEIF X > Y THEN
  CABS = X * SQR(1! + SQR(Y / X))
```

```
ELSE
  CABS = Y * SQR(1! + SQR(X / Y))
END IF
END FUNCTION


FUNCTION CDIV1 (A1, A2, B1, B2)          Real part of complex quotient.
IF ABS(B1) >= ABS(B2) THEN
  R = B2 / B1
  DEN = B1 + R * B2
  CDIV1 = (A1 + A2 * R) / DEN
ELSE
  R = B1 / B2
  DEN = B2 + R * B1
  CDIV1 = (A1 * R + A2) / DEN
END IF
END FUNCTION


FUNCTION CDIV2 (A1, A2, B1, B2)          Imaginary part of complex quotient.
IF ABS(B1) >= ABS(B2) THEN
  R = B2 / B1
  DEN = B1 + R * B2
  CDIV2 = (A2 - A1 * R) / DEN
ELSE
  R = B1 / B2
  DEN = B2 + R * B1
  CDIV2 = (A2 * R - A1) / DEN
END IF
END FUNCTION


FUNCTION CSQR1 (X, Y)          Real part of complex square root.
IF X = 0! AND Y = 0! THEN
  U = 0!
ELSE
  IF ABS(X) >= ABS(Y) THEN
    W = SQR(ABS(X)) * SQR(.5 * (1! + SQR(1! + SQR(ABS(Y / X)))))
  ELSE
    R = ABS(X / Y)
    W = SQR(ABS(Y)) * SQR(.5 * (R + SQR(1! + SQR(R))))
  END IF
  IF X >= 0! THEN
    U = W
    V = Y / (2! * U)
  ELSE
    IF Y >= 0! THEN V = W ELSE V = -W
    U = Y / (2! * V)
  END IF
END IF
CSQR1 = U
END FUNCTION


FUNCTION CSQR2 (X, Y)          Imaginary part of complex square root.
IF X = 0! AND Y = 0! THEN
  V = 0!
ELSE
  IF ABS(X) >= ABS(Y) THEN
```

```
      W = SQR(ABS(X)) * SQR(.5 * (1! + SQR(1! + SQR(ABS(Y / X)))))
   ELSE
      R = ABS(X / Y)
      W = SQR(ABS(Y)) * SQR(.5 * (R + SQR(1! + SQR(R))))
   END IF
   IF X >= 0! THEN
      U = W
      V = Y / (2! * U)
   ELSE
      IF Y >= 0! THEN V = W ELSE V = -W
      U = Y / (2! * V)
   END IF
END IF
CSQR2 = V
END FUNCTION


SUB LAGUER (A(), M, X(), EPS, POLISH%)
```
Given the degree M and the M+1 complex coefficients A of the polynomial $\sum_{i=1}^{M+1} A(i)x^{i-1}$, and given EPS the desired fractional accuracy, and given a complex value X, this routine improves X by Laguerre's method until it converges to a root of the given polynomial. For normal use POLISH% should be input as 0. When POLISH% is input as -1, the routine ignores EPS and instead attempts to improve X (assumed to be a good initial guess) to the achievable roundoff limit.

```
DIM ZERO(2), B(2), D(2), F(2), G(2), H(2)
DIM G2(2), SQ(2), GP(2), GM(2), DX(2), X1(2)
ZERO(1) = 0!
ZERO(2) = 0!
EPSS = 6E-08
MAXIT = 100
DXOLD = CABS(X(1), X(2))
FOR ITER = 1 TO MAXIT            Loop over iterations up to allowed maximum.
   B(1) = A(1, M + 1)
   B(2) = A(2, M + 1)
   ERQ = CABS(B(1), B(2))
   D(1) = ZERO(1)
   D(2) = ZERO(2)
   F(1) = ZERO(1)
   F(2) = ZERO(2)
   ABX = CABS(X(1), X(2))
   FOR J = M TO 1 STEP -1                   Efficient computation of the polynomial and
      DUM = X(1) * F(1) - X(2) * F(2) + D(1)   its first two derivatives.
      F(2) = X(2) * F(1) + X(1) * F(2) + D(2)
      F(1) = DUM
      DUM = X(1) * D(1) - X(2) * D(2) + B(1)
      D(2) = X(2) * D(1) + X(1) * D(2) + B(2)
      D(1) = DUM
      DUM = X(1) * B(1) - X(2) * B(2) + A(1, J)
      B(2) = X(2) * B(1) + X(1) * B(2) + A(2, J)
      B(1) = DUM
      ERQ = CABS(B(1), B(2)) + ABX * ERQ
   NEXT J
   ERQ = EPSS * ERQ                 Estimate of roundoff error in evaluating polynomial.
   IF CABS(B(1), B(2)) <= ERQ THEN         We are on the root.
      ERASE X1, DX, GM, GP, SQ, G2, H, G, F, D, B, ZERO
      EXIT SUB
```

```
ELSE                            The generic case: use Laguerre's formula.
  G(1) = CDIV1(D(1), D(2), B(1), B(2))
  G(2) = CDIV2(D(1), D(2), B(1), B(2))
  G2(1) = G(1) * G(1) - G(2) * G(2)
  G2(2) = 2! * G(1) * G(2)
  H(1) = G2(1) - 2! * CDIV1(F(1), F(2), B(1), B(2))
  H(2) = G2(2) - 2! * CDIV2(F(1), F(2), B(1), B(2))
  DUM1 = (M - 1) * (M * H(1) - G2(1))
  DUM2 = (M - 1) * (M * H(2) - G2(2))
  SQ(1) = CSQR1(DUM1, DUM2)
  SQ(2) = CSQR2(DUM1, DUM2)
  GP(1) = G(1) + SQ(1)
  GP(2) = G(2) + SQ(2)
  GM(1) = G(1) - SQ(1)
  GM(2) = G(2) - SQ(2)
  IF CABS(GP(1), GP(2)) < CABS(GM(1), GM(2)) THEN
    GP(1) = GM(1)
    GP(2) = GM(2)
  END IF
  DX(1) = CDIV1(M, 0, GP(1), GP(2))
  DX(2) = CDIV2(M, 0, GP(1), GP(2))
END IF
X1(1) = X(1) - DX(1)
X1(2) = X(2) - DX(2)
IF X(1) = X1(1) AND X(2) = X1(2) THEN              Converged.
  ERASE X1, DX, GM, GP, SQ, G2, H, G, F, D, B, ZERO
  EXIT SUB
END IF
X(1) = X1(1)
X(2) = X1(2)
CDX = CABS(DX(1), DX(2))
DXOLD = CDX
IF NOT POLISH% THEN
  IF CDX <= EPS * CABS(X(1), X(2)) THEN            Converged.
    ERASE X1, DX, GM, GP, SQ, G2, H, G, F, D, B, ZERO
    EXIT SUB
  END IF
END IF
NEXT ITER
PRINT "too many iterations"      Very unusual — can only occur for complex roots.
END SUB
```

A sample program using **LAGUER** is the following:

```
DECLARE SUB LAGUER (A!(), M!, X!(), EPS!, POLISH%)
DECLARE FUNCTION CABS! (R!, I!)

'PROGRAM D9R10
'Driver for routine LAGUER
CLS
M = 4
MP1 = M + 1
NTRY = 21
EPS = .000001
DIM A(2, MP1), Y(2, NTRY), X(2)
```

```
FOR J = 1 TO MP1
  FOR I = 1 TO 2
    READ A(I, J)
  NEXT I
NEXT J
DATA 0.0,2.0,0.0,0.0,-1.0,-2.0,0.0,0.0,1.0,0.0
PRINT "Roots of polynomial x^4-(1+2i)*x^2+2i"
PRINT
PRINT "              Real        Complex"
PRINT
N = 0
POLISH% = 0
FOR I = 1 TO NTRY
  X(1) = (I - 11!) / 10!
  X(2) = (I - 11!) / 10!
  CALL LAGUER(A(), M, X(), EPS, POLISH%)
  IF N = 0 THEN
    N = 1
    Y(1, 1) = X(1)
    Y(2, 1) = X(2)
    PRINT USING "#####"; N;
    PRINT USING "########.######"; X(1); X(2)
  ELSE
    IFLAG = 0
    FOR J = 1 TO N
      IF CABS(X(1) - Y(1, J), X(2) - Y(2, J)) <= EPS * CABS(X(1), X(2)) THEN
        IFLAG = 1
      END IF
    NEXT J
    IF IFLAG = 0 THEN
      N = N + 1
      Y(1, N) = X(1)
      Y(2, N) = X(2)
      PRINT USING "#####"; N;
      PRINT USING "########.######"; X(1); X(2)
    END IF
  END IF
NEXT I
END
```

ZROOTS is a driver for LAGUER. Sample program D9R11 exercises ZROOTS, using the same polynomial as the previous routine. First it finds the four roots. Then it corrupts each one by multiplying by 1.01. Finally it uses ZROOTS again to polish the corrupted roots by setting the logical parameter POLISH% to -1.

Here is the recipe ZROOTS:

```
DECLARE SUB LAGUER (A!(), M!, X!(), EPS!, POLISH%)
```

```
SUB ZROOTS (A(), M, ROOTS(), POLISH%)
```
> Given the degree M and the M+1 complex coefficients A of the polynomial $\sum_{i=1}^{M+1} A(i)x^{i-1}$, this routine successively calls LAGUER and finds all M complex ROOTS. The logical variable

POLISH% should be input as -1 if polishing (also by Laguerre's method) is desired, 0 if the roots will be subsequently polished by other means.

```
EPS = .000001                          Desired accuracy and maximum anticipated value of M+1.
MAXM = 101
DIM AD(2, MAXM), X(2), B(2), C(2), DUM(2)
FOR J = 1 TO M + 1                     Copy of coefficients for successive deflation.
  AD(1, J) = A(1, J)
  AD(2, J) = A(2, J)
NEXT J
FOR J = M TO 1 STEP -1                  Loop over each root to be found.
  X(1) = 0!                            Start at zero to favor convergence to smallest remaining root.
  X(2) = 0!
  CALL LAGUER(AD(), J, X(), EPS, 0)       Find the root.
  IF ABS(X(2)) <= 2! * EPS ^ 2 * ABS(X(1)) THEN X(2) = 0!
  ROOTS(1, J) = X(1)
  ROOTS(2, J) = X(2)
  B(1) = AD(1, J + 1)                  Forward deflation.
  B(2) = AD(2, J + 1)
  FOR JJ = J TO 1 STEP -1
    C(1) = AD(1, JJ)
    C(2) = AD(2, JJ)
    AD(1, JJ) = B(1)
    AD(2, JJ) = B(2)
    DUM = B(1)
    B(1) = X(1) * DUM - X(2) * B(2) + C(1)
    B(2) = X(2) * DUM + X(1) * B(2) + C(2)
  NEXT JJ
NEXT J
IF POLISH% THEN
  FOR J = 1 TO M                       Polish the roots using the undeflated coefficients.
    DUM(1) = ROOTS(1, J)
    DUM(2) = ROOTS(2, J)
    CALL LAGUER(A(), M, DUM(), EPS, -1)
  NEXT J
END IF
FOR J = 2 TO M                         Sort roots by their real parts by straight insertion.
  X(1) = ROOTS(1, J)
  X(2) = ROOTS(2, J)
  FOR I = J - 1 TO 1 STEP -1
    IF ROOTS(1, I) <= X(1) THEN EXIT FOR
    ROOTS(1, I + 1) = ROOTS(1, I)
    ROOTS(2, I + 1) = ROOTS(2, I)
  NEXT I
  IF ROOTS(1, I) > X(1) THEN I = 0
  ROOTS(1, I + 1) = X(1)
  ROOTS(2, I + 1) = X(2)
NEXT J
ERASE DUM, C, B, X, AD
END SUB
```

A sample program using **ZROOTS** is the following:

```
DECLARE SUB ZROOTS (A!(), M!, ROOTS!(), POLISH%)

'PROGRAM D9R11
'Driver for routine ZROOTS
CLS
M = 4
M1 = M + 1
DIM A(2, M1), X(2), ROOTS(2, M)
FOR J = 1 TO M1
  FOR I = 1 TO 2
    READ A(I, J)
  NEXT I
NEXT J
DATA 0.0,2.0,0.0,0.0,-1.0,-2.0,0.0,0.0,1.0,0.0
PRINT "Roots of the polynomial x^4-(1+2i)*x^2+2i"
PRINT
POLISH% = 0
CALL ZROOTS(A(), M, ROOTS(), POLISH%)
PRINT "Unpolished roots:"
PRINT "      Root #       Real        Imag."
FOR I = 1 TO M
  PRINT USING "##########"; I;
  PRINT USING "#########.#####"; ROOTS(1, I);
  PRINT USING "#####.#####"; ROOTS(2, I)
NEXT I
PRINT
PRINT "Corrupted roots:"
FOR I = 1 TO M
  ROOTS(1, I) = ROOTS(1, I) * (1! + .01 * I)
  ROOTS(2, I) = ROOTS(2, I) * (1! + .01 * I)
NEXT I
PRINT "      Root #       Real        Imag."
FOR I = 1 TO M
  PRINT USING "##########"; I;
  PRINT USING "#########.#####"; ROOTS(1, I);
  PRINT USING "#####.#####"; ROOTS(2, I)
NEXT I
PRINT
POLISH% = -1
CALL ZROOTS(A(), M, ROOTS(), POLISH%)
PRINT "Polished roots:"
PRINT "      Root #       Real        Imag."
FOR I = 1 TO M
  PRINT USING "##########"; I;
  PRINT USING "#########.#####"; ROOTS(1, I);
  PRINT USING "#####.#####"; ROOTS(2, I)
NEXT I
END
```

QROOT is used for finding quadratic factors of polynomials with real coefficients. In the case of sample program D9R12, the polynomial is

$$P(x) = x^6 - 6x^5 + 16x^4 - 24x^3 + 25x^2 - 18x + 10.$$

The program proceeds like that of LAGUER. Successive trial values for quadratic factors $x^2 + Bx + C$ (in the form of guesses for B and C) are made, and for each trial, QROOT converges on correct values. If the B and C which are found are unlike any previous values, then they are printed. By this means, all three quadratic factors are located. You can, of course, compare their product to the polynomial above.

Here is the recipe QROOT:

```
DECLARE SUB POLDIV (U!(), N!, V!(), NV!, Q!(), R!())

SUB QROOT (P(), N, B, C, EPS)
      Given N coefficients P of a polynomial of degree N-1, and trial values for the coefficients of
      a quadratic factor X*X+B*X+C, improve the solution until the coefficients B,C change by less
      than EPS. The routine POLDIV §5.3 is used.
ITMAX = 20                        At most ITMAX iterations.
TINY = .000001
DIM Q(N), D(3), REQ(N), QQ(N)
D(3) = 1!
FOR ITER = 1 TO ITMAX
  D(2) = B
  D(1) = C
  CALL POLDIV(P(), N, D(), 3, Q(), REQ())
  S = REQ(1)                      First division R,S.
  R = REQ(2)
  CALL POLDIV(Q(), N - 1, D(), 3, QQ(), REQ())
  SC = -REQ(1)                    Second division partial R,S with respect to C.
  RC = -REQ(2)
  FOR I = N - 1 TO 1 STEP -1
    Q(I + 1) = Q(I)
  NEXT I
  Q(1) = 0!
  CALL POLDIV(Q(), N, D(), 3, QQ(), REQ())
  SB = -REQ(1)                    Third division partial R,S with respect to B.
  RB = -REQ(2)
  DIV = 1! / (SB * RC - SC * RB)  Solve 2x2 equation.
  DELB = (R * SC - S * RC) * DIV
  DELC = (-R * SB + S * RB) * DIV
  B = B + DELB
  C = C + DELC
  DB = ABS(DELB) - EPS * ABS(B)
  DC = ABS(DELC) - EPS * ABS(C)
  IF (DB <= 0 OR ABS(B) < TINY) AND (DC <= 0 OR ABS(C) < TINY) THEN
    ERASE QQ, REQ, D, Q
    EXIT SUB                      Coefficients converged.
  END IF
NEXT ITER
PRINT "too many iterations in QROOT"
END SUB
```

A sample program using QROOT is the following:

```
DECLARE SUB QROOT (P!(), N!, B!, C!, EPS!)

'PROGRAM D9R12
'Driver for routine QROOT
CLS
N = 7
EPS = .000001
NTRY = 10
TINY = .00001
DIM P(N), B(NTRY), C(NTRY)
FOR I = 1 TO N
  READ P(I)
NEXT I
DATA 10.0,-18.0,25.0,-24.0,16.0,-6.0,1.0
PRINT "P(x)=x^6-6x^5+16x^4-24x^3+25x^2-18x+10"
PRINT "Quadratic factors x^2+Bx+C"
PRINT
PRINT "Factor        B            C"
PRINT
NROOT = 0
FOR I = 1 TO NTRY
  C(I) = .5 * I
  B(I) = -.5 * I
  CALL QROOT(P(), N, B(I), C(I), EPS)
  IF NROOT = 0 THEN
    PRINT USING "###"; NROOT;
    PRINT USING "#######.#####"; B(I);
    PRINT USING "#####.#####"; C(I)
    NROOT = 1
  ELSE
    NFLAG = 0
    FOR J = 1 TO NROOT
      IF ABS(B(I) - B(J)) < TINY AND ABS(C(I) - C(J)) < TINY THEN NFLAG = 1
    NEXT J
    IF NFLAG = 0 THEN
      PRINT USING "###"; NROOT;
      PRINT USING "#######.#####"; B(I);
      PRINT USING "#####.#####"; C(I)
      NROOT = NROOT + 1
    END IF
  END IF
NEXT I
END
```

Finally, MNEWT looks for roots of multiple nonlinear equations. In order to run a sample program D9R13 we supply a subroutine USRFUN that returns the matrix ALPHA of partial derivatives of the functions with respect to each of the variables, and vector BETA, containing the negatives of the function values. The sample program tries to find sets of variables that solve the four equations

$$-x_1^2 - x_2^2 - x_3^2 + x_4 = 0$$

$$x_1^2 + x_2^2 + x_3^2 + x_4^2 - 1 = 0$$

$$x_1 - x_2 = 0$$

$$x_2 - x_3 = 0$$

You will probably be able to find the two solutions to this set even without MNEWT, noting that $x_1 = x_2$ and $x_2 = x_3$. If not, simply take the output from MNEWT and plug it into these equations for verification. The output from MNEWT should convince you of the need for good starting values.

Here is the recipe MNEWT:

```
DECLARE SUB USRFUN (X!(), ALPHA!(), BETA!())
DECLARE SUB LUDCMP (A!(), N!, NP!, INDX!(), D!)
DECLARE SUB LUBKSB (A!(), N!, NP!, INDX!(), B!())

SUB MNEWT (NTRIAL, X(), N, TOLX, TOLF)
          Given an initial guess X for a root in N dimensions, take NTRIAL Newton-Raphson steps to
          improve the root. Stop if the root converges in either summed absolute variable increments
          TOLX or summed absolute function values TOLF.
DIM ALPHA(N, N), BETA(N), INDX(N)
FOR K = 1 TO NTRIAL
  CALL USRFUN(X(), ALPHA(), BETA())          User subroutine supplies matrix coefficients.
  ERRF = 0!                                  Check function convergence.
  FOR I = 1 TO N
    ERRF = ERRF + ABS(BETA(I))
  NEXT I
  IF ERRF <= TOLF THEN EXIT FOR
  CALL LUDCMP(ALPHA(), N, N, INDX(), D)      Solve linear equations using LU decomposition.
  CALL LUBKSB(ALPHA(), N, N, INDX(), BETA())
  ERRX = 0!                                  Check root convergence.
  FOR I = 1 TO N                             Update solution.
    ERRX = ERRX + ABS(BETA(I))
    X(I) = X(I) + BETA(I)
  NEXT I
  IF ERRX <= TOLX THEN EXIT FOR
NEXT K
ERASE INDX, BETA, ALPHA
END SUB
```

A sample program using MNEWT is the following:

```
DECLARE SUB USRFUN (X!(), ALPHA!(), BETA!())
DECLARE SUB MNEWT (NTRIAL!, X!(), N!, TOLX!, TOLF!)


'PROGRAM D9R13
'Driver for routine MNEWT
CLS
NTRIAL = 5
TOLX = .000001
N = 4
```

```
TOLF = .000001
NP = 15
DIM X(NP), ALPHA(NP, NP), BETA(NP)
FOR KK = -1 TO 1 STEP 2
  FOR K = 1 TO 3
    XX = .2001 * K * KK
    PRINT "Starting vector number"; K
    FOR I = 1 TO 4
      X(I) = XX + .2 * I
      PRINT "    X(";
      PRINT USING "#"; I;
      PRINT ") =";
      PRINT USING "###.##"; X(I)
    NEXT I
    PRINT
    FOR J = 1 TO NTRIAL
      CALL MNEWT(1, X(), N, TOLX, TOLF)
      CALL USRFUN(X(), ALPHA(), BETA())
      PRINT "    I          X(I)               F"
      PRINT
      FOR I = 1 TO N
        PRINT USING "####"; I;
        PRINT "    ";
        PRINT USING "#.######^^^^"; X(I);
        PRINT "    ";
        PRINT USING "#.######^^^^"; -BETA(I)
      NEXT I
      PRINT
      PRINT "press RETURN to continue..."
      LINE INPUT DUM$
    NEXT J
  NEXT K
NEXT KK
END

SUB USRFUN (X(), ALPHA(), BETA())
ALPHA(1, 1) = -2! * X(1)
ALPHA(1, 2) = -2! * X(2)
ALPHA(1, 3) = -2! * X(3)
ALPHA(1, 4) = 1!
ALPHA(2, 1) = 2! * X(1)
ALPHA(2, 2) = 2! * X(2)
ALPHA(2, 3) = 2! * X(3)
ALPHA(2, 4) = 2! * X(4)
ALPHA(3, 1) = 1!
ALPHA(3, 2) = -1!
ALPHA(3, 3) = 0!
ALPHA(3, 4) = 0!
ALPHA(4, 1) = 0!
ALPHA(4, 2) = 1!
ALPHA(4, 3) = -1!
ALPHA(4, 4) = 0!
BETA(1) = X(1) ^ 2 + X(2) ^ 2 + X(3) ^ 2 - X(4)
BETA(2) = -X(1) ^ 2 - X(2) ^ 2 - X(3) ^ 2 - X(4) ^ 2 + 1!
BETA(3) = -X(1) + X(2)
```

```
BETA(4) = -X(2) + X(3)
END SUB
```

Chapter 10: Minimization and Maximization of Functions

Chapter 10 of *Numerical Recipes* deals with finding the maxima and minima of functions. The task has two parts, first the discovery of one or more bracketing intervals, and then the convergence to an extremum. MNBRAK begins with two specified abscissas of a function and searches in the "downhill" direction for brackets of a minimum. GOLDEN can then take a bracketing triplet and perform a golden section search to a specified precision, for the minimum itself. When you are not concerned with worst-case examples, but only very efficient average-case performance, Brent's method (routine BRENT) is recommended. In the event that means are at hand for calculating the function's derivative as well as its value, consider DBRENT.

Multidimensional minimization strategies may be based on the one-dimensional algorithms. Our single example of an algorithm that is *not* so based is AMOEBA, which utilizes the downhill simplex method. Among the ones that *do* use one-dimensional methods are POWELL, FRPRMN, and DFPMIN. These three all make calls to LINMIN, a subroutine that minimizes a function along a given direction in space. LINMIN in turn uses the one-dimensional algorithm BRENT, if derivatives are not known, or DBRENT if they are. POWELL uses only function values and minimizes along an artfully chosen set of favorable directions. FRPRMN uses a Fletcher-Reeves-Polak-Ribiere minimization and requires the calculation of derivatives for the function. DFPMIN uses a variant of the Davidon-Fletcher-Powell variable metric method. This, too, requires calculation of derivatives.

The chapter ends with two topics of somewhat different nature. The first is linear programming, which deals with the maximization of a linear combination of variables, subject to linear constraints. This problem is dealt with by the simplex method in routine SIMPLX. The second is the subject of large scale optimization, which is illustrated with the method of simulated annealing, and applied particularly to the "traveling salesman" problem in routine ANNEAL.

⋆ ⋆ ⋆ ⋆

MNBRAK searches a given function for a minimum. Given two values AX and BX of abscissa, it searches in the downward direction until it can find three new values AX,BX,CX that bracket a minimum. FA,FB,FC are the values of the function at these points. Sample program D10R1 is a simple application of MNBRAK applied to the Bessel function J_0. It tries a series of starting values AX,BX each encompassing an interval of length 1.0. MNBRAK then finds several bracketing intervals of various minima of J_0.

Here is the recipe MNBRAK:

```
DECLARE FUNCTION FUNC! (X!)

SUB MNBRAK (AX, BX, CX, FA, FB, FC, DUM)
```
Given a user-supplied function FUNC, and given distinct initial points AX and BX, this routine searches in the downhill direction (defined by the function as evaluated at the initial points) and returns new points AX, BX, CX which bracket a minimum of the function. Also returned are the function values at the three points, FA, FB, and FC.
```
GOLD = 1.618034
GLIMIT = 100!
```
GOLD is the default ratio by which successive intervals are magnified; GLIMIT is the maximum magnification allowed for a parabolic-fit step.
```
TINY = 1E-20
FA = FUNC(AX)
FB = FUNC(BX)
IF FB > FA THEN                  Switch roles of A and B so that we can go downhill in the
  DUM = AX                       direction from A to B.
  AX = BX
  BX = DUM
  DUM = FB
  FB = FA
  FA = DUM
END IF
CX = BX + GOLD * (BX - AX)       First guess for C.
FC = FUNC(CX)
DO                               "DO...LOOP": keep returning here until we bracket.
  IF FB < FC THEN EXIT DO
  DONE% = -1
  R = (BX - AX) * (FB - FC)      Compute U by parabolic extrapolation from A,B,C. TINY is
  Q = (BX - CX) * (FB - FA)      used to prevent any possible division by zero.
  DUM = Q - R
  IF ABS(DUM) < TINY THEN DUM = TINY
  U = BX - ((BX - CX) * Q - (BX - AX) * R) / (2! * DUM)
  ULIM = BX + GLIMIT * (CX - BX)    We won't go further than this. Now to test possibilities:
  IF (BX - U) * (U - CX) > 0! THEN   Parabolic U is between B and C: try it.
    FU = FUNC(U)
    IF FU < FC THEN              Got a minimum between B and C.
      AX = BX
      FA = FB
      BX = U
      FB = FU
      EXIT SUB
    ELSEIF FU > FB THEN          Got a minimum between between A and U.
      CX = U
      FC = FU
      EXIT SUB
    END IF
    U = CX + GOLD * (CX - BX)    Parabolic fit was no use. Use default magnification.
    FU = FUNC(U)
  ELSEIF (CX - U) * (U - ULIM) > 0! THEN   Parabolic fit is between C and its al-
    FU = FUNC(U)                           lowed limit.
    IF FU < FC THEN
      BX = CX
      CX = U
      U = CX + GOLD * (CX - BX)
```

```
        FB = FC
        FC = FU
        FU = FUNC(U)
      END IF
    ELSEIF (U - ULIM) * (ULIM - CX) >= 0! THEN          Limit parabolic U to maximum allowed
        U = ULIM                                        value.
        FU = FUNC(U)
    ELSE                                                Reject parabolic U, use default magnification.
        U = CX + GOLD * (CX - BX)
        FU = FUNC(U)
    END IF
    IF DONE% THEN
      AX = BX                                           Eliminate oldest point and continue.
      BX = CX
      CX = U
      FA = FB
      FB = FC
      FC = FU
    ELSE
      DONE% = 0
    END IF
LOOP WHILE NOT DONE%
END SUB
```

A sample program using MNBRAK is the following:

```
DECLARE FUNCTION FUNC! (U!)
DECLARE SUB MNBRAK (AX!, BX!, CX!, FA!, FB!, FC!, DUM!)
DECLARE FUNCTION BESSJ0! (X!)

'PROGRAM D10R1
'Driver for routine MNBRAK
CLS
FOR I = 1 TO 10
  AX = I * .5
  BX = (I + 1!) * .5
  CALL MNBRAK(AX, BX, CX, FA, FB, FC, DUM)
  PRINT "            A            B            C"
  PRINT "  X";
  PRINT USING "#####.######"; AX; BX; CX
  PRINT "  F";
  PRINT USING "#####.######"; FA; FB; FC
NEXT I
END

FUNCTION FUNC (U)
FUNC = BESSJ0(U)
END FUNCTION
```

Routine GOLDEN continues the minimization process by taking a bracketing triplet AX,BX,CX and performing a golden section search to isolate the contained minimum to a stated precision TOL. Sample program D10R2 again uses J_0 as the test function. Using intervals (AX,BX) of length 1.0 it uses MNBRAK to bracket all minima between $x = 0.0$ and $x = 100.0$. Some minima are bracketed more than once. On each pass, the bracketed solution is tracked down by GOLDEN. It is then compared to all previously located minima, and if different it is added to the collection by incrementing NMIN (number of minima found) and adding the location XMIN of the minima to the list in array AMIN. As a check of GOLDEN, the routine prints out the value of J_0 at the minimum, and also the value of J_1, which ought to be zero at extrema of J_0.

Here is the recipe GOLDEN:

```
DECLARE FUNCTION GOLDEN! (AX!, BX!, CX!, DUM!, TOL!, XMIN!)
DECLARE FUNCTION FUNC! (X!)

FUNCTION GOLDEN (AX, BX, CX, DUM, TOL, XMIN)
```
       Given a function FUNC, and given a bracketing triplet of abscissas AX, BX, CX (such that BX is between AX and CX, and FUNC(BX) is less than both FUNC(AX) and FUNC(CX)), this routine performs a golden section search for the minimum, isolating it to a fractional precision of about TOL. The abscissa of the minimum is returned as XMIN, and the minimum function value is returned as GOLDEN, the returned function value.

```
R = .61803399#                     Golden ratios.
C = .38196602#
X0 = AX                            At any given time we will keep track of four points, X0,X1,X2,X3.
X3 = CX
IF ABS(CX - BX) > ABS(BX - AX) THEN    Make X0 to X1 the smaller segment,
  X1 = BX
  X2 = BX + C * (CX - BX)           and fill in the new point to be tried.
ELSE
  X2 = BX
  X1 = BX - C * (BX - AX)
END IF
F1 = FUNC(X1)                      The initial function evaluations. Note that we never need to
F2 = FUNC(X2)                      evaluate the function at the original endpoints.
WHILE ABS(X3 - X0) > TOL * (ABS(X1) + ABS(X2))    WHILE...WEND loop: we keep re-
  IF F2 < F1 THEN                  One possible outcome,           turning here.
    X0 = X1                        its housekeeping,
    X1 = X2
    X2 = R * X1 + C * X3
    F1 = F2
    F2 = FUNC(X2)                  and a new function evaluation.
  ELSE                             The other outcome,
    X3 = X2
    X2 = X1
    X1 = R * X2 + C * X0
    F2 = F1
    F1 = FUNC(X1)                  and its new function evaluation.
  END IF
WEND                               Back to see if we are done.
IF F1 < F2 THEN                    We are done. Output the best of the two current values.
  GOLDEN = F1
  XMIN = X1
ELSE
  GOLDEN = F2
```

```
  XMIN = X2
END IF
END FUNCTION
```

A sample program using GOLDEN is the following:

```
DECLARE FUNCTION FUNC! (X!)
DECLARE SUB MNBRAK (AX!, BX!, CX!, FA!, FB!, FC!, DUM!)
DECLARE FUNCTION GOLDEN! (AX!, BX!, CX!, DUM!, TOL!, XMIN!)
DECLARE FUNCTION BESSJ0! (X!)
DECLARE FUNCTION BESSJ1! (X!)

'PROGRAM D10R2
'Driver for routine GOLDEN
CLS
TOL = .000001
EQL = .001
DIM AMIN(20)
NMIN = 0
PRINT "Minima of the function BESSJ0"
PRINT
PRINT "    Min. #        X        BESSJ0(X)     BESSJ1(X)"
PRINT
FOR I = 1 TO 100
  AX = I
  BX = I + 1!
  CALL MNBRAK(AX, BX, CX, FA, FB, FC, DUM)
  G = GOLDEN(AX, BX, CX, DUM, TOL, XMIN)
  IF NMIN = 0 THEN
    AMIN(1) = XMIN
    NMIN = 1
    PRINT USING "#######"; NMIN;
    PRINT "    ";
    PRINT USING "#####.######"; XMIN; BESSJ0(XMIN); BESSJ1(XMIN)
  ELSE
    IFLAG = 0
    FOR J = 1 TO NMIN
      IF ABS(XMIN - AMIN(J)) <= EQL * XMIN THEN IFLAG = 1
    NEXT J
    IF IFLAG = 0 THEN
      NMIN = NMIN + 1
      AMIN(NMIN) = XMIN
      PRINT USING "#######"; NMIN;
      PRINT "    ";
      PRINT USING "#####.######"; XMIN; BESSJ0(XMIN); BESSJ1(XMIN)
    END IF
  END IF
NEXT I
END

FUNCTION FUNC (X)
FUNC = BESSJ0(X)
END FUNCTION
```

There are two other routines presented which also take the bracketing triplet AX,BX,CX from MNBRAK and find the contained minimum. They are BRENT and DBRENT. The sample programs for these two, D10R3 and D10R4, are virtually identical to that used on GOLDEN. In all of these programs, the function to be minimized is called FUNC(X) and cannot be changed in the argument of the function call. The trivial FUNCTION subroutine is provided with FUNC(X) set equal to BESSJ0(X). One advantage of this method is that if you are planning to deal with several different functions, you can avoid multiple compilations of the program; simply recompile the very short function subroutines and link each in turn to the same program. In program D10R4 an additional function DF(X), which is the derivative of FUNC(X), is set equal to -BESSJ1(X). Note that DBRENT is only used when the derivative can be calculated conveniently.

Here is the recipe BRENT:

```
DECLARE FUNCTION BRENT! (AX!, BX!, CX!, DUM!, TOL!, XMIN!)
DECLARE FUNCTION FUNC! (X!)

FUNCTION BRENT (AX, BX, CX, DUM, TOL, XMIN)
```
> Given a user-supplied function FUNC, and given a bracketing triplet of abscissas AX, BX, CX (such that BX is between AX and CX, and FUNC(BX) is less than both FUNC(AX) and FUNC(CX)), this routine isolates the minimum to a fractional precision of about TOL using Brent's method. The abscissa of the minimum is returned as XMIN, and the minimum function value is returned as BRENT, the returned function value.

`ITMAX = 100`	Maximum allowed number of iterations; golden ratio; and a
`CGOLD = .381966#`	small number which protects against trying to achieve frac-
`ZEPS = 1E-10`	tional accuracy for a minimum that happens to be exactly zero.
`A = AX`	A and B must be in ascending order, though the input abscis-
`IF CX < AX THEN A = CX`	sas need not be.
`B = AX`	
`IF CX > AX THEN B = CX`	
`V = BX`	Initializations...
`W = V`	
`X = V`	
`E = 0!`	This will be the distance moved on the step before last.
`FX = FUNC(X)`	
`FV = FX`	
`FW = FX`	
`FOR ITER = 1 TO ITMAX`	Main program loop.
`  XM = .5 * (A + B)`	
`  TOL1 = TOL * ABS(X) + ZEPS`	
`  TOL2 = 2! * TOL1`	
`  IF ABS(X - XM) <= TOL2 - .5 * (B - A) THEN EXIT FOR`	Test for done here.
`  DONE% = -1`	
`  IF ABS(E) > TOL1 THEN`	Construct a trial parabolic fit.
`    R = (X - W) * (FX - FV)`	
`    Q = (X - V) * (FX - FW)`	
`    P = (X - V) * Q - (X - W) * R`	
`    Q = 2! * (Q - R)`	
`    IF Q > 0! THEN P = -P`	
`    Q = ABS(Q)`	
`    ETEMP = E`	
`    E = D`	
`    DUM = ABS(.5 * Q * ETEMP)`	
`    IF ABS(P) < DUM AND P > Q * (A - X) AND P < Q * (B - X) THEN`	

The above conditions determine the acceptability of the parabolic fit. Here it is o.k.:

```
        D = P / Q                                  Take the parabolic step.
        U = X + D
        IF U - A < TOL2 OR B - U < TOL2 THEN D = ABS(TOL1) * SGN(XM - X)
        DONE% = 0                                  A signal to skip golden section step.
      END IF
    END IF
    IF DONE% THEN                                  Don't skip the golden section step.
      IF X >= XM THEN                              We arrive here for a golden section step, which we take into
        E = A - X                                  the larger of the two segments.
      ELSE
        E = B - X
      END IF
      D = CGOLD * E                                Take the golden section step.
    END IF
    IF ABS(D) >= TOL1 THEN                         Arrive here with D computed either from parabolic fit, or else
      U = X + D                                    from golden section.
    ELSE
      U = X + ABS(TOL1) * SGN(D)
    END IF
    FU = FUNC(U)                                   This is the one function evaluation per iteration,
    IF FU <= FX THEN                               and now we have to decide what to do with our function
      IF U >= X THEN                               evaluation. Housekeeping follows:
        A = X
      ELSE
        B = X
      END IF
      V = W
      FV = FW
      W = X
      FW = FX
      X = U
      FX = FU
    ELSE
      IF U < X THEN
        A = U
      ELSE
        B = U
      END IF
      IF FU <= FW OR W = X THEN
        V = W
        FV = FW
        W = U
        FW = FU
      ELSEIF FU <= FV OR V = X OR V = W THEN
        V = U
        FV = FU
      END IF
    END IF                                         Done with housekeeping. Back for another iteration.
  NEXT ITER
  IF ITER > ITMAX THEN PRINT "Brent exceed maximum iterations.": END
  XMIN = X                                         Arrive here ready to exit with best values.
  BRENT = FX
END FUNCTION
```

A sample program using BRENT is the following:

```
DECLARE FUNCTION FUNC! (X!)
DECLARE SUB MNBRAK (AX!, BX!, CX!, FA!, FB!, FC!, DUM!)
DECLARE FUNCTION BRENT! (AX!, BX!, CX!, DUM!, TOL!, XMIN!)
DECLARE FUNCTION BESSJ0! (X!)
DECLARE FUNCTION BESSJ1! (X!)
DECLARE FUNCTION BRENT! (AX!, BX!, CX!, DUM!, TOL!, XMIN!)

'PROGRAM D10R3
'Driver for routine BRENT
CLS
TOL = .000001
EQL = .0001
DIM AMIN(20)
NMIN = 0
PRINT "Minima of the function BESSJ0"
PRINT
PRINT "    Min. #        X           BESSJ0(X)    BESSJ1(X)"
PRINT
FOR I = 1 TO 100
  AX = I
  BX = I + 1!
  CALL MNBRAK(AX, BX, CX, FA, FB, FC, DUM)
  B = BRENT(AX, BX, CX, DUM, TOL, XMIN)
  IF NMIN = 0 THEN
    AMIN(1) = XMIN
    NMIN = 1
    PRINT USING "#######"; NMIN;
    PRINT "   ";
    PRINT USING "#####.######"; XMIN; BESSJ0(XMIN); BESSJ1(XMIN)
  ELSE
    IFLAG = 0
    FOR J = 1 TO NMIN
      IF ABS(XMIN - AMIN(J)) < EQL * XMIN THEN IFLAG = 1
    NEXT J
    IF IFLAG = 0 THEN
      NMIN = NMIN + 1
      AMIN(NMIN) = XMIN
      PRINT USING "#######"; NMIN;
      PRINT "   ";
      PRINT USING "#####.######"; XMIN; BESSJ0(XMIN); BESSJ1(XMIN)
    END IF
  END IF
NEXT I
END

FUNCTION FUNC (X)
FUNC = BESSJ0(X)
END FUNCTION
```

Here is the recipe DBRENT:

```
DECLARE FUNCTION DBRENT! (AX!, BX!, CX!, DUM!, DUM!, TOL!, XMIN!)
DECLARE FUNCTION FUNC! (X!)
DECLARE FUNCTION DF! (X!)

FUNCTION DBRENT (AX, BX, CX, DUM1, DUM2, TOL, XMIN)
```
   Given a user-supplied function FUNC and its derivative function DF (also a user-supplied function), and given a bracketing triplet of abscissas AX, BX, CX [such that BX is between AX and CX, and FUNC(BX) is less than both FUNC(AX) and FUNC(CX)], this routine isolates the minimum to a fractional precision of about TOL using a modification of Brent's method that uses derivatives. The abscissa of the minimum is returned as XMIN, and the minimum function value is returned as DBRENT, the returned function value.
```
ITMAX = 100
ZEPS = 1E-10
```
   Comments following will point out only differences from the routine BRENT. Read that routine first.
```
A = AX
IF CX < AX THEN A = CX
B = AX
IF CX > AX THEN B = CX
V = BX
W = V
X = V
E = 0!
FX = FUNC(X)
FV = FX
FW = FX
DX = DF(X)          All our housekeeping chores are doubled by the necessity of
DV = DX             moving derivative values around as well as function values.
DW = DX
FOR ITER = 1 TO ITMAX
  XM = .5 * (A + B)
  TOL1 = TOL * ABS(X) + ZEPS
  TOL2 = 2! * TOL1
  IF ABS(X - XM) <= TOL2 - .5 * (B - A) THEN    A normal exit.
    DONE% = -1
    EXIT FOR
  ELSE
    DONE% = 0
  END IF
  IF ABS(E) > TOL1 THEN
    D1 = 2! * (B - A)          Initialize these D's to an out-of-bracket value.
    D2 = D1
    IF DW <> DX THEN D1 = (W - X) * DX / (DX - DW)    Secant method.
    IF DV <> DX THEN D2 = (V - X) * DX / (DX - DV)    Other secant method.
```
   Which of these two estimates of D shall we take? We will insist that they be within the bracket, and on the side pointed to by the derivative at X:
```
    U1 = X + D1
    U2 = X + D2
    OK1% = ((A - U1) * (U1 - B) > 0!) AND (DX * D1 <= 0!)    Flags for whether pro-
    OK2% = ((A - U2) * (U2 - B) > 0!) AND (DX * D2 <= 0!)    posed steps are accept-
    OLDE = E                   Movement on the step before last.    able.
    E = D
    IF OK1% OR OK2% THEN       Take only an acceptable D, and if both are acceptable, then
                              take the smallest one.
```

```
IF OK1% AND OK2% THEN
   IF ABS(D1) < ABS(D2) THEN
      D = D1
   ELSE
      D = D2
   END IF
ELSEIF OK1% THEN
   D = D1
ELSE
   D = D2
END IF
IF ABS(D) <= ABS(.5 * OLDE) THEN
   U = X + D
   IF U - A < TOL2 OR B - U < TOL2 THEN D = ABS(TOL1) * SGN(XM - X)
END IF
END IF
END IF
IF DX >= 0! THEN              Decide which segment by the sign of the derivative.
   E = A - X
ELSE
   E = B - X
END IF
D = .5 * E                    Bisect, not golden section.
IF ABS(D) >= TOL1 THEN
   U = X + D
   FU = FUNC(U)
ELSE
   U = X + ABS(TOL1) * SGN(D)
   FU = FUNC(U)
   IF FU > FX THEN            If the minimum step in the downhill direction takes us uphill,
      DONE% = -1             then we are done.
      EXIT FOR
   ELSE
      DONE% = 0
   END IF
END IF
DU = DF(U)                    Now all the housekeeping, sigh.
IF FU <= FX THEN
   IF U >= X THEN
      A = X
   ELSE
      B = X
   END IF
   V = W
   FV = FW
   DV = DW
   W = X
   FW = FX
   DW = DX
   X = U
   FX = FU
   DX = DU
ELSE
   IF U < X THEN
      A = U
```

```
      ELSE
        B = U
      END IF
      IF FU <= FW OR W = X THEN
        V = W
        FV = FW
        DV = DW
        W = U
        FW = FU
        DW = DU
      ELSEIF FU <= FV OR V = X OR V = W THEN
        V = U
        FV = FU
        DV = DU
      END IF
    END IF
  NEXT ITER
  IF NOT DONE% THEN
    PRINT "DBRENT exceeded maximum iterations."
  ELSE
    XMIN = X
    DBRENT = FX
  END IF
END FUNCTION
```

A sample program using **DBRENT** is the following:

```
DECLARE FUNCTION FUNC! (X!)
DECLARE FUNCTION DF! (X!)
DECLARE SUB MNBRAK (AX!, BX!, CX!, FA!, FB!, FC!, DUM!)
DECLARE FUNCTION BESSJ0! (X!)
DECLARE FUNCTION BESSJ1! (X!)
DECLARE FUNCTION DBRENT! (AX!, BX!, CX!, DUM!, DUM!, TOL!, XIMIN!)

'PROGRAM D10R4
'Driver for routine DBRENT
CLS
TOL = .000001
EQL = .0001
DIM AMIN(20)
NMIN = 0
PRINT "Minima of the function BESSJ0"
PRINT
PRINT "    Min. #        X        BESSJ0(X)     BESSJ1(X)        DBRENT"
PRINT
FOR I = 1 TO 100
  AX = I
  BX = I + 1!
  CALL MNBRAK(AX, BX, CX, FA, FB, FC, DUM)
  DBR = DBRENT(AX, BX, CX, DUM, DERIV, TOL, XMIN)
  IF NMIN = 0 THEN
    AMIN(1) = XMIN
    NMIN = 1
    PRINT USING "#######"; NMIN;
    PRINT "    ";
```

```
    PRINT USING "#####.######"; XMIN; BESSJ0(XMIN); DF(XMIN); DBR
  ELSE
    IFLAG = 0
    FOR J = 1 TO NMIN
      IF ABS(XMIN - AMIN(J)) <= EQL * XMIN THEN IFLAG = 1
    NEXT J
    IF IFLAG = 0 THEN
      NMIN = NMIN + 1
      AMIN(NMIN) = XMIN
      PRINT USING "#######"; NMIN;
      PRINT "   ";
      PRINT USING "#####.######"; XMIN; BESSJ0(XMIN); DF(XMIN); DBR
    END IF
  END IF
NEXT I
END


FUNCTION DF (X)
DF = -BESSJ1(X)
END FUNCTION


FUNCTION FUNC (X)
FUNC = BESSJ0(X)
END FUNCTION
```

Numerical Recipes presents several methods for minimization in multiple dimensions. Among these, the downhill simplex method carried out by AMOEBA was the only one that did not treat the problem as a series of one-dimensional minimizations. As input, AMOEBA requires the coordinates of $N + 1$ vertices of a starting simplex in N-dimensional space, and the values Y of the function at each of these vertices. Sample program D10R5 tries the method out on the exotic function

$$\text{AMOEB} = 0.6 - J_0[(x - 0.5)^2 + (y - 0.6)^2 + (z - 0.7)^2]$$

which has a minimum at $(x, y, z) = (0.5, 0.6, 0.7)$. As vertices of the starting simplex, specified by P in a DATA statement, we used $(0, 0, 0)$, $(1, 0, 0)$, $(0, 1, 0)$, and $(0, 0, 1)$. A vector X(I) is set successively to each vertex to allow the evaluation of function values Y. These data are submitted to AMOEBA along with FTOL=1.0E-6 to specify the tolerance on the function value. The vertices and corresponding function values of the final simplex are printed out, and you can easily check whether the specified tolerance is met.

Here is the recipe AMOEBA:

DECLARE FUNCTION AMOEB! (X!(), NDIM!)

SUB AMOEBA (P(), Y(), MP, NP, NDIM, FTOL, DUM, ITER)
    Multidimensional minimization of the user-supplied function AMOEB(X) where X is an NDIM-dimensional vector, by the downhill simplex method of Nelder and Mead. Input is a matrix P whose NDIM+1 rows are NDIM-dimensional vectors which are the vertices of the starting simplex. [Logical dimensions of P are P(NDIM+1,NDIM); physical dimensions are input as P(MP,NP).] Also input is the vector Y of length NDIM+1, whose components must be pre-initialized to the values of AMOEB evaluated at the NDIM+1 vertices (rows) of P; and FTOL the fractional convergence tolerance to be achieved in the function value (n.b.!). On output,

P and Y will have been reset to NDIM+1 new points all within FTOL of a minimum function value, and ITER gives the number of iterations taken.

```
ALPHA = 1!
BETA = .5
GAMMA = 2!
ITMAX = 500
DIM PR(NDIM), PRR(NDIM), PBAR(NDIM)
MPTS = NDIM + 1
ITER = 0
DO
  ILO = 1
  IF Y(1) > Y(2) THEN
    IHI = 1
    INHI = 2
  ELSE
    IHI = 2
    INHI = 1
  END IF
  FOR I = 1 TO MPTS
    IF Y(I) < Y(ILO) THEN ILO = I
    IF Y(I) > Y(IHI) THEN
      INHI = IHI
      IHI = I
    ELSEIF Y(I) > Y(INHI) THEN
      IF I <> IHI THEN INHI = I
    END IF
  NEXT I
```

Three parameters which define the expansions and contractions, and maximum allowed number of iterations.

Note that MP is the physical dimension corresponding to the logical dimension MPTS, NP to NDIM.

First we must determine which point is the highest (worst), next-highest, and lowest (best),

by looping over the points in the simplex.

Compute the fractional range from highest to lowest and return if satisfactory.

```
RTOL = 2! * ABS(Y(IHI) - Y(ILO)) / (ABS(Y(IHI)) + ABS(Y(ILO)))
IF RTOL < FTOL THEN ERASE PBAR, PRR, PR: EXIT SUB
IF ITER = ITMAX THEN PRINT "Amoeba exceeding maximum iterations.": EXIT SUB
ITER = ITER + 1
FOR J = 1 TO NDIM
  PBAR(J) = 0!
NEXT J
FOR I = 1 TO MPTS
  IF I <> IHI THEN
    FOR J = 1 TO NDIM
      PBAR(J) = PBAR(J) + P(I, J)
    NEXT J
  END IF
NEXT I
FOR J = 1 TO NDIM
  PBAR(J) = PBAR(J) / NDIM
  PR(J) = (1! + ALPHA) * PBAR(J) - ALPHA * P(IHI, J)
NEXT J
YPR = AMOEB(PR(), NDIM)
IF YPR <= Y(ILO) THEN
  FOR J = 1 TO NDIM
    PRR(J) = GAMMA * PR(J) + (1! - GAMMA) * PBAR(J)
  NEXT J
  YPRR = AMOEB(PRR(), NDIM)
  IF YPRR < Y(ILO) THEN
    FOR J = 1 TO NDIM
```

Begin a new iteration. Compute the vector average of all points except the highest, i.e. the center of the "face" of the simplex across from the high point. We will subsequently explore along the ray from the high point through that center.

Extrapolate by a factor ALPHA through the face, i.e. reflect the simplex from the high point.

Evaluate the function at the reflected point. Gives a result better than the best point, so try an additional extrapolation by a factor GAMMA,

and check out the function there. The additional extrapolation succeeded, and replaces the high point.

```
      P(IHI, J) = PRR(J)
    NEXT J
    Y(IHI) = YPRR
  ELSE                              The additional extrapolation failed,
    FOR J = 1 TO NDIM               but we can still use the reflected point.
      P(IHI, J) = PR(J)
    NEXT J
    Y(IHI) = YPR
  END IF
ELSEIF YPR >= Y(INHI) THEN          The reflected point is worse than the second-highest.
  IF YPR < Y(IHI) THEN              If it's better than the highest, then replace the highest,
    FOR J = 1 TO NDIM
      P(IHI, J) = PR(J)
    NEXT J
    Y(IHI) = YPR
  END IF
  FOR J = 1 TO NDIM                 but look for an intermediate lower point,
    PRR(J) = BETA * P(IHI, J) +     (1! - BETA) * PBAR(J)
  NEXT J                            in other words, perform a contraction of the simplex along
  YPRR = AMOEB(PRR(), NDIM)         one dimension. Then evaluate the function.
  IF YPRR < Y(IHI) THEN             Contraction gives an improvement,
    FOR J = 1 TO NDIM               so accept it.
      P(IHI, J) = PRR(J)
    NEXT J
    Y(IHI) = YPRR
  ELSE                              Can't seem to get rid of that high point. Better contract
    FOR I = 1 TO MPTS               around the lowest (best) point.
      IF I <> ILO THEN
        FOR J = 1 TO NDIM
          PR(J) = .5 * (P(I, J) + P(ILO, J))
          P(I, J) = PR(J)
        NEXT J
        Y(I) = AMOEB(PR(), NDIM)
      END IF
    NEXT I
  END IF
ELSE                                We arrive here if the original reflection gives a middling point.
  FOR J = 1 TO NDIM                 Replace the old high point and continue
    P(IHI, J) = PR(J)
  NEXT J
  Y(IHI) = YPR
END IF
LOOP                                for the test of doneness and the next iteration.
END SUB
```

A sample program using `AMOEBA` is the following:

```
DECLARE SUB AMOEBA (P!(), Y!(), MP!, NP!, NDIM!, FTOL!, DUM!, ITER!)
DECLARE FUNCTION AMOEB! (X!(), NDIM!)
DECLARE FUNCTION BESSJ0! (X!)

'PROGRAM D10R5
'Driver for routine AMOEBA
CLS
NP = 3
```

```
MP = 4
FTOL = .000001
DIM P(MP, NP), X(NP), Y(MP)
FOR J = 1 TO NP
  FOR I = 1 TO MP
    READ P(I, J)
  NEXT I
NEXT J
DATA 0.0,1.0,0.0,0.0,0.0,0.0,1.0,0.0,0.0,0.0,0.0,1.0
NDIM = NP
FOR I = 1 TO MP
  FOR J = 1 TO NP
    X(J) = P(I, J)
  NEXT J
  Y(I) = AMOEB(X(), NP)
NEXT I
CALL AMOEBA(P(), Y(), MP, NP, NDIM, FTOL, DUM, ITER)
PRINT "Iterations: "; ITER
PRINT
PRINT "Vertices of final 3-D simplex and"
PRINT "function values at the vertices:"
PRINT
PRINT "  I        X(I)         Y(I)         Z(I)        FUNCTION"
PRINT
FOR I = 1 TO MP
  PRINT USING "###"; I;
  FOR J = 1 TO NP
    PRINT USING "#####.######"; P(I, J);
  NEXT J
  PRINT USING "#####.######"; Y(I)
NEXT I
PRINT
PRINT "True minimum is at (0.5,0.6,0.7)"
END

FUNCTION AMOEB (X(), NDIM)
AMOEB = .6 - BESSJ0((X(1) - .5) ^ 2 + (X(2) - .6) ^ 2 + (X(3) - .7) ^ 2)
END FUNCTION
```

POWELL carries out one-dimensional minimizations along favorable directions in N-dimensional space. The function minimized must be called FUNC2, and in sample program D10R6 a function subroutine is defined for

$$\text{FUNC2}(x, y, z) = \tfrac{1}{2} - J_0[(x - 1)^2 + (y - 2)^2 + (z - 3)^2].$$

The program provides POWELL with a starting point P of $(3/2, 3/2, 5/2)$ and a set of initial directions, here chosen to be the unit directions $(1, 0, 0)$, $(0, 1, 0)$, and $(0, 0, 1)$. POWELL performs its one-dimensional minimizations with LINMIN, which is discussed next.

Here is the recipe POWELL:

```
DECLARE SUB LINMIN (P!(), XI!(), N!, FRET!)
DECLARE FUNCTION FUNC2! (X!(), N!)

SUB POWELL (P(), XI(), N, NP, FTOL, ITER, FRET)
```
Minimization of a function FUNC2 of N variables. (FUNC2 is not an argument, it is a fixed function name.) Input consists of an initial starting point P that is a vector of length N; an initial matrix XI whose logical dimensions are N by N, physical dimensions NP by NP, and whose columns contain the initial set of directions (usually the N unit vectors); and FTOL, the fractional tolerance in the function value such that failure to decrease by more than this amount on one iteration signals doneness. On output, P is set to the best point found, XI is the then-current direction set, FRET is the returned function value at P, and ITER is the number of iterations taken. The routine LINMIN is used.

```
ITMAX = 200                           Maximum allowed iterations.
DIM PT(N), PTT(N), XIT(N)
FRET = FUNC2(P(), N)
FOR J = 1 TO N                        Save the initial point.
  PT(J) = P(J)
NEXT J
ITER = 0
DO
  DO
    DO
      ITER = ITER + 1
      FP = FRET
      IBIG = 0
      DEL = 0!                        Will be the biggest function decrease.
      FOR I = 1 TO N                  In each iteration, loop over all directions in the set.
        FOR J = 1 TO N                Copy the direction,
          XIT(J) = XI(J, I)
        NEXT J
        FPTT = FRET
        CALL LINMIN(P(), XIT(), N, FRET)   minimize along it,
        IF ABS(FPTT - FRET) > DEL THEN     and record it if it is the largest decrease so far.
          DEL = ABS(FPTT - FRET)
          IBIG = I
        END IF
      NEXT I
      IF 2! * ABS(FP - FRET) <= FTOL * (ABS(FP) + ABS(FRET)) THEN   Termination
        ERASE XIT, PTT, PT                                          criterion.
        EXIT SUB
      END IF
      IF ITER = ITMAX THEN
        PRINT "Powell exceeding maximum iterations."
        EXIT SUB
      END IF
      FOR J = 1 TO N                  Construct extrapolated point and average direction moved.
        PTT(J) = 2! * P(J) - PT(J)    Save the old starting point.
        XIT(J) = P(J) - PT(J)
        PT(J) = P(J)
      NEXT J
      FPTT = FUNC2(PTT(), N)          Function value at extrapolated point.
    LOOP WHILE FPTT >= FP             One reason not to use new direction.
    DUM = FP - 2! * FRET + FPTT
    T = 2! * DUM * (FP - FRET - DEL) ^ 2 - DEL * (FP - FPTT) ^ 2
  LOOP WHILE T >= 0!                  Other reason not to use new direction.
```

```
CALL LINMIN(P(), XIT(), N, FRET)
FOR J = 1 TO N
   XI(J, IBIG) = XIT(J)
NEXT J
LOOP
END SUB
```

Move to the minimum of the new direction, and save the new direction.

Back for another iteration.

A sample program using POWELL is the following:

```
DECLARE FUNCTION FUNC2! (X!(), N!)
DECLARE FUNCTION FUNC! (X!)
DECLARE FUNCTION F1DIM! (X!)
DECLARE SUB POWELL (P!(), XI!(), N!, NP!, FTOL!, ITER!, FRET!)
DECLARE FUNCTION BESSJ0! (X!)
COMMON NCOM, PCOM(), XICOM()

'PROGRAM D10R6
'Driver for routine POWELL
CLS
NDIM = 3
FTOL = .000001
DIM P(NDIM), XI(NDIM, NDIM), PCOM(50), XICOM(50)
NP = NDIM
FOR J = 1 TO NDIM
  FOR I = 1 TO NDIM
    READ XI(I, J)
  NEXT I
NEXT J
DATA 1.0,0.0,0.0,0.0,1.0,0.0,0.0,0.0,1.0
FOR I = 1 TO NDIM
  READ P(I)
NEXT I
DATA 1.5,1.5,2.5
CALL POWELL(P(), XI(), NDIM, NP, FTOL, ITER, FRET)
PRINT "Iterations:"; ITER
PRINT
PRINT "Minimum found at:"
FOR I = 1 TO NDIM
  PRINT USING "#####.######"; P(I);
NEXT I
PRINT
PRINT
PRINT "Minimum function value =";
PRINT USING "#####.######"; FRET
PRINT
PRINT "True minimum of function is at:"
PRINT USING "#####.######"; 1!; 2!; 3!
END

FUNCTION FUNC (X)
FUNC = F1DIM(X)
END FUNCTION

FUNCTION FUNC2 (X(), N)
FUNC2 = .5 - BESSJ0((X(1) - 1!) ^ 2 + (X(2) - 2!) ^ 2 + (X(3) - 3!) ^ 2)
```

END FUNCTION

LINMIN, as we have said, finds the minimum of a function FUNC2 along a direction in N-dimensional space. To use it we specify a point P and a direction vector XI, both in N-space. LINMIN then does the bookkeeping required to treat the function as a function of position along this line, and minimizes the function with a conventional one-dimensional minimization routine. Sample program D10R7 feeds LINMIN the function

$$\text{FUNC2}(x, y, z) = (x - 1)^2 + (y - 1)^2 + (z - 1)^2$$

which has a minimum at $(x, y, z) = (1, 1, 1)$. It also chooses point P to be the origin $(0, 0, 0)$, and tries a series of directions

$$\left(\sqrt{2} \cos\left(\frac{\pi}{2}\frac{I}{10.0}\right), \quad \sqrt{2} \sin\left(\frac{\pi}{2}\frac{I}{10.0}\right), \quad 1.0 \right) \qquad I = 1, \dots, 10$$

For each pass, the location of the minimum, and the value of the function at the minimum, are printed. Among the directions searched is the direction $(1, 1, 1)$. Along this direction, of course, the minimum function value should be zero and should occur at $(1, 1, 1)$.

Here is the recipe LINMIN:

```
DECLARE SUB MNBRAK (AX!, BX!, CX!, FA!, FB!, FC!, DUM!)
DECLARE FUNCTION BRENT! (AX!, BX!, CX!, DUM!, TOL!, XMIN!)
COMMON SHARED NCOM, PCOM(), XICOM()

SUB LINMIN (P(), XI(), N, FRET)
```
> Given an N dimensional point P and an N dimensional direction XI, moves and resets P to where the function FUNC2(P) takes on a minimum along the direction XI from P, and replaces XI by the actual vector displacement that P was moved. Also returns as FRET the value of FUNC2 at the returned location P. This is actually all accomplished by calling the routines MNBRAK and BRENT.

```
TOL = .0001                    TOL passed to BRENT.
NCOM = N                       Set up the common block.
FOR J = 1 TO N
  PCOM(J) = P(J)
  XICOM(J) = XI(J)
NEXT J
AX = 0!                        Initial guess for brackets.
XX = 1!
CALL MNBRAK(AX, XX, BX, FA, FX, FB, DUM)
FRET = BRENT(AX, XX, BX, DUM, TOL, XMIN)
FOR J = 1 TO N                 Construct the vector results to return.
  XI(J) = XMIN * XI(J)
  P(J) = P(J) + XI(J)
NEXT J
END SUB
```

A sample program using LINMIN is the following:

```
DECLARE FUNCTION FUNC! (X!)
DECLARE FUNCTION FUNC2! (X!(), N!)
DECLARE FUNCTION F1DIM! (X!)
DECLARE SUB LINMIN (P!(), XI!(), N!, FRET!)
COMMON NCOM, PCOM(), XICOM()

'PROGRAM D10R7
'Driver for routine LINMIN
CLS
NDIM = 3
PIO2 = 1.5707963#
DIM P(NDIM), XI(NDIM), PCOM(50), XICOM(50)
PRINT "Minimum of a 3-D quadratic centered"
PRINT "at (1.0,1.0,1.0). Minimum is found"
PRINT "along a series of radials."
PRINT
PRINT "        x            y            z        minimum"
PRINT
FOR I = 0 TO 10
  X = PIO2 * I / 10!
  SR2 = SQR(2!)
  XI(1) = SR2 * COS(X)
  XI(2) = SR2 * SIN(X)
  XI(3) = 1!
  P(1) = 0!
  P(2) = 0!
  P(3) = 0!
  CALL LINMIN(P(), XI(), NDIM, FRET)
  FOR J = 1 TO 3
    PRINT USING "#####.######"; P(J);
  NEXT J
  PRINT USING "#####.######"; FRET
NEXT I
END


FUNCTION FUNC (X)
FUNC = F1DIM(X)
END FUNCTION


FUNCTION FUNC2 (X(), N)
F = 0!
F = F + (X(1) - 1!) ^ 2
F = F + (X(2) - 1!) ^ 2
F = F + (X(3) - 1!) ^ 2
FUNC2 = F
END FUNCTION
```

F1DIM accompanies LINMIN and is the routine that makes the N-dimensional function FUNC effectively a one-dimensional function along a given line in N-space. There is little to check here, and our perfunctory demonstration of its use, in sample program D10R8, simply plots F1DIM as a one-dimensional function, given the function

$$\text{FUNC2}(x,y,z) = (x-1)^2 + (y-1)^2 + (z-1)^2.$$

You get to choose the direction; then SCRSHO plots the function along this direction. Try the direction $(1,1,1)$ along which you should find a minimum value of FUNC=0 at position $(1,1,1)$.

Here is the recipe F1DIM:

```
DECLARE FUNCTION F1DIM! (X!)
DECLARE FUNCTION FUNC2! (X!(), NCOM!)
COMMON SHARED NCOM, PCOM(), XICOM()

FUNCTION F1DIM (X)
        Must accompany LINMIN.
DIM XT(NCOM)
FOR J = 1 TO NCOM
  XT(J) = PCOM(J) + X * XICOM(J)
NEXT J
F1DIM = FUNC2(XT(), NCOM)
ERASE XT
END FUNCTION
```

A sample program using F1DIM is the following:

```
DECLARE FUNCTION FUNC! (X!)
DECLARE FUNCTION FUNC2! (X!(), N!)
DECLARE FUNCTION F1DIM! (X!)
DECLARE SUB SCRSHO (DUM!)
COMMON NCOM, PCOM(), XICOM()

'PROGRAM D10R8
'Driver for routine F1DIM
CLS
NDIM = 3
DIM P(NDIM), XI(NDIM), PCOM(50), XICOM(50)
FOR I = 1 TO NDIM
  READ P(I)
NEXT I
DATA 0.0,0.0,0.0
NCOM = NDIM
PRINT "Enter vector direction along which to"
PRINT "plot the function. Minimum is in the"
PRINT "direction 1.0,1.0,1.0 - Enter X,Y,Z:"
INPUT XI(1), XI(2), XI(3)
FOR J = 1 TO NDIM
  PCOM(J) = P(J)
  XICOM(J) = XI(J)
NEXT J
CALL SCRSHO(DUM)
END
```

```
FUNCTION FUNC (X)
FUNC = F1DIM(X)
END FUNCTION

FUNCTION FUNC2 (X(), N)
F = 0!
F = F + (X(1) - 1!) ^ 2
F = F + (X(2) - 1!) ^ 2
F = F + (X(3) - 1!) ^ 2
FUNC2 = F
END FUNCTION
```

FRPRMN is another multidimensional minimizer that relies on the one-dimensional minimizations of LINMIN. It works, however, via the Fletcher-Reeves-Polak-Ribiere method and requires that routines be supplied for calculating both the function and its gradient. Sample program D10R9, for example, uses

$$\text{FUNC2}(x, y, z) = 1.0 - J_0(x - \tfrac{1}{2})J_0(y - \tfrac{1}{2})J_0(z - \tfrac{1}{2})$$

and

$$\text{DFUNC} = \frac{\partial \text{FUNC2}}{\partial x} = J_1(x - \tfrac{1}{2})J_0(y - \tfrac{1}{2})J_0(z - \tfrac{1}{2})$$

etc. A number of trial starting vectors are used, and each time, FRPRMN manages to find the minimum at $(1/2, 1/2, 1/2)$.

Here is the recipe FRPRMN:

```
DECLARE SUB DFUNC (X!(), DF!())
DECLARE SUB LINMIN (P!(), XI!(), N!, FRET!)
DECLARE FUNCTION FUNC2! (P!(), N!)

SUB FRPRMN (P(), N, FTOL, ITER, FRET)
```
     Given a starting point P that is a vector of length N, Fletcher-Reeves-Polak-Ribiere minimization is performed on a function FUNC2, using its gradient as calculated by a routine DFUNC. The convergence tolerance on the function value is input as FTOL. Returned quantities are P (the location of the minimum), ITER (the number of iterations that were performed), and FRET (the minimum value of the function). The routine LINMIN is called to perform line minimizations.
```
ITMAX = 200
EPS = 1E-10
```
     Maximum allowed number of iterations; small number to rectify special case of converging to exactly zero function value.
```
DIM G(N), H(N), XI(N)
FP = FUNC2(P(), N)                  Initializations.
CALL DFUNC(P(), XI())
FOR J = 1 TO N
  G(J) = -XI(J)
  H(J) = G(J)
  XI(J) = H(J)
NEXT J
FOR ITS = 1 TO ITMAX                Loop over iterations.
  ITER = ITS
  CALL LINMIN(P(), XI(), N, FRET)            Next statement is the normal return:
  IF 2! * ABS(FRET - FP) <= FTOL * (ABS(FRET) + ABS(FP) + EPS) THEN EXIT FOR
```

```
FP = FUNC2(P(), N)
CALL DFUNC(P(), XI())
GG = 0!
DGG = 0!
FOR J = 1 TO N
  GG = GG + G(J) ^ 2
  DGG = DGG + XI(J) ^ 2          This statement for Fletcher-Reeves.
  DGG = DGG + (XI(J) + G(J)) * XI(J)     This statement for Polak-Ribiere.
NEXT J
IF GG = 0! THEN EXIT FOR          Unlikely. If gradient is exactly zero then we are already done.
GAM = DGG / GG
FOR J = 1 TO N
  G(J) = -XI(J)
  H(J) = G(J) + GAM * H(J)
  XI(J) = H(J)
NEXT J
NEXT ITS
IF ITS > ITMAX THEN PRINT "FRPR maximum iterations exceeded"
ERASE XI, H, G
END SUB
```

A sample program using FRPRMN is the following:

```
DECLARE SUB FRPRMN (P!(), N!, FTOL!, ITER!, FRET!)
DECLARE FUNCTION BESSJ0! (X!)
DECLARE FUNCTION BESSJ1! (X!)
DECLARE FUNCTION F1DIM! (X!)
DECLARE FUNCTION FUNC2! (X!(), N!)
DECLARE FUNCTION FUNC! (X!)
COMMON NCOM, PCOM(), XICOM()

'PROGRAM D10R9
'Driver for routine FRPRMN
CLS
NDIM = 3
FTOL = .000001
PIO2 = 1.5707963#
DIM P(NDIM), PCOM(50), XICOM(50)
PRINT "Program finds the minimum of a function"
PRINT "with different trial starting vectors."
PRINT "True minimum is (0.5,0.5,0.5)"
PRINT
FOR K = 0 TO 4
  ANGL = PIO2 * K / 4!
  P(1) = 2! * COS(ANGL)
  P(2) = 2! * SIN(ANGL)
  P(3) = 0!
  PRINT "Starting vector: (";
  PRINT USING "#.####"; P(1);
  PRINT ",";
  PRINT USING "#.####"; P(2);
  PRINT ",";
  PRINT USING "#.####"; P(3);
  PRINT ")"
  CALL FRPRMN(P(), NDIM, FTOL, ITER, FRET)
```

```
PRINT "Iterations:"; ITER
PRINT "Solution vector: (";
PRINT USING "#.####"; P(1);
PRINT ",";
PRINT USING "#.####"; P(2);
PRINT ",";
PRINT USING "#.####"; P(3);
PRINT ")"
PRINT "Func. value at solution   ";
PRINT USING ".######^^^^"; FRET
PRINT
NEXT K
END


SUB DFUNC (X(), DF())
DF(1) = BESSJ1(X(1) - .5) * BESSJ0(X(2) - .5) * BESSJ0(X(3) - .5)
DF(2) = BESSJ0(X(1) - .5) * BESSJ1(X(2) - .5) * BESSJ0(X(3) - .5)
DF(3) = BESSJ0(X(1) - .5) * BESSJ0(X(2) - .5) * BESSJ1(X(3) - .5)
END SUB

FUNCTION FUNC (X)
FUNC = F1DIM(X)
END FUNCTION

FUNCTION FUNC2 (X(), N)
FUNC2 = 1! - BESSJ0(X(1) - .5) * BESSJ0(X(2) - .5) * BESSJ0(X(3) - .5)
END FUNCTION
```

Completeness requires that we provide a sample program for DF1DIM, which is presented in *Numerical Recipes* as a routine for converting the N-dimensional gradient subroutine to one that provides the first derivative of the function along a specified line in N-dimensional space. It is exactly analogous to F1DIM and the program D10R10 is the same.

Here is the recipe DF1DIM:

```
DECLARE SUB DFUNC (X!(), DF!())
DECLARE FUNCTION DF1DIM! (X!)
COMMON SHARED NCOM, PCOM(), XICOM()

FUNCTION DF1DIM (X)
DIM XT(NCOM), DF(NCOM)
FOR J = 1 TO NCOM
  XT(J) = PCOM(J) + X * XICOM(J)
NEXT J
CALL DFUNC(XT(), DF())
DF1 = 0!
FOR J = 1 TO NCOM
  DF1 = DF1 + DF(J) * XICOM(J)
NEXT J
DF1DIM = DF1
ERASE DF, XT
END FUNCTION
```

A sample program using DF1DIM is the following:

```
DECLARE FUNCTION FUNC! (X!)
DECLARE SUB SCRSHO (FUNC!)
DECLARE FUNCTION DF1DIM! (X!)
COMMON NCOM, PCOM(), XICOM()

'PROGRAM D10R10
'Driver for routine DF1DIM
CLS
NDIM = 3
DIM P(NDIM), XI(NDIM), PCOM(50), XICOM(50)
FOR I = 1 TO NDIM
  READ P(I)
NEXT I
DATA 0.0,0.0,0.0
NCOM = NDIM
PRINT "Enter vector direction along which to"
PRINT "plot the function. Minimum is in the"
PRINT "direction 1.0,1.0,1.0 - Enter X,Y,Z:"
INPUT XI(1), XI(2), XI(3)
FOR J = 1 TO NDIM
  PCOM(J) = P(J)
  XICOM(J) = XI(J)
NEXT J
CALL SCRSHO(DUM)
END

SUB DFUNC (X(), DF())
FOR I = 1 TO 3
  DF(I) = (X(I) - 1!) ^ 2
NEXT I
END SUB

FUNCTION FUNC (X)
FUNC = DF1DIM(X)
END FUNCTION
```

DFPMIN implements the Broyden-Fletcher-Goldfarb-Shanno variant of the Davidon-Fletcher-Powell minimization by variable metric methods. It requires somewhat more intermediate storage than the preceding routine and is not considered superior in other ways. However, it is a popular method. Sample program **D10R11** works just as did the program for **FRPRMN**, including the fact that it requires a subroutine for calculation of the derivative.

Here is the recipe **DFPMIN**:

```
DECLARE SUB DFUNC (X!(), DF!())
DECLARE SUB LINMIN (P!(), XI!(), N!, FRET!)
DECLARE FUNCTION FUNC2! (P!(), N!)

SUB DFPMIN (P(), N, FTOL, ITER, FRET)
```
    Given a starting point P that is a vector of length N, the Broyden-Fletcher-Goldfarb-Shanno vari-
    ant of Davidon-Fletcher-Powell minimization is performed on a user-supplied function FUNC2,

using its gradient as calculated by a routine DFUNC. The convergence requirement on the function value is input as FTOL. Returned quantities are P (the location of the minimum), ITER (the number of iterations that were performed), and FRET (the minimum value of the function). The routine LINMIN is called to perform line minimizations.

```
ITMAX = 200                              Maximum allowed number of iterations; small number to rec-
EPS = 1E-10                              tify special case of converging to exactly zero function value.
DIM HESSIN(N, N), XI(N), G(N), DG(N), HDG(N)
FP = FUNC2(P(), N)                       Calculate starting function value and gradient,
CALL DFUNC(P(), G())
FOR I = 1 TO N                           and initialize the inverse Hessian to the unit matrix.
  FOR J = 1 TO N
    HESSIN(I, J) = 0!
  NEXT J
  HESSIN(I, I) = 1!
  XI(I) = -G(I)                          Initial line direction.
NEXT I
FOR ITS = 1 TO ITMAX                     Main loop over the iterations.
  ITER = ITS
  CALL LINMIN(P(), XI(), N, FRET)        Next statement is the normal return:
  IF 2! * ABS(FRET - FP) <= FTOL * (ABS(FRET) + ABS(FP) + EPS) THEN
    ERASE HDG, DG, G, XI, HESSIN
    EXIT SUB
  END IF
  FP = FRET                              Save the old function value
  FOR I = 1 TO N                         and the old gradient.
    DG(I) = G(I)
  NEXT I
  FRET = FUNC2(P(), N)                   Get new function value and gradient.
  CALL DFUNC(P(), G())
  FOR I = 1 TO N                         Compute difference of gradients,
    DG(I) = G(I) - DG(I)
  NEXT I
  FOR I = 1 TO N                         and difference times current matrix.
    HDG(I) = 0!
    FOR J = 1 TO N
      HDG(I) = HDG(I) + HESSIN(I, J) * DG(J)
    NEXT J
  NEXT I
  FAC = 0!                               Calculate dot products for the denominators,
  FAE = 0!
  FOR I = 1 TO N
    FAC = FAC + DG(I) * XI(I)
    FAE = FAE + DG(I) * HDG(I)
  NEXT I
  FAC = 1! / FAC                         and make the denominators multiplicative.
  FAD = 1! / FAE
  FOR I = 1 TO N                         The vector which makes BFGS different from DFP:
    DG(I) = FAC * XI(I) - FAD * HDG(I)
  NEXT I
  FOR I = 1 TO N
    FOR J = 1 TO N                       The BFGS updating formula:
      DUM = FAC * XI(I) * XI(J) - FAD * HDG(I) * HDG(J) + FAE * DG(I) * DG(J)
      HESSIN(I, J) = HESSIN(I, J) + DUM
    NEXT J
  NEXT I
```

```
FOR I = 1 TO N                    Now calculate the next direction to go,
  XI(I) = 0!
  FOR J = 1 TO N
    XI(I) = XI(I) - HESSIN(I, J) * G(J)
  NEXT J
NEXT I
NEXT ITS                          and go back for another iteration.
PRINT "too many iterations in DFPMIN"
END SUB
```

A sample program using **DFPMIN** is the following:

```
DECLARE FUNCTION FUNC! (X!)
DECLARE FUNCTION FUNC2! (X! (), N)
DECLARE SUB DFPMIN (P! (), N!, FTOL!, ITER!, FRET!)
DECLARE FUNCTION F1DIM! (X!)
DECLARE FUNCTION BESSJ0! (X!)
DECLARE FUNCTION BESSJ1! (X!)
COMMON NCOM, PCOM(), XICOM()

'PROGRAM D10R11
'Driver for routine DFPMIN
CLS
NDIM = 3
FTOL = .000001
PIO2 = 1.5707963#
DIM P(NDIM), PCOM(50), XICOM(50)
PRINT "Program finds the minimum of a function"
PRINT "with different trial starting vectors."
PRINT "True minimum is (0.5,0.5,0.5)"
PRINT
FOR K = 0 TO 4
  ANGL = PIO2 * K / 4!
  P(1) = 2! * COS(ANGL)
  P(2) = 2! * SIN(ANGL)
  P(3) = 0!
  PRINT "Starting vector: (";
  PRINT USING "#.####"; P(1);
  PRINT ",";
  PRINT USING "#.####"; P(2);
  PRINT ",";
  PRINT USING "#.####"; P(3);
  PRINT ")"
  CALL DFPMIN(P(), NDIM, FTOL, ITER, FRET)
  PRINT "Iterations:"; ITER
  PRINT "Solution vector: (";
  PRINT USING "#.####"; P(1);
  PRINT ",";
  PRINT USING "#.####"; P(2);
  PRINT ",";
  PRINT USING "#.####"; P(3);
  PRINT ")"
  PRINT "Func. value at solution   ";
  PRINT USING ".######^^^^"; FRET
  PRINT
```

```
NEXT K
END


SUB DFUNC (X(), DF())
DF(1) = BESSJ1(X(1) - .5) * BESSJ0(X(2) - .5) * BESSJ0(X(3) - .5)
DF(2) = BESSJ0(X(1) - .5) * BESSJ1(X(2) - .5) * BESSJ0(X(3) - .5)
DF(3) = BESSJ0(X(1) - .5) * BESSJ0(X(2) - .5) * BESSJ1(X(3) - .5)
END SUB


FUNCTION FUNC (X)
FUNC = F1DIM(X)
END FUNCTION


FUNCTION FUNC2 (X(), N)
FUNC2 = 1! - BESSJ0(X(1) - .5) * BESSJ0(X(2) - .5) * BESSJ0(X(3) - .5)
END FUNCTION
```

SIMPLX is a subroutine for dealing with problems in linear programming. In these problems the goal is to maximize a linear combination of N variables, subject to the constraint that none be negative, and that as a group they satisfy a number of other constraints. In order to clarify the subject, *Numerical Recipes* presents a sample problem in equations (10.8.6) and (10.8.7), translating the problem into tableau format in (10.8.18), and presenting a solution in equation (10.8.19). Sample program D10R12 carries out the analysis that leads to this solution.

Here is the recipe SIMPLX:

```
DECLARE SUB SIMP1 (A!(), MP!, NP!, MM!, LL!(), NLL!, IABF!, KP!, BMAX!)
DECLARE SUB SIMP2 (A!(), M!, N!, MP!, NP!, L2!(), NL2!, IP!, KP!, Q1!)
DECLARE SUB SIMP3 (A!(), MP!, NP!, I1!, K1!, IP!, KP!)

SUB SIMPLX (A(), M, N, MP, NP, M1, M2, M3, ICASE, IZROV(), IPOSV())
        Simplex method for linear programming. Input parameters A, M, N, MP, NP, M1, M2, and M3,
        and output parameters A, ICASE, IZROV, and IPOSV are described in the main book.
EPS = .000001
DIM L1(M), L2(M), L3(M)
IF M <> M1 + M2 + M3 THEN         M is the maximum number of constraints expected.
  PRINT "Bad input constraint counts."
  EXIT SUB
END IF
NL1 = N                           Initially make all variables right-hand.
FOR K = 1 TO N                    Initialize index lists.
  L1(K) = K
  IZROV(K) = K
NEXT K
NL2 = M                           Make all artificial variables left-hand,
FOR I = 1 TO M                    and initialize those lists.
  IF A(I + 1, 1) < 0! THEN        Constants b_i must be nonnegative.
    PRINT "Bad input tableau."
    EXIT SUB
  END IF
  L2(I) = I
  IPOSV(I) = N + I
NEXT I
FOR I = 1 TO M2                   Used later, but initialized here.
```

```
   L3(I) = 1
NEXT I
IR = 0                               This flag setting means we are in phase two, i.e have a feasible
IF M2 + M3 = 0 THEN GOTO 3           starting solution. GOTO 3 if origin is a feasible solution.
IR = 1                               Flag meaning that we must start out in phase one.
FOR K = 1 TO N + 1                   Compute the auxiliary objective function.
  Q1 = 0!
  FOR I = M1 + 1 TO M
    Q1 = Q1 + A(I + 1, K)
  NEXT I
  A(M + 2, K) = -Q1
NEXT K
DO
  CALL SIMP1(A(), MP, NP, M + 1, L1(), NL1, 0, KP, BMAX)     Find max. coeff. of aux-
  IF BMAX <= EPS AND A(M + 2, 1) < -EPS THEN                 iliary objective fn.
    ICASE = -1                       Auxiliary objective function is still negative and can't be im-
    ERASE L3, L2, L1                 proved, hence no feasible solution exists.
    EXIT SUB
  ELSEIF BMAX <= EPS AND A(M + 2, 1) <= EPS THEN
    M12 = M1 + M2 + 1                Auxiliary objective function is zero and can't be improved.
    IF M12 <= M THEN                 This signals that we have a feasible starting vector. Clean
      FOR IP = M12 TO M              out the artificial variables and then move on to phase two by
        IF IPOSV(IP) = IP + N THEN   EXIT DO.
          CALL SIMP1(A(), MP, NP, IP, L1(), NL1, 1, KP, BMAX)
          IF BMAX > 0! THEN GOTO 1
        END IF
      NEXT IP
    END IF
    IR = 0                           Set flag indicating we have reached phase two.
    M12 = M12 - 1
    IF M1 + 1 > M12 THEN EXIT DO
    FOR I = M1 + 1 TO M12
      IF L3(I - M1) = 1 THEN
        FOR K = 1 TO N + 1
          A(I + 1, K) = -A(I + 1, K)
        NEXT K
      END IF
    NEXT I
    EXIT DO                          Go to phase two.
  END IF
  CALL SIMP2(A(), M, N, MP, NP, L2(), NL2, IP, KP, Q1)     Locate a pivot element.
  IF IP = 0 THEN                     Maximum of auxiliary objective function is unbounded, so no
    ICASE = -1                       feasible solution exists.
    ERASE L3, L2, L1
    EXIT SUB
  END IF
1 CALL SIMP3(A(), MP, NP, M + 1, N, IP, KP)     Exchange a left- and right-hand variable
  IF IPOSV(IP) >= N + M1 + M2 + 1 THEN          (phase one), then update lists.
    FOR K = 1 TO NL1
      IF L1(K) = KP THEN EXIT FOR
    NEXT K
    NL1 = NL1 - 1
    FOR IQ = K TO NL1
      L1(IQ) = L1(IQ + 1)
    NEXT IQ
```

```
      ELSE
        IF IPOSV(IP) < N + M1 + 1 THEN GOTO 2
        KH = IPOSV(IP) - M1 - N
        IF L3(KH) = 0 THEN GOTO 2
        L3(KH) = 0
      END IF
      A(M + 2, KP + 1) = A(M + 2, KP + 1) + 1!
      FOR I = 1 TO M + 2
        A(I, KP + 1) = -A(I, KP + 1)
      NEXT I
2   IQ = IZROV(KP)
    IZROV(KP) = IPOSV(IP)
    IPOSV(IP) = IQ
  LOOP WHILE IR <> 0                    If still in phase one, loop again.
        End of phase one code for finding an initial feasible solution. Now, in phase two, optimize it.
3 CALL SIMP1(A(), MP, NP, 0, L1(), NL1, 0, KP, BMAX)   Test the z-row for doneness.
  IF BMAX <= 0! THEN                    Done. Solution found. Return with the good news.
    ICASE = 0
    ERASE L3, L2, L1
    EXIT SUB
  END IF
  CALL SIMP2(A(), M, N, MP, NP, L2(), NL2, IP, KP, Q1)   Locate a pivot element.
  IF IP = 0 THEN                        Objective function is unbounded. Report and return.
    ICASE = 1
    ERASE L3, L2, L1
    EXIT SUB
  END IF
  CALL SIMP3(A(), MP, NP, M, N, IP, KP)   Exchange a left- and a right-hand variable,
  GOTO 2                                and return for another iteration.
  END SUB
```

Here is the recipe SIMP1:

```
SUB SIMP1 (A(), MP, NP, MM, LL(), NLL, IABF, KP, BMAX)
        Determines the maximum of those elements whose index is contained in the supplied list LL,
        either with or without taking the absolute value, as flagged by IABF.
KP = LL(1)
BMAX = A(MM + 1, KP + 1)
FOR K = 2 TO NLL
  IF IABF = 0 THEN
    TEST = A(MM + 1, LL(K) + 1) - BMAX
  ELSE
    TEST = ABS(A(MM + 1, LL(K) + 1)) - ABS(BMAX)
  END IF
  IF TEST > 0! THEN
    BMAX = A(MM + 1, LL(K) + 1)
    KP = LL(K)
  END IF
NEXT K
END SUB
```

Here is the recipe SIMP2:

```
SUB SIMP2 (A(), M, N, MP, NP, L2(), NL2, IP, KP, Q1)
     Locate a pivot element, taking degeneracy into account.
EPS = .000001
IP = 0
FLAG = 0
FOR I = 1 TO NL2
  IF A(L2(I) + 1, KP + 1) < -EPS THEN FLAG = 1
  IF FLAG = 1 THEN EXIT FOR
NEXT I
IF FLAG = 0 THEN EXIT SUB          No possible pivots. Return.
Q1 = -A(L2(I) + 1, 1) / A(L2(I) + 1, KP + 1)
IP = L2(I)
FOR I = I + 1 TO NL2
  II = L2(I)
  IF A(II + 1, KP + 1) < -EPS THEN
    Q = -A(II + 1, 1) / A(II + 1, KP + 1)
    IF Q < Q1 THEN
      IP = II
      Q1 = Q
    ELSEIF Q = Q1 THEN            We have a degeneracy.
      FOR K = 1 TO N
        QP = -A(IP + 1, K + 1) / A(IP + 1, KP + 1)
        QO = -A(II + 1, K + 1) / A(II + 1, KP + 1)
        IF QO <> QP THEN EXIT FOR
      NEXT K
      IF QO < QP THEN IP = II
    END IF
  END IF
NEXT I
END SUB
```

Here is the recipe SIMP3:

```
SUB SIMP3 (A(), MP, NP, I1, K1, IP, KP)
     Matrix operations to exchange a left-hand and right-hand variable (see text).
PIV = 1! / A(IP + 1, KP + 1)
FOR II = 1 TO I1 + 1
  IF II - 1 <> IP THEN
    A(II, KP + 1) = A(II, KP + 1) * PIV
    FOR KK = 1 TO K1 + 1
      IF KK - 1 <> KP THEN
        A(II, KK) = A(II, KK) - A(IP + 1, KK) * A(II, KP + 1)
      END IF
    NEXT KK
  END IF
NEXT II
FOR KK = 1 TO K1 + 1
  IF KK - 1 <> KP THEN A(IP + 1, KK) = -A(IP + 1, KK) * PIV
NEXT KK
A(IP + 1, KP + 1) = PIV
END SUB
```

A sample program using **SIMPLX**, **SIMP1**, **SIMP2**, and **SIMP3** is the following:

```
DECLARE SUB SIMPLX (A!(), M!, N!, MP!, NP!, M1!, M2!, M3!, ICASE!, IZROV!(), IP

'PROGRAM D10R12
'Driver for routine SIMPLX
'Incorporates examples discussed in text
CLS
N = 4
M = 4
NP = 5
MP = 6
M1 = 2
M2 = 1
M3 = 1
NM1M2 = N + M1 + M2
DIM A(MP, NP), IZROV(N), IPOSV(M), ANUM(NP), TXT$(NM1M2), ALPHA$(NP)
FOR I = 1 TO NM1M2
  READ TXT$(I)
NEXT I
DATA x1,x2,x3,x4,y1,y2,y3
FOR J = 1 TO NP
  FOR I = 1 TO MP
    READ A(I, J)
  NEXT I
NEXT J
DATA 0.0,740.0,0.0,0.5,9.0,0.0,1.0,-1.0,0.0,0.0,-1.0,0.0,1.0,0.0,-2.0,-1.0
DATA -1.0,0.0,3.0,-2.0,0.0,1.0,-1.0,0.0,-0.5,0.0,7.0,-2.0,-1.0,0.0
CALL SIMPLX(A(), M, N, MP, NP, M1, M2, M3, ICASE, IZROV(), IPOSV())
IF ICASE = 1 THEN
  PRINT "Unbounded objective function"
ELSEIF ICASE = -1 THEN
  PRINT "No solutions satisfy constraints given"
ELSE
  JJ = 1
  FOR I = 1 TO N
    IF IZROV(I) <= NM1M2 THEN
      ALPHA$(JJ) = TXT$(IZROV(I))
      JJ = JJ + 1
    END IF
  NEXT I
  JMAX = JJ - 1
  PRINT "                ";
  FOR JJ = 1 TO JMAX
    PRINT "          ";
    PRINT ALPHA$(JJ);
  NEXT JJ
  PRINT
  FOR I = 1 TO M + 1
    IF I = 1 THEN
      ALPHA$(1) = "  "
      RITE% = -1
    ELSEIF IPOSV(I - 1) <= NM1M2 THEN
      ALPHA$(1) = TXT$(IPOSV(I - 1))
      RITE% = -1
```

```
   ELSE
     RITE% = 0
   END IF
   IF RITE% THEN
     ANUM(1) = A(I, 1)
     JJ = 2
     FOR J = 2 TO N + 1
       IF IZROV(J - 1) <= NM1M2 THEN
         ANUM(JJ) = A(I, J)
         JJ = JJ + 1
       END IF
     NEXT J
     JMAX = JJ - 1
     PRINT ALPHA$(1);
     FOR JJ = 1 TO JMAX
       PRINT USING "#######.##"; ANUM(JJ);
     NEXT JJ
     PRINT
   END IF
 NEXT I
END IF
END
```

ANNEAL is a subroutine for solving the traveling salesman problem—a problem that is included as a demonstration of the use of simulated annealing. Sample program D10R13 has the function of setting up the initial route for the salesman and printing final results. For each of NCITY=10 cities, it chooses random coordinates X(I),Y(I) using routine RAN3, and puts an entry for each city in the array IPTR(I). The array indicates the order in which the cities will be visited. On the originally specified path, the cities are in the order I=1,..,10 so the sample program initially takes IPTR(I)=I. (It is assumed that the salesman will return to the first city after visiting the last.) A call is then made to ANNEAL, which attempts to find the shortest alternative route, which is recorded in the array. After finding a path that resists further improvement, the driver lists the modified itinerary.

Here is the recipe ANNEAL:

```
DECLARE SUB TRNCST (X!(), Y!(), IORDER!(), NCITY!, N!(), DE!)
DECLARE SUB METROP (DE!, T!, ANS%)
DECLARE SUB TRNSPT (IORDER!(), NCITY!, N!())
DECLARE SUB REVCST (X!(), Y!(), IORDER!(), NCITY!, N!(), DE!)
DECLARE SUB REVERS (IORDER!(), NCITY!, N!())
DECLARE FUNCTION ALEN! (X1!, X2!, Y1!, Y2!)
DECLARE FUNCTION RAN3! (IDUM&)
DECLARE FUNCTION IRBIT1! (ISEED!)

FUNCTION ALEN (X1, X2, Y1, Y2)
ALEN = SQR((X2 - X1) ^ 2 + (Y2 - Y1) ^ 2)
END FUNCTION

SUB ANNEAL (X(), Y(), IORDER(), NCITY)
```
   This algorithm finds the shortest round-trip path to NCITY cities whose coordinates are in the arrays X(I),Y(I). The array IORDER(I) specifies the order in which the cities are visited.

On input, the elements of IORDER may be set to any permutation of the numbers 1 to NCITY. This routine will return the best alternative path it can find.

```
DIM N(6)
NOVER = 100 * NCITY                 Maximum number of paths tried at any temperature.
NLIMIT = 10 * NCITY                 Maximum number of successful path changes before continuing.
TFACTR = .9                         Annealing schedule – T is reduced by this factor on each step.
PATH = 0!
T = .5
FOR I = 1 TO NCITY - 1              Calculate initial path length.
  I1 = IORDER(I)
  I2 = IORDER(I + 1)
  PATH = PATH + ALEN(X(I1), X(I2), Y(I1), Y(I2))
NEXT I
I1 = IORDER(NCITY)                  Close the loop by tying path ends together.
I2 = IORDER(1)
PATH = PATH + ALEN(X(I1), X(I2), Y(I1), Y(I2))
IDUM& = -1
ISEED = 111
FOR J = 1 TO 100                    Try up to 100 temperature steps.
  NSUCC = 0
  FOR K = 1 TO NOVER
    DO
      N(1) = 1 + INT(NCITY * RAN3(IDUM&))        Choose beginning of segment
      N(2) = 1 + INT((NCITY - 1) * RAN3(IDUM&))  and end of segment.
      IF N(2) >= N(1) THEN N(2) = N(2) + 1
      IDEC = IRBIT1(ISEED)          Decide whether to do a segment reversal or transport.
      NN = 1 + (N(1) - N(2) + NCITY - 1) MOD NCITY      Cities not on segment.
    LOOP WHILE NN < 3
    IF IDEC = 0 THEN                                    Do a transport.
      N(3) = N(2) + INT(ABS(NN - 2) * RAN3(IDUM&)) + 1
      N(3) = 1 + N(3) - 1 - NCITY * INT((N(3) - 1) / NCITY)
      CALL TRNCST(X(), Y(), IORDER(), NCITY, N(), DE)  Calculate cost.
      CALL METROP(DE, T, ANS%)                          Consult the oracle.
      IF ANS% THEN
        NSUCC = NSUCC + 1
        PATH = PATH + DE
        CALL TRNSPT(IORDER(), NCITY, N())              Carry out the transport.
      END IF
    ELSE                                                Do a path reversal
      CALL REVCST(X(), Y(), IORDER(), NCITY, N(), DE)  Calculate cost.
      CALL METROP(DE, T, ANS%)                          Consult the oracle.
      IF ANS% THEN
        NSUCC = NSUCC + 1
        PATH = PATH + DE
        CALL REVERS(IORDER(), NCITY, N())             Carry out the reversal.
      END IF
    END IF
    IF NSUCC >= NLIMIT THEN EXIT FOR      Finish early if we have enough successful changes.
  NEXT K
  PRINT
  PRINT "T =";
  PRINT USING "####.######"; T;
  PRINT "   Path Length =";
  PRINT USING "####.######"; PATH
  PRINT "Successful Moves: "; NSUCC
```

```
T = T * TFACTR                    Annealing schedule.
IF NSUCC = 0 THEN EXIT FOR        If no success, we are done.
NEXT J
ERASE N
END SUB
```

Here is the recipe LINK, which contains five subroutines called by ANNEAL:

```
DECLARE FUNCTION ALEN! (X1!, X2!, X3!, X4!)
DECLARE FUNCTION RAN3! (IDUM&)

SUB METROP (DE, T, ANS%)
```
   Metropolis algorithm. ANS% is a logical variable which issues a verdict on whether to accept a reconfiguration which leads to a change DE in the objective function E. If DE<0, ANS%=-1, while if DE>0, ANS% is only -1 with probability exp(-DE/T), where T is a temperature determined by the annealing schedule.
```
JDUM& = 1
ANS% = (DE < 0!) OR (RAN3(JDUM&) < EXP(-DE / T))
END SUB

SUB REVCST (X(), Y(), IORDER(), NCITY, N(), DE)
```
   This subroutine returns the value of the cost function for a proposed path reversal. NCITY is the number of cities, and arrays X(I),Y(I) give the coordinates of these cities. IORDER(I) holds the present itinerary. The first two values N(1) and N(2) of array N give the starting and ending cities along the path segment which is to be reversed. On output, DE is the cost of making the reversal. The actual reversal is not performed by this routine.
```
DIM XX(4), YY(4)
N(3) = 1 + N(1) + NCITY - 2 - NCITY * INT((N(1) + NCITY - 2) / NCITY)
N(4) = 1 + N(2) - NCITY * INT(N(2) / NCITY)       Find the city before N(1) and the
FOR J = 1 TO 4                                     city after N(2)
   II = IORDER(N(J))                               Find coordinates for the four cities
   XX(J) = X(II)                                   involved.
   YY(J) = Y(II)
NEXT J
DE = -ALEN(XX(1), XX(3), YY(1), YY(3))            Calculate cost of disconnecting the
DE = DE - ALEN(XX(2), XX(4), YY(2), YY(4))        segment at both ends and recon-
DE = DE + ALEN(XX(1), XX(4), YY(1), YY(4))        necting in the opposite order.
DE = DE + ALEN(XX(2), XX(3), YY(2), YY(3))
ERASE YY, XX
END SUB

SUB REVERS (IORDER(), NCITY, N())
```
   This routine performs a path segment reversal. IORDER(I) is an input array giving the present itinerary. The vector N has as its first four elements the first and last cities N(1),N(2) of the path segment to be reversed, and the two cities N(3) and N(4) which immediately precede and follow this segment. N(3) and N(4) are found by subroutine REVCST. On output, IORDER(I) contains the segment from N(1) to N(2) in reversed order.
```
NN = (1 + (N(2) - N(1) + NCITY) MOD NCITY) / 2     This many cities must be swapped
FOR J = 1 TO NN                                    to effect the reversal.
   K = 1 + N(1) + J - 2 - NCITY * INT((N(1) + J - 2) / NCITY)
   L = 1 + (N(2) - J + NCITY) MOD NCITY            Start at the ends of the segment
   ITMP = IORDER(K)                                and swap pairs of cities, moving to-
   IORDER(K) = IORDER(L)                           ward the center.
   IORDER(L) = ITMP
```

```
NEXT J
END SUB


SUB TRNCST (X(), Y(), IORDER(), NCITY, N(), DE)
```
This subroutine returns the value of the cost function for a proposed path segment transport. NCITY is the number of cities, and arrays X(I) and Y(I) give the city coordinates. IORDER is an array giving the present itinerary. The first three elements of array N give the starting and ending cities of the path to be transported, and the point among the remaining cities after which it is to be inserted. On output, DE is the cost of the change. The actual transport is not performed by this routine.

```
DIM XX(6), YY(6)
N(4) = 1 + N(3) - NCITY * INT(N(3) / NCITY)        Find the city following N(3)..
N(5) = 1 + N(1) + NCITY - 2 - NCITY * INT((N(1) + NCITY - 2) / NCITY)
N(6) = 1 + N(2) - NCITY * INT(N(2) / NCITY)        .. the one preceding N(1), and the
FOR J = 1 TO 6                                       one following N(2).
  II = IORDER(N(J))          Determine coordinates for the six cities involved.
  XX(J) = X(II)
  YY(J) = Y(II)
NEXT J
DE = -ALEN(XX(2), XX(6), YY(2), YY(6))             Calculate the cost of disconnect-
DE = DE - ALEN(XX(1), XX(5), YY(1), YY(5))         ing the path segment from N(1)
DE = DE - ALEN(XX(3), XX(4), YY(3), YY(4))         to N(2), opening a space between
DE = DE + ALEN(XX(1), XX(3), YY(1), YY(3))         N(3) and N(4), connecting the seg-
DE = DE + ALEN(XX(2), XX(4), YY(2), YY(4))         ment in the space, and connecting
DE = DE + ALEN(XX(5), XX(6), YY(5), YY(6))         N(5) to N(6).
ERASE YY, XX
END SUB


SUB TRNSPT (IORDER(), NCITY, N())
```
This routine does the actual path transport, once METROP has approved. IORDER is an input array of length NCITY giving the present itinerary. The array N has as its six elements the beginning N(1) and end N(2) of the path to be transported, the adjacent cities N(3) and N(4) between which the path is to be placed, and the cities N(5) and N(6) which precede and follow the path. N(4), N(5) and N(6) are calculated by subroutine TRNCST. On output, IORDER is modified to reflect the movement of the path segment.

```
DIM JORDER(NCITY)                Number of cities.
M1 = 1 + N(2) - N(1) + NCITY - NCITY * INT((N(2) - N(1) + NCITY) / NCITY)
M2 = 1 + N(5) - N(4) + NCITY - NCITY * INT((N(5) - N(4) + NCITY) / NCITY)
M3 = 1 + N(3) - N(6) + NCITY - NCITY * INT((N(3) - N(6) + NCITY) / NCITY)
NN = 1
FOR J = 1 TO M1
  JJ = 1 + J + N(1) - 2 - NCITY * INT((J + N(1) - 2) / NCITY)       Copy the cho-
  JORDER(NN) = IORDER(JJ)                                           sen segment.
  NN = NN + 1
NEXT J
IF M2 > 0 THEN
  FOR J = 1 TO M2             Then copy the segment from N(4) to N(5).
    JJ = 1 + J + N(4) - 2 - NCITY * INT((J + N(4) - 2) / NCITY)
    JORDER(NN) = IORDER(JJ)
    NN = NN + 1
  NEXT J
END IF
IF M3 > 0 THEN
  FOR J = 1 TO M3            Finally, the segment from N(6) to N(3).
    JJ = 1 + J + N(6) - 2 - NCITY * INT((J + N(6) - 2) / NCITY)
```

```
    JORDER(NN) = IORDER(JJ)
    NN = NN + 1
  NEXT J
END IF
FOR J = 1 TO NCITY
  IORDER(J) = JORDER(J)          Copy JORDER back into IORDER.
NEXT J
ERASE JORDER
END SUB
```

A sample program using **ANNEAL** is the following:

```
DECLARE SUB ANNEAL (X!(), Y!(), IORDER!(), NCITY!)
DECLARE FUNCTION RAN3! (IDUM&)

'PROGRAM D10R13
CLS
NCITY = 10
DIM X(NCITY), Y(NCITY), IORDER(NCITY)
'Create points of sale
IDUM& = -111
FOR I = 1 TO NCITY
  X(I) = RAN3(IDUM&)
  Y(I) = RAN3(IDUM&)
  IORDER(I) = I
NEXT I
CALL ANNEAL(X(), Y(), IORDER(), NCITY)
PRINT "*** System Frozen ***"
PRINT "Final path:"
PRINT " city        x           y"
FOR I = 1 TO NCITY
  II = IORDER(I)
  PRINT USING "####"; II;
  PRINT USING "#####.####"; X(II); Y(II)
NEXT I
END
```

Chapter 11: Eigensystems

In Chapter 11 of *Numerical Recipes*, we deal with the problem of finding eigenvectors and eigenvalues of matrices, first dealing with symmetric matrices, and then with more general cases. For real symmetric matrices of small-to-moderate size, the routine JACOBI is recommended as a simple and foolproof scheme of finding eigenvalues and eigenvectors. Routine EIGSRT may be used to reorder the output of JACOBI into ascending order of eigenvalue. A more efficient (but operationally more complicated) procedure is to reduce the symmetric matrix to tridiagonal form before doing the eigenvalue analysis. TRED2 uses the Householder scheme to perform this reduction and is used in conjunction with TQLI. TQLI determines the eigenvalues and eigenvectors of a real, symmetric, tridiagonal matrix.

For nonsymmetric matrices, we offer only routines for finding eigenvalues, and not eigenvectors. To ameliorate problems with roundoff error, BALANC makes the corresponding rows and columns of the matrix have comparable norms while leaving eigenvalues unchanged. Then the matrix is reduced to Hessenberg form by Gaussian elimination using ELMHES. Finally HQR applies the QR algorithm to find the eigenvalues of the Hessenberg matrix.

★   ★   ★   ★

JACOBI is a reliable scheme for finding both the eigenvalues and eigenvectors of a symmetric matrix. It is not the most efficient scheme available, but it is simple and trustworthy, and it is recommended for problems of small-to-moderate order. Sample program D11R1 defines three matrices A,B,C for use by JACOBI. They are of order 3, 5, and 10 respectively. To check the index handling in JACOBI, they are, each in turn, loaded into a 10×10 matrix E, and then sent to JACOBI with NP=10 (the physical size of the matrix) and N=3, 5 or 10 (the logical array dimension). For each matrix, the eigenvalues and eigenvectors are reported. Then, an eigenvector test takes place in which the original matrix is applied to the purported eigenvector, and the ratio of the result to the vector itself is found. The ratio should, of course, be the eigenvalue.

Here is the recipe JACOBI:

```
SUB JACOBI (A(), N, NP, D(), V(), NROT)
        Computes all eigenvalues and eigenvectors of a real symmetric matrix A, which is of size N by N,
        stored in a physical NP by NP array. On output, elements of A above the diagonal are destroyed.
        D returns the eigenvalues of A in its first N elements. V is a matrix with the same logical and
        physical dimensions as A whose columns contain, on output, the normalized eigenvectors of A.
        NROT returns the number of Jacobi rotations which were required.
DIM B(N), Z(N)
FOR IP = 1 TO N                         Initialize to the identity matrix.
```

```
FOR IQ = 1 TO N
  V(IP, IQ) = 0!
NEXT IQ
V(IP, IP) = 1!
NEXT IP
FOR IP = 1 TO N
  B(IP) = A(IP, IP)                    Initialize B and D to the diagonal of A.
  D(IP) = B(IP)
  Z(IP) = 0!                           This vector will accumulate terms of the form $ta_{pq}$ as in equa-
NEXT IP                                tion (11.1.14).
NROT = 0
FOR I = 1 TO 50
  SM = 0!
  FOR IP = 1 TO N - 1                  Sum off-diagonal elements.
    FOR IQ = IP + 1 TO N
      SM = SM + ABS(A(IP, IQ))
    NEXT IQ
  NEXT IP
  IF SM = 0 THEN ERASE Z, B: EXIT SUB  The normal return, which relies on quadratic conver-
  IF I < 4 THEN                        gence to machine underflow.
    TRESH = .2 * SM / N ^ 2            ...on the first three sweeps.
  ELSE
    TRESH = 0!                         ...thereafter.
  END IF
  FOR IP = 1 TO N - 1
    FOR IQ = IP + 1 TO N
      G = 100! * ABS(A(IP, IQ))
      After four sweeps, skip the rotation if the off-diagonal element is small.
      DUM = ABS(D(IP))
      IF I > 4 AND DUM + G = DUM AND ABS(D(IQ)) + G = ABS(D(IQ)) THEN
        A(IP, IQ) = 0!
      ELSEIF ABS(A(IP, IQ)) > TRESH THEN
        H = D(IQ) - D(IP)
        IF ABS(H) + G = ABS(H) THEN
          T = A(IP, IQ) / H            $t = 1/(2\theta)$
        ELSE
          THETA = .5 * H / A(IP, IQ)   Equation (11.1.10).
          T = 1! / (ABS(THETA) + SQR(1! + THETA ^ 2))
          IF THETA < 0! THEN T = -T
        END IF
        C = 1! / SQR(1 + T ^ 2)
        S = T * C
        TAU = S / (1! + C)
        H = T * A(IP, IQ)
        Z(IP) = Z(IP) - H
        Z(IQ) = Z(IQ) + H
        D(IP) = D(IP) - H
        D(IQ) = D(IQ) + H
        A(IP, IQ) = 0!
        FOR J = 1 TO IP - 1            Case of rotations $1 \leq j < p$.
          G = A(J, IP)
          H = A(J, IQ)
          A(J, IP) = G - S * (H + G * TAU)
          A(J, IQ) = H + S * (G - H * TAU)
```

```
        NEXT J
        FOR J = IP + 1 TO IQ - 1   Case of rotations p < j < q.
          G = A(IP, J)
          H = A(J, IQ)
          A(IP, J) = G - S * (H + G * TAU)
          A(J, IQ) = H + S * (G - H * TAU)
        NEXT J
        FOR J = IQ + 1 TO N         Case of rotations q < j ≤n.
          G = A(IP, J)
          H = A(IQ, J)
          A(IP, J) = G - S * (H + G * TAU)
          A(IQ, J) = H + S * (G - H * TAU)
        NEXT J
        FOR J = 1 TO N
          G = V(J, IP)
          H = V(J, IQ)
          V(J, IP) = G - S * (H + G * TAU)
          V(J, IQ) = H + S * (G - H * TAU)
        NEXT J
        NROT = NROT + 1
      END IF
    NEXT IQ
  NEXT IP
  FOR IP = 1 TO N
    B(IP) = B(IP) + Z(IP)
    D(IP) = B(IP)                    Update D with the sum of ta_{pq},
    Z(IP) = 0!                       and reinitialize Z.
  NEXT IP
NEXT I
PRINT "50 iterations should never happen"
END SUB
```

A sample program using **JACOBI** is the following:

```
DECLARE SUB JACOBI (A!(), N!, NP!, D!(), V!(), NROT!)

'PROGRAM D11R1
'Driver for routine JACOBI
CLS
NP = 10
NMAT = 3
DIM D(NP), V(NP, NP), R(NP)
DIM A(3, 3), B(5, 5), C(10, 10), E(NP, NP), NUM(3)
FOR I = 1 TO 3
  READ NUM(I)
NEXT I
DATA 3,5,10
FOR J = 1 TO 3
  FOR I = 1 TO 3
    READ A(I, J)
  NEXT I
NEXT J
DATA 1.0,2.0,3.0,2.0,2.0,3.0,3.0,3.0,3.0
FOR J = 1 TO 5
  FOR I = 1 TO 5
```

```
      READ B(I, J)
    NEXT I
  NEXT J
  DATA -2.0,-1.0,0.0,1.0,2.0,-1.0,-1.0,0.0,1.0,2.0,0.0,0.0,0.0,1.0,2.0,1.0,1.0
  DATA 1.0,1.0,2.0,2.0,2.0,2.0,2.0,2.0
  FOR J = 1 TO 10
    FOR I = 1 TO 10
      READ C(I, J)
    NEXT I
  NEXT J
  DATA 5.0,4.0,3.0,2.0,1.0,0.0,-1.0,-2.0,-3.0,-4.0,4.0,5.0,4.0,3.0,2.0,1.0,0.0
  DATA -1.0,-2.0,-3.0,3.0,4.0,5.0,4.0,3.0,2.0,1.0,0.0,-1.0,-2.0,2.0,3.0,4.0,5.0
  DATA 4.0,3.0,2.0,1.0,0.0,-1.0,1.0,2.0,3.0,4.0,5.0,4.0,3.0,2.0,1.0,0.0,0.0,1.0
  DATA 2.0,3.0,4.0,5.0,4.0,3.0,2.0,1.0,-1.0,0.0,1.0,2.0,3.0,4.0,5.0,4.0,3.0,2.0
  DATA -2.0,-1.0,0.0,1.0,2.0,3.0,4.0,5.0,4.0,3.0,-3.0,-2.0,-1.0,0.0,1.0,2.0,3.0
  DATA 4.0,5.0,4.0,-4.0,-3.0,-2.0,-1.0,0.0,1.0,2.0,3.0,4.0,5.0
  FOR I = 1 TO NMAT
    IF I = 1 THEN
      FOR II = 1 TO 3
        FOR JJ = 1 TO 3
          E(II, JJ) = A(II, JJ)
        NEXT JJ
      NEXT II
      CALL JACOBI(E(), 3, NP, D(), V(), NROT)
    ELSEIF I = 2 THEN
      FOR II = 1 TO 5
        FOR JJ = 1 TO 5
          E(II, JJ) = B(II, JJ)
        NEXT JJ
      NEXT II
      CALL JACOBI(E(), 5, NP, D(), V(), NROT)
    ELSEIF I = 3 THEN
      FOR II = 1 TO 10
        FOR JJ = 1 TO 10
          E(II, JJ) = C(II, JJ)
        NEXT JJ
      NEXT II
      CALL JACOBI(E(), 10, NP, D(), V(), NROT)
    END IF
    PRINT "Matrix Number", I
    PRINT "Number of JACOBI rotations:", NROT
    PRINT
    PRINT "Eigenvalues:"
    FOR J = 1 TO NUM(I)
      PRINT USING "#####.######"; D(J)
    NEXT J
    PRINT
    PRINT "Eigenvectors:"
    FOR J = 1 TO NUM(I)
      PRINT "Number"; J
      FOR K = 1 TO NUM(I)
        PRINT USING "#####.######"; V(K, J);
        IF (K MOD 5) = 0 OR K = NUM(I) THEN PRINT
      NEXT K
    NEXT J
```

```
PRINT
'Eigenvector test
PRINT "Eigenvector Test"
PRINT
FOR J = 1 TO NUM(I)
  FOR L = 1 TO NUM(I)
    R(L) = 0!
    FOR K = 1 TO NUM(I)
      IF K > L THEN
        KK = L
        LL = K
      ELSE
        KK = K
        LL = L
      END IF
      IF I = 1 THEN
        R(L) = R(L) + A(LL, KK) * V(K, J)
      ELSEIF I = 2 THEN
        R(L) = R(L) + B(LL, KK) * V(K, J)
      ELSEIF I = 3 THEN
        R(L) = R(L) + C(LL, KK) * V(K, J)
      END IF
    NEXT K
  NEXT L
  PRINT "Vector Number", J
  PRINT
  PRINT "     Vector     Mtrx*Vec.     Ratio"
  FOR L = 1 TO NUM(I)
    RATIO = R(L) / V(L, J)
    PRINT USING "#####.######"; V(L, J); R(L); RATIO
  NEXT L
  PRINT
NEXT J
PRINT "press RETURN to continue..."
LINE INPUT DUM$
PRINT
NEXT I
END
```

EIGSRT reorders the output of **JACOBI** so that the eigenvectors are in the order of decreasing eigenvalue. Sample program **D11R2** uses matrix **C** from the previous program to illustrate. This 10×10 matrix is passed to **JACOBI** and the ten eigenvectors are found. They are printed, along with their eigenvalues, in the order that **JACOBI** returns them. Then the matrices **D** and **V** from **JACOBI**, which contain the eigenvalues and eigenvectors, are passed to **EIGSRT**, and ought to return in descending order of eigenvalue. The result is printed for inspection.

Here is the recipe **EIGSRT**:

```
SUB EIGSRT (D(), V(), N, NP)
```
> Given the eigenvalues D and eigenvectors V as output from JACOBI (§11.1) or TQLI (§11.3), this routine sorts the eigenvalues into descending order, and rearranges the columns of V correspondingly. The method is straight insertion.

```
FOR I = 1 TO N - 1
  K = I
  P = D(I)
  FOR J = I + 1 TO N
    IF D(J) >= P THEN
      K = J
      P = D(J)
    END IF
  NEXT J
  IF K <> I THEN
    D(K) = D(I)
    D(I) = P
    FOR J = 1 TO N
      P = V(J, I)
      V(J, I) = V(J, K)
      V(J, K) = P
    NEXT J
  END IF
NEXT I
END SUB
```

A sample program using **EIGSRT** is the following:

```
DECLARE SUB EIGSRT (D!(), V!(), N!, NP!)
DECLARE SUB JACOBI (A!(), N!, NP!, D!(), V!(), NROT!)

'PROGRAM D11R2
'Driver for routine EIGSRT
CLS
NP = 10
DIM D(NP), V(NP, NP), C(NP, NP)
FOR J = 1 TO NP
  FOR I = 1 TO NP
    READ C(I, J)
  NEXT I
NEXT J
DATA 5.0,4.0,3.0,2.0,1.0,0.0,-1.0,-2.0,-3.0,-4.0,4.0,5.0,4.0,3.0,2.0,1.0,0.0
DATA -1.0,-2.0,-3.0,3.0,4.0,5.0,4.0,3.0,2.0,1.0,0.0,-1.0,-2.0,2.0,3.0,4.0,5.0
DATA 4.0,3.0,2.0,1.0,0.0,-1.0,1.0,2.0,3.0,4.0,5.0,4.0,3.0,2.0,1.0,0.0,0.0,1.0
DATA 2.0,3.0,4.0,5.0,4.0,3.0,2.0,1.0,-1.0,0.0,1.0,2.0,3.0,4.0,5.0,4.0,3.0,2.0
DATA -2.0,-1.0,0.0,1.0,2.0,3.0,4.0,5.0,4.0,3.0,-3.0,-2.0,-1.0,0.0,1.0,2.0,3.0
DATA 4.0,5.0,4.0,-4.0,-3.0,-2.0,-1.0,0.0,1.0,2.0,3.0,4.0,5.0
CALL JACOBI(C(), NP, NP, D(), V(), NROT)
PRINT "Unsorted Eigenvectors:"
PRINT
FOR I = 1 TO NP
  PRINT
  PRINT "Eigenvalue"; I; " =";
  PRINT USING "#####.######"; D(I)
  PRINT "Eigenvector:"
```

```
    FOR J = 1 TO NP
      PRINT USING "#####.######"; V(J, I);
      IF (J MOD 5) = 0 OR J = NP THEN PRINT
    NEXT J
NEXT I
PRINT
PRINT "****** sorting ******"
PRINT
CALL EIGSRT(D(), V(), NP, NP)
PRINT "Sorted Eigenvectors:"
PRINT
FOR I = 1 TO NP
  PRINT
  PRINT "Eigenvalue"; I; " =";
  PRINT USING "#####.######"; D(I)
  PRINT "Eigenvector:"
  FOR J = 1 TO NP
    PRINT USING "#####.######"; V(J, I);
    IF (J MOD 5) = 0 OR J = NP THEN PRINT
  NEXT J
NEXT I
END
```

TRED2 reduces a real symmetric matrix to tridiagonal form. Sample program TRED2 again uses matrix C from the earlier programs, and copies it into matrix A. Matrix A is sent to TRED2, while C is saved for a check of the transformation matrix that TRED2 returns in A. The program prints the diagonal and off-diagonal elements of the reduced matrix. It then forms the matrix F defined by $F = A^T C A$ to prove that F is tridiagonal and that the listed diagonal and off-diagonal elements are correct.

Here is the recipe TRED2:

```
SUB TRED2 (A(), N, NP, D(), E())
        Householder reduction of a real, symmetric, N by N matrix A, stored in an NP by NP physical
        array. On output, A is replaced by the orthogonal matrix Q effecting the transformation. D
        returns the diagonal elements of the tridiagonal matrix, and E the off-diagonal elements, with
        E(1)=0. Several statements, as noted in comments, can be omitted if only eigenvalues are to
        be found, in which case A contains no useful information on output. Otherwise they are to be
        included.
FOR I = N TO 2 STEP -1
  L = I - 1
  H = 0!
  SCALE = 0!
  IF L > 1 THEN
    FOR K = 1 TO L
      SCALE = SCALE + ABS(A(I, K))
    NEXT K
    IF SCALE = 0! THEN              Skip transformation.
      E(I) = A(I, L)
    ELSE
      FOR K = 1 TO L
        A(I, K) = A(I, K) / SCALE      Use scaled a's for transformation.
        H = H + A(I, K) ^ 2            Form σ in H.
```

```
         NEXT K
         F = A(I, L)
         G = -ABS(SQR(H)) * SGN(F)
         E(I) = SCALE * G
         H = H - F * G                        Now H is equation (11.2.4).
         A(I, L) = F - G                      Store u in the Ith row of A.
         F = 0!
         FOR J = 1 TO L
' Omit following line if finding only eigenvalues
           A(J, I) = A(I, J) / H              Store u/H in Ith column of A.
           G = 0!                             Form an element of A · u in G.
           FOR K = 1 TO J
             G = G + A(J, K) * A(I, K)
           NEXT K
           FOR K = J + 1 TO L
             G = G + A(K, J) * A(I, K)
           NEXT K
           E(J) = G / H                       Form element of p in temporarily unused element of E.
           F = F + E(J) * A(I, J)
         NEXT J
         HH = F / (H + H)                     Form K, equation (11.2.11).
         FOR J = 1 TO L                       Form q and store in E overwriting p.
           F = A(I, J)                        Note that E(L)=E(I-1) survives.
           G = E(J) - HH * F
           E(J) = G
           FOR K = 1 TO J                     Reduce A, equation (11.2.13).
             A(J, K) = A(J, K) - F * E(K) - G * A(I, K)
           NEXT K
         NEXT J
       END IF
     ELSE
       E(I) = A(I, L)
     END IF
     D(I) = H
NEXT I
' Omit following line if finding only eigenvalues
D(1) = 0!
E(1) = 0!
FOR I = 1 TO N                               Begin accumulation of transformation matrices.
' Delete lines from here ...
   L = I - 1
   IF D(I) <> 0! THEN                        This block skipped when I=1.
     FOR J = 1 TO L
       G = 0!
       FOR K = 1 TO L                        Use u and u/H stored in A to form P · Q.
         G = G + A(I, K) * A(K, J)
       NEXT K
       FOR K = 1 TO L
         A(K, J) = A(K, J) - G * A(K, I)
       NEXT K
     NEXT J
   END IF
' ... to here when finding only eigenvalues.
   D(I) = A(I, I)                            This statement remains.
' Also delete lines from here ...
```

```
    A(I, I) = 1!                    Reset row and column of A to identity matrix for next itera-
    FOR J = 1 TO L                  tion.
      A(I, J) = 0!
      A(J, I) = 0!
    NEXT J
'  ... to here when finding only eigenvalues.
  NEXT I
END SUB
```

A sample program using TRED2 is the following:

```
DECLARE SUB TRED2 (A! (), N!, NP!, D! (), E! ())

'PROGRAM D11R3
'Driver for routine TRED2
CLS
NP = 10
DIM A(NP, NP), C(NP, NP), D(NP), E(NP), F(NP, NP)
FOR J = 1 TO NP
  FOR I = 1 TO NP
    READ C(I, J)
  NEXT I
NEXT J
DATA 5.0,4.0,3.0,2.0,1.0,0.0,-1.0,-2.0,-3.0,-4.0,4.0,5.0,4.0,3.0,2.0,1.0,0.0
DATA -1.0,-2.0,-3.0,3.0,4.0,5.0,4.0,3.0,2.0,1.0,0.0,-1.0,-2.0,2.0,3.0,4.0,5.0
DATA 4.0,3.0,2.0,1.0,0.0,-1.0,1.0,2.0,3.0,4.0,5.0,4.0,3.0,2.0,1.0,0.0,0.0,1.0
DATA 2.0,3.0,4.0,5.0,4.0,3.0,2.0,1.0,-1.0,0.0,1.0,2.0,3.0,4.0,5.0,4.0,3.0,2.0
DATA -2.0,-1.0,0.0,1.0,2.0,3.0,4.0,5.0,4.0,3.0,-3.0,-2.0,-1.0,0.0,1.0,2.0,3.0
DATA 4.0,5.0,4.0,-4.0,-3.0,-2.0,-1.0,0.0,1.0,2.0,3.0,4.0,5.0
FOR I = 1 TO NP
  FOR J = 1 TO NP
    A(I, J) = C(I, J)
  NEXT J
NEXT I
PRINT
CALL TRED2(A(), NP, NP, D(), E())
PRINT "Diagonal elements"
FOR I = 1 TO NP
  PRINT USING "#####.######"; D(I);
  IF (I MOD 5) = 0 OR I = NP THEN PRINT
NEXT I
PRINT
PRINT "Off-diagonal elements"
FOR I = 2 TO NP
  PRINT USING "#####.######"; E(I);
  IF (I MOD 5) = 1 OR I = NP THEN PRINT
NEXT I
'Check transformation matrix
FOR J = 1 TO NP
  FOR K = 1 TO NP
    F(J, K) = 0!
    FOR L = 1 TO NP
      FOR M = 1 TO NP
        F(J, K) = F(J, K) + A(L, J) * C(L, M) * A(M, K)
      NEXT M
```

```
    NEXT L
   NEXT K
 NEXT J
 PRINT
 'How does it look?
 PRINT "Tridiagonal matrix"
 FOR I = 1 TO NP
   FOR J = 1 TO NP
     PRINT USING "####.##"; F(I, J);
   NEXT J
   PRINT
 NEXT I
 END
```

TQLI finds the eigenvectors and eigenvalues for a real, symmetric, tridiagonal matrix. Sample program D11R4 operates with matrix C again, and uses TRED2 to reduce it to tridiagonal form as before. More specifically, C is copied into matrix A, which is sent to TRED2. From TRED2 come two vectors D,E which are the diagonal and subdiagonal elements of the tridiagonal matrix. D and E are made arguments of TQLI, as is A, the returned transformation matrix from TRED2. On output from TQLI, D is replaced with eigenvalues, and A with corresponding eigenvectors. These are checked as in the program for JACOBI. That is, the original matrix C is applied to each eigenvector, and the result is divided (element by element) by the eigenvector. Look for a result equal to the eigenvalue. (Note: in some cases, the vector element is zero or nearly so. These cases are flagged with the words "div. by zero".)

Here is the recipe TQLI:

```
SUB TQLI (D(), E(), N, NP, Z())
```
> QL algorithm with implicit shifts, to determine the eigenvalues and eigenvectors of a real, symmetric, tridiagonal matrix, or of a real, symmetric matrix previously reduced by TRED2 §11.2. D is a vector of length NP. On input, its first N elements are the diagonal elements of the tridiagonal matrix. On output, it returns the eigenvalues. The vector E inputs the subdiagonal elements of the tridiagonal matrix, with E(1) arbitrary. On output E is destroyed. When finding only the eigenvalues, several lines may be omitted, as noted in the comments. If the eigenvectors of a tridiagonal matrix are desired, the matrix Z (N by N matrix stored in NP by NP array) is input as the identity matrix. If the eigenvectors of a matrix that has been reduced by TRED2 are required, then Z is input as the matrix output by TRED2. In either case, the Kth column of Z returns the normalized eigenvector corresponding to D(K).

```
FOR I = 2 TO N                              Convenient to renumber the elements of E.
  E(I - 1) = E(I)
NEXT I
E(N) = 0!
FOR L = 1 TO N
  ITER = 0
  DO
    DONE% = -1
    FOR M = L TO N - 1                      Look for a single small subdiagonal element
      DD = ABS(D(M)) + ABS(D(M + 1))        to split the matrix.
      IF ABS(E(M)) + DD = DD THEN EXIT FOR
    NEXT M
    IF ABS(E(M)) + DD <> DD THEN M = N
    IF M <> L THEN
      IF ITER = 30 THEN PRINT "too many iterations": EXIT SUB
```

```
      ITER = ITER + 1
      G = (D(L + 1) - D(L)) / (2! * E(L))        Form shift.
      R = SQR(G ^ 2 + 1!)
      G = D(M) - D(L) + E(L) / (G + ABS(R) * SGN(G))        This is dₘ - kₛ.
      S = 1!
      C = 1!
      P = 0!
      FOR I = M - 1 TO L STEP -1          A plane rotation as in the original QL, followed by Givens
        F = S * E(I)                      rotations to restore tridiagonal form.
        B = C * E(I)
        IF ABS(F) >= ABS(G) THEN
          C = G / F
          R = SQR(C ^ 2 + 1!)
          E(I + 1) = F * R
          S = 1! / R
          C = C * S
        ELSE
          S = F / G
          R = SQR(S ^ 2 + 1!)
          E(I + 1) = G * R
          C = 1! / R
          S = S * C
        END IF
        G = D(I + 1) - P
        R = (D(I) - G) * S + 2! * C * B
        P = S * R
        D(I + 1) = G + P
        G = C * R - B
' Omit lines from here ...
        FOR K = 1 TO N                    Form eigenvectors.
          F = Z(K, I + 1)
          Z(K, I + 1) = S * Z(K, I) + C * F
          Z(K, I) = C * Z(K, I) - S * F
        NEXT K
' ... to here when finding only eigenvalues.
      NEXT I
      D(L) = D(L) - P
      E(L) = G
      E(M) = 0!
      DONE% = 0
    END IF
  LOOP WHILE NOT DONE%
NEXT L
END SUB
```

A sample program using TQLI is the following:

```
DECLARE SUB TQLI (D!(), E!(), N!, NP!, Z!())
DECLARE SUB TRED2 (A!(), N!, NP!, D!(), E!())

'PROGRAM D11R4
'Driver for routine TQLI
CLS
NP = 10
TINY = .000001
```

```
DIM A(NP, NP), C(NP, NP), D(NP), E(NP), F(NP)
FOR J = 1 TO NP
  FOR I = 1 TO NP
    READ C(I, J)
  NEXT I
NEXT J
DATA 5.0,4.0,3.0,2.0,1.0,0.0,-1.0,-2.0,-3.0,-4.0,4.0,5.0,4.0,3.0,2.0,1.0,0.0
DATA -1.0,-2.0,-3.0,3.0,4.0,5.0,4.0,3.0,2.0,1.0,0.0,-1.0,-2.0,2.0,3.0,4.0,5.0
DATA 4.0,3.0,2.0,1.0,0.0,-1.0,1.0,2.0,3.0,4.0,5.0,4.0,3.0,2.0,1.0,0.0,0.0,1.0
DATA 2.0,3.0,4.0,5.0,4.0,3.0,2.0,1.0,-1.0,0.0,1.0,2.0,3.0,4.0,5.0,4.0,3.0,2.0
DATA -2.0,-1.0,0.0,1.0,2.0,3.0,4.0,5.0,4.0,3.0,-3.0,-2.0,-1.0,0.0,1.0,2.0,3.0
DATA 4.0,5.0,4.0,-4.0,-3.0,-2.0,-1.0,0.0,1.0,2.0,3.0,4.0,5.0
FOR I = 1 TO NP
  FOR J = 1 TO NP
    A(I, J) = C(I, J)
  NEXT J
NEXT I
CALL TRED2(A(), NP, NP, D(), E())
CALL TQLI(D(), E(), NP, NP, A())
PRINT "Eigenvectors for a real symmetric matrix"
FOR I = 1 TO NP
  FOR J = 1 TO NP
    F(J) = 0!
    FOR K = 1 TO NP
      F(J) = F(J) + C(J, K) * A(K, I)
    NEXT K
  NEXT J
  PRINT
  PRINT "Eigenvalue"; I; " =";
  PRINT USING "#####.######"; D(I)
  PRINT
  PRINT "    Vector    Mtrx*Vect.    Ratio"
  FOR J = 1 TO NP
    IF ABS(A(J, I)) < TINY THEN
      PRINT USING "#####.######"; A(J, I); F(J);
      PRINT "      div. by 0"
    ELSE
      PRINT USING "#####.######"; A(J, I); F(J);
      PRINT "    ";
      PRINT USING ".######^^^^"; F(J) / A(J, I)
    END IF
  NEXT J
  PRINT
  PRINT "press ENTER to continue..."
  LINE INPUT DUM$
NEXT I
END
```

BALANC reduces error in eigenvalue problems involving nonsymmetric matrices. It does this by adjusting corresponding rows and columns to have comparable norms, without changing eigenvalues. Sample program **D11R5** prepares the following array **A** for **BALANC**

$$
\begin{pmatrix}
1 & 100 & 1 & 100 & 1 \\
1 & 1 & 1 & 1 & 1 \\
1 & 100 & 1 & 100 & 1 \\
1 & 1 & 1 & 1 & 1 \\
1 & 100 & 1 & 100 & 1
\end{pmatrix}
$$

The norms of the five rows and five columns are printed out. It is clear from the array that three of the rows and two of the columns have much larger norms than the others. After balancing with BALANC, the norms are recalculated, and this time the row, column pairs should be much more nearly equal.

Here is the recipe BALANC:

```
SUB BALANC (A(), N, NP)
```
Given an N by N matrix A stored in an array of physical dimensions NP by NP, this routine replaces it by a balanced matrix with identical eigenvalues. A symmetric matrix is already balanced and is unaffected by this procedure. The parameter RADIX should be the machine's floating point radix.

```
RADIX = 2!
SQRDX = 4!
DO
  LAST = 1
  FOR I = 1 TO N
    C = 0!
    R = 0!
    FOR J = 1 TO N
      IF J <> I THEN
        C = C + ABS(A(J, I))
        R = R + ABS(A(I, J))
      END IF
    NEXT J
    IF C <> 0! AND R <> 0! THEN
      G = R / RADIX
      F = 1!
      S = C + R
      WHILE C < G
        F = F * RADIX
        C = C * SQRDX
      WEND
      G = R * RADIX
      WHILE C > G
        F = F / RADIX
        C = C / SQRDX
      WEND
      IF (C + R) / F < .95 * S THEN
        LAST = 0
        G = 1! / F
        FOR J = 1 TO N
          A(I, J) = A(I, J) * G
        NEXT J
        FOR J = 1 TO N
          A(J, I) = A(J, I) * F
        NEXT J
```

```
      END IF
    END IF
  NEXT I
LOOP WHILE LAST = 0
END SUB
```

A sample program using **BALANC** is the following:

```
DECLARE SUB BALANC (A!(), N!, NP!)

'PROGRAM D11R5
'Driver for routine BALANC
CLS
NP = 5
DIM A(NP, NP), R(NP), C(NP)
FOR J = 1 TO NP
  FOR I = 1 TO NP
    READ A(I, J)
  NEXT I
NEXT J
DATA 1.0,1.0,1.0,1.0,1.0,100.0,1.0,100.0,1.0,100.0,1.0,1.0,1.0,1.0,1.0,100.0
DATA 1.0,100.0,1.0,100.0,1.0,1.0,1.0,1.0,1.0
'Print norms
FOR I = 1 TO NP
  R(I) = 0!
  C(I) = 0!
  FOR J = 1 TO NP
    R(I) = R(I) + ABS(A(I, J))
    C(I) = C(I) + ABS(A(J, I))
  NEXT J
NEXT I
PRINT "Rows:"
FOR I = 1 TO NP
  PRINT USING "########.######"; R(I);
NEXT I
PRINT
PRINT "Columns:"
FOR I = 1 TO NP
  PRINT USING "########.######"; C(I);
NEXT I
PRINT
PRINT
PRINT "***** Balancing Matrix *****"
PRINT
CALL BALANC(A(), NP, NP)
'Print norms
FOR I = 1 TO NP
  R(I) = 0!
  C(I) = 0!
  FOR J = 1 TO NP
    R(I) = R(I) + ABS(A(I, J))
    C(I) = C(I) + ABS(A(J, I))
  NEXT J
NEXT I
PRINT "Rows:"
```

```
FOR I = 1 TO NP
  PRINT USING "########.######"; R(I);
NEXT I
PRINT
PRINT "Columns:"
FOR I = 1 TO NP
  PRINT USING "########.######"; C(I);
NEXT I
PRINT
END
```

ELMHES reduces a general matrix to Hessenberg form using Gaussian elimination. It is particularly valuable for real, nonsymmetric matrices. Sample program D11R6 employs BALANC and ELMHES to get a nonsymmetric and grossly unbalanced matrix into Hessenberg form. The matrix A is

$$\begin{pmatrix} 1 & 2 & 300 & 4 & 5 \\ 2 & 3 & 400 & 5 & 6 \\ 3 & 4 & 5 & 6 & 7 \\ 4 & 5 & 600 & 7 & 8 \\ 5 & 6 & 700 & 8 & 9 \end{pmatrix}$$

After printing the original matrix, the program feeds it to BALANC and prints the balanced version. This is submitted to ELMHES and the result is printed. Notice that the elements of A with I>J+1 are all set to zero by the program, because ELMHES returns random values in this part of the matrix. Therefore, you should not attach any importance to the fact that the printed output of the program has Hessenberg form. More important are the contents of the nonzero entries. We include here the expected results for comparison.

Balanced Matrix:

1.00	2.00	37.50	4.00	5.00
2.00	3.00	50.00	5.00	6.00
24.00	32.00	5.00	48.00	56.00
4.00	5.00	75.00	7.00	8.00
5.00	6.00	87.50	8.00	9.00

Reduced to Hessenberg Form:

.1000E+01	.3938E+02	.9618E+01	.3333E+01	.4000E+01
.2400E+02	.2733E+02	.1161E+03	.4800E+02	.4800E+02
.0000E+00	.8551E+02	-.4780E+01	-.1333E+01	-.2000E+01
.0000E+00	.0000E+00	.5188E+01	.1447E+01	.2171E+01
.0000E+00	.0000E+00	.0000E+00	-.9155E-07	.7874E-07

Here is the recipe ELMHES:

```
SUB ELMHES (A(), N, NP)
```
   Reduction to Hessenberg form by the elimination method. The real, nonsymmetric, N by N matrix A, stored in an array of physical dimensions NP by NP, is replaced by an upper Hessenberg matrix with identical eigenvalues. Recommended, but not required, is that this routine be preceded by BALANC. On output, the Hessenberg matrix is in elements A(I,J) with I ≤ J+1. Elements with I > J+1 are to be thought of as zero, but are returned with random values.
```
FOR M = 2 TO N - 1            M is called r + 1 in the text.
  X = 0!
  I = M
  FOR J = M TO N              Find the pivot.
    IF ABS(A(J, M - 1)) > ABS(X) THEN
```

```
      X = A(J, M - 1)
      I = J
   END IF
 NEXT J
 IF I <> M THEN                    Interchange rows and columns.
   FOR J = M - 1 TO N
     Y = A(I, J)
     A(I, J) = A(M, J)
     A(M, J) = Y
   NEXT J
   FOR J = 1 TO N
     Y = A(J, I)
     A(J, I) = A(J, M)
     A(J, M) = Y
   NEXT J
 END IF
 IF X <> 0! THEN                   Carry out the elimination.
   FOR I = M + 1 TO N
     Y = A(I, M - 1)
     IF Y <> 0! THEN
       Y = Y / X
       A(I, M - 1) = Y
       FOR J = M TO N
         A(I, J) = A(I, J) - Y * A(M, J)
       NEXT J
       FOR J = 1 TO N
         A(J, M) = A(J, M) + Y * A(J, I)
       NEXT J
     END IF
   NEXT I
 END IF
NEXT M
END SUB
```

A sample program using **ELMHES** is the following:

```
DECLARE SUB ELMHES (A!(), N!, NP!)
DECLARE SUB BALANC (A!(), N!, NP!)

'PROGRAM D11R6
'Driver for ELMHES
CLS
NP = 5
DIM A(NP, NP), R(NP), C(NP)
FOR J = 1 TO NP
  FOR I = 1 TO NP
    READ A(I, J)
  NEXT I
NEXT J
DATA 1.0,2.0,3.0,4.0,5.0,2.0,3.0,4.0,5.0,6.0,300.0,400.0,5.0,600.0,700.0,4.0
DATA 5.0,6.0,7.0,8.0,5.0,6.0,7.0,8.0,9.0
PRINT "***** Original Matrix *****"
PRINT
FOR I = 1 TO NP
  FOR J = 1 TO NP
```

```
      PRINT USING "#########.##"; A(I, J);
    NEXT J
    PRINT
  NEXT I
  PRINT
  PRINT "***** Balance Matrix *****"
  CALL BALANC(A(), NP, NP)
  FOR I = 1 TO NP
    FOR J = 1 TO NP
      PRINT USING "#########.##"; A(I, J);
    NEXT J
    PRINT
  NEXT I
  PRINT
  PRINT "***** Reduce to Hessenberg Form *****"
  PRINT
  CALL ELMHES(A(), NP, NP)
  FOR J = 1 TO NP - 2
    FOR I = J + 2 TO NP
      A(I, J) = 0!
    NEXT I
  NEXT J
  FOR I = 1 TO NP
    FOR J = 1 TO NP
      PRINT "   ";
      PRINT USING "#.####^^^^"; A(I, J);
    NEXT J
    PRINT
  NEXT I
  END
```

HQR, finally, is a routine for finding the eigenvalues of a Hessenberg matrix using the QR algorithm. The 5×5 matrix A specified in the DATA statement is treated just as you would expect to treat any general real nonsymmetric matrix. It is fed to BALANC for balancing, to ELMHES for reduction to Hessenberg form, and to HQR for eigenvalue determination. The eigenvalues may be complex-valued, and both real and imaginary parts are given. The original matrix has enough strategically placed zeros in it that you should have no trouble finding the eigenvalues by hand. Alternatively, you may check them against the list below:

Matrix:

1.00	2.00	.00	.00	.00
-2.00	3.00	.00	.00	.00
3.00	4.00	50.00	.00	.00
-4.00	5.00	-60.00	7.00	.00
-5.00	6.00	-70.00	8.00	-9.00

Eigenvalues:

#	Real	Imag.
1	.500000E+02	.000000E+00
2	.200000E+01	-.173205E+01
3	.200000E+01	.173205E+01
4	.700000E+01	.000000E+00
5	-.900000E+01	.000000E+00

Here is the recipe HQR:

```
SUB HQR (A(), N, NP, WR(), WI())
        Finds all eigenvalues of an N by N upper Hessenberg matrix A that is stored in an NP by NP
        array. On input A can be exactly as output from ELMHES §11.5; on output it is destroyed. The
        real and imaginary parts of the eigenvalues are returned in WR and WI respectively.
ANORM = ABS(A(1, 1))                      Compute matrix norm for possible use in locating single small
FOR I = 2 TO N                            subdiagonal element.
  FOR J = I - 1 TO N
    ANORM = ANORM + ABS(A(I, J))
  NEXT J
NEXT I
NN = N
T = 0!                                    ...gets changed only by an exceptional shift.
WHILE NN >= 1                             Begin search for next eigenvalue.
  ITS = 0
  DO
    DONE% = -1
    FOR L = NN TO 2 STEP -1                       Begin iteration: look for single small
      S = ABS(A(L - 1, L - 1)) + ABS(A(L, L))     subdiagonal element.
      IF S = 0! THEN S = ANORM
      IF ABS(A(L, L - 1)) + S = S THEN EXIT FOR
    NEXT L
    IF ABS(A(L, L - 1)) + S <> S THEN L = 1
    X = A(NN, NN)
    IF L = NN THEN                        One root found.
      WR(NN) = X + T
      WI(NN) = 0!
      NN = NN - 1
    ELSE
      Y = A(NN - 1, NN - 1)
      W = A(NN, NN - 1) * A(NN - 1, NN)
      IF L = NN - 1 THEN                  Two roots found...
        P = .5 * (Y - X)
        Q = P ^ 2 + W
        Z = SQR(ABS(Q))
        X = X + T
        IF Q >= 0! THEN                   ...a real pair.
          Z = P + ABS(Z) * SGN(P)
          WR(NN) = X + Z
          WR(NN - 1) = WR(NN)
          IF Z <> 0! THEN WR(NN) = X - W / Z
          WI(NN) = 0!
          WI(NN - 1) = 0!
        ELSE                              ...a complex pair.
          WR(NN) = X + P
          WR(NN - 1) = WR(NN)
          WI(NN) = Z
          WI(NN - 1) = -Z
        END IF
        NN = NN - 2
      ELSE                       No roots found. Continue iteration.
        IF ITS = 30 THEN PRINT "too many iterations": EXIT SUB
        IF ITS = 10 OR ITS = 20 THEN      Form exceptional shift.
          T = T + X
          FOR I = 1 TO NN
            A(I, I) = A(I, I) - X
```

```
      NEXT I
      S = ABS(A(NN, NN - 1)) + ABS(A(NN - 1, NN - 2))
      X = .75 * S
      Y = X
      W = -.4375 * S ^ 2
   END IF
   ITS = ITS + 1
   FOR M = NN - 2 TO L STEP -1          Form shift and then look for 2 consecutive small
      Z = A(M, M)                       subdiagonal elements.
      R = X - Z
      S = Y - Z
      P = (R * S - W) / A(M + 1, M) + A(M, M + 1)      Equation 11.6.23.
      Q = A(M + 1, M + 1) - Z - R - S
      R = A(M + 2, M + 1)
      S = ABS(P) + ABS(Q) + ABS(R)          Scale to prevent overflow or underflow.
      P = P / S
      Q = Q / S
      R = R / S
      IF M = L THEN EXIT FOR
      U = ABS(A(M, M - 1)) * (ABS(Q) + ABS(R))
      V = ABS(P) * (ABS(A(M - 1, M - 1)) + ABS(Z) + ABS(A(M + 1, M + 1)))
      IF U + V = V THEN EXIT FOR          Equation 11.6.26.
   NEXT M
   FOR I = M + 2 TO NN
      A(I, I - 2) = 0!
      IF I <> M + 2 THEN A(I, I - 3) = 0!
   NEXT I
   FOR K = M TO NN - 1          Double QR step on rows L to NN and columns M to NN.
      IF K <> M THEN
         P = A(K, K - 1)          Begin setup of Householder vector.
         Q = A(K + 1, K - 1)
         R = 0!
         IF K <> NN - 1 THEN R = A(K + 2, K - 1)
         X = ABS(P) + ABS(Q) + ABS(R)
         IF X <> 0! THEN
            P = P / X               Scale to prevent overflow or underflow.
            Q = Q / X
            R = R / X
         END IF
      END IF
      S = ABS(SQR(P ^ 2 + Q ^ 2 + R ^ 2)) * SGN(P)
      IF S <> 0! THEN
         IF K = M THEN
            IF L <> M THEN A(K, K - 1) = -A(K, K - 1)
         ELSE
            A(K, K - 1) = -S * X
         END IF
         P = P + S               Equations 11.6.24.
         X = P / S
         Y = Q / S
         Z = R / S
         Q = Q / P
         R = R / P
         FOR J = K TO NN          Row modification.
            P = A(K, J) + Q * A(K + 1, J)
```

```
             IF K <> NN - 1 THEN
                P = P + R * A(K + 2, J)
                A(K + 2, J) = A(K + 2, J) - P * Z
             END IF
             A(K + 1, J) = A(K + 1, J) - P * Y
             A(K, J) = A(K, J) - P * X
           NEXT J
           NTMP = NN
           IF K + 3 < NN THEN NTMP = K + 3
           FOR I = L TO NTMP        Column modification.
             P = X * A(I, K) + Y * A(I, K + 1)
             IF K <> NN - 1 THEN
                P = P + Z * A(I, K + 2)
                A(I, K + 2) = A(I, K + 2) - P * R
             END IF
             A(I, K + 1) = A(I, K + 1) - P * Q
             A(I, K) = A(I, K) - P
           NEXT I
         END IF
       NEXT K                       ...for next iteration on current eigenvalue.
       DONE% = 0
     END IF
   END IF
 LOOP WHILE NOT DONE%               ...for next eigenvalue.
WEND
END SUB
```

A sample program using HQR is the following:

```
DECLARE SUB HQR (A!(), N!, NP!, WR!(), WI!())
DECLARE SUB BALANC (A!(), N!, NP!)
DECLARE SUB ELMHES (A!(), N!, NP!)

'PROGRAM D11R7
'Driver for routine HQR
CLS
NP = 5
DIM A(NP, NP), WR(NP), WI(NP)
FOR J = 1 TO NP
  FOR I = 1 TO NP
    READ A(I, J)
  NEXT I
NEXT J
DATA 1.0,-2.0,3.0,-4.0,-5.0,2.0,3.0,4.0,5.0,6.0,0.0,0.0,50.0,-60.0,-70.0,0.0
DATA 0.0,0.0,7.0,8.0,0.0,0.0,0.0,0.0,-9.0
PRINT "Matrix:"
FOR I = 1 TO NP
  FOR J = 1 TO NP
    PRINT USING "#########.##"; A(I, J);
  NEXT J
  PRINT
NEXT I
PRINT
CALL BALANC(A(), NP, NP)
CALL ELMHES(A(), NP, NP)
```

```
CALL HQR(A(), NP, NP, WR(), WI())
PRINT "Eigenvalues:"
PRINT
PRINT "        Real            Imag."
PRINT
FOR I = 1 TO NP
  PRINT "    ";
  PRINT USING "#.######^^^^"; WR(I);
  PRINT "   ";
  PRINT USING "#.######^^^^"; WI(I)
NEXT I
END
```

Chapter 12: Fourier Methods

Chapter 12 of *Numerical Recipes* covers Fourier transform spectral methods, particularly the transform of discretely sampled data. Central to the chapter is the fast Fourier transform (FFT). Routine FOUR1 performs the FFT on a complex data array. TWOFFT does the same transform on two real-valued data arrays (at the same time) and returns two complex-valued transforms. Finally, REALFT finds the Fourier transform of a single real-valued array. Two related transforms are the sine transform and the cosine transform, given by SINFT and COSFT.

Two common uses of the Fourier transform are the convolution of data with a response function, and the computation of the correlation of two data sets. These operations are carried out by CONVLV and CORREL respectively. Other applications of Fourier methods include data filtering, power spectrum estimation (SPCTRM, or EVLMEM with MEMCOF), and linear prediction (PREDIC with FIXRTS). All of these applications assume data in one dimension. For FFTs in two or more dimensions the routine FOURN is supplied.

$$\star \quad \star \quad \star \quad \star$$

Routine FOUR1 performs the fast Fourier transform on a complex-valued array of data points. Example program D12R1 has five tests for this transform. First, it checks the following four symmetries (where $h(t)$ is the data and $H(n)$ is the transform):
1. If $h(t)$ is real-valued and even, then $H(n) = H(N - n)$ and H is real.
2. If $h(t)$ is imaginary-valued and even, then $H(n) = H(N - n)$ and H is imaginary.
3. If $h(t)$ is real-valued and odd, then $H(n) = -H(N - n)$ and H is imaginary.
4. If $h(t)$ is imaginary-valued and odd, then $H(n) = -H(N - n)$ and H is real.

The fifth test is that if a data array is Fourier transformed twice in succession, the resulting array should be identical to the original.

Here is the recipe FOUR1:

```
SUB FOUR1 (DATQ(), NN, ISIGN)
        Replaces DATQ by its discrete Fourier transform, if ISIGN is input as 1; or replaces DATQ by
        NN times its inverse discrete Fourier transform, if ISIGN is input as −1. DATQ is a real array
        of length 2*NN. NN MUST be an integer power of 2 (this is not checked for!). Use double
        precision for the trigonometric recurrences.
N = 2 * NN
J = 1
FOR I = 1 TO N STEP 2                    This is the bit-reversal section of the routine.
  IF J > I THEN
     TEMPR = DATQ(J)                      Exchange the two complex numbers.
     TEMPI = DATQ(J + 1)
     DATQ(J) = DATQ(I)
     DATQ(J + 1) = DATQ(I + 1)
```

```
      DATQ(I) = TEMPR
      DATQ(I + 1) = TEMPI
    END IF
    M = INT(N / 2)
    WHILE M >= 2 AND J > M
      J = J - M
      M = INT(M / 2)
    WEND
    J = J + M
  NEXT I
  MMAX = 2                         Here begins the Danielson-Lanczos section of the routine.
  WHILE N > MMAX                   Outer loop executed log₂ NN times.
    ISTEP = 2 * MMAX
    THETA# = 6.28318530717959# / (ISIGN * MMAX)    Initialize for the trigonometric recurrence.
    WPR# = -2# * SIN(.5# * THETA#) ^ 2
    WPI# = SIN(THETA#)
    WR# = 1#
    WI# = 0#
    FOR M = 1 TO MMAX STEP 2       Here are the two nested inner loops.
      FOR I = M TO N STEP ISTEP
        J = I + MMAX               This is the Danielson-Lanczos formula:
        TEMPR = CSNG(WR#) * DATQ(J) - CSNG(WI#) * DATQ(J + 1)
        TEMPI = CSNG(WR#) * DATQ(J + 1) + CSNG(WI#) * DATQ(J)
        DATQ(J) = DATQ(I) - TEMPR
        DATQ(J + 1) = DATQ(I + 1) - TEMPI
        DATQ(I) = DATQ(I) + TEMPR
        DATQ(I + 1) = DATQ(I + 1) + TEMPI
      NEXT I
      WTEMP# = WR#                 Trigonometric recurrence.
      WR# = WR# * WPR# - WI# * WPI# + WR#
      WI# = WI# * WPR# + WTEMP# * WPI# + WI#
    NEXT M
    MMAX = ISTEP
  WEND
END SUB
```

A sample program using `FOUR1` is the following:

```
DECLARE SUB PRNTFT (DATQ!(), NN2!)
DECLARE SUB FOUR1 (DATQ!(), NN!, ISIGN!)

'PROGRAM D12R1
'Driver for routine FOUR1
CLS
NN = 32
NN2 = 2 * NN
DIM DATQ(NN2), DCMP(NN2)
PRINT "h(t)=real-valued even-function"
PRINT "H(n)=H(N-n) and real?"
PRINT
FOR I = 1 TO 2 * NN - 1 STEP 2
  DATQ(I) = 1! / (((I - NN - 1!) / NN) ^ 2 + 1!)
  DATQ(I + 1) = 0!
NEXT I
ISIGN = 1
```

```
CALL FOUR1 (DATQ (), NN, ISIGN)
CALL PRNTFT (DATQ (), NN2)
PRINT "h (t) =imaginary-valued even-function"
PRINT "H (n) =H (N-n) and imaginary?"
PRINT
FOR I = 1 TO 2 * NN - 1 STEP 2
  DATQ (I + 1) = 1! / (((I - NN - 1!) / NN) ^ 2 + 1!)
  DATQ (I) = 0!
NEXT I
ISIGN = 1
CALL FOUR1 (DATQ (), NN, ISIGN)
CALL PRNTFT (DATQ (), NN2)
PRINT "h (t) =real-valued odd-function"
PRINT "H (n) =-H (N-n) and imaginary?"
PRINT
FOR I = 1 TO 2 * NN - 1 STEP 2
  DATQ (I) = (I - NN - 1!) / NN / (((I - NN - 1!) / NN) ^ 2 + 1!)
  DATQ (I + 1) = 0!
NEXT I
DATQ (1) = 0!
ISIGN = 1
CALL FOUR1 (DATQ (), NN, ISIGN)
CALL PRNTFT (DATQ (), NN2)
PRINT "h (t) =imaginary-valued odd-function"
PRINT "H (n) =-H (N-n) and real?"
PRINT
FOR I = 1 TO 2 * NN - 1 STEP 2
  DATQ (I + 1) = (I - NN - 1!) / NN / (((I - NN - 1!) / NN) ^ 2 + 1!)
  DATQ (I) = 0!
NEXT I
DATQ (2) = 0!
ISIGN = 1
CALL FOUR1 (DATQ (), NN, ISIGN)
CALL PRNTFT (DATQ (), NN2)
'Transform, inverse-transform test
FOR I = 1 TO 2 * NN - 1 STEP 2
  DATQ (I) = 1! / ((.5 * (I - NN - 1) / NN) ^ 2 + 1!)
  DCMP (I) = DATQ (I)
  DATQ (I + 1) = (.25 * (I - NN - 1) / NN) * EXP (- (.5 * (I - NN - 1!) / NN) ^ 2)
  DCMP (I + 1) = DATQ (I + 1)
NEXT I
ISIGN = 1
CALL FOUR1 (DATQ (), NN, ISIGN)
ISIGN = -1
CALL FOUR1 (DATQ (), NN, ISIGN)
PRINT "        Original Data:              Double Fourier Transform:"
PRINT
PRINT "   k     Real h (k)      Imag h (k)      Real h (k)      Imag h (k)"
PRINT
FOR I = 1 TO NN STEP 2
  J = (I + 1) / 2
  PRINT USING "####"; J;
  PRINT USING "#######.######"; DCMP (I); DCMP (I + 1); DATQ (I) / NN; DATQ (I + 1)
/ NN
NEXT I
```

```
END

SUB PRNTFT (DATQ(), NN2)
PRINT "   n        Real H(n)      Imag H(n)      Real H(N-n)      Imag H(N-n)"
PRINT USING "####"; 0;
PRINT USING "#######.######"; DATQ(1); DATQ(2); DATQ(1); DATQ(2)
FOR N = 3 TO NN2 / 2 + 1 STEP 2
  M = (N - 1) / 2
  MM = NN2 + 2 - N
  PRINT USING "####"; M;
  PRINT USING "#######.######"; DATQ(N); DATQ(N + 1); DATQ(MM); DATQ(MM + 1)
NEXT N
PRINT
PRINT " press RETURN to continue ..."
LINE INPUT DUM$
END SUB
```

TWOFFT is a routine that performs an efficient FFT of two real arrays at once by packing them into a complex array and transforming with **FOUR1**. Sample program **D12R2** generates two periodic data sets, out of phase with one another, and performs a transform and an inverse transform on each. It will be difficult to judge whether the transform itself gives the right answer, but if the inverse transform gets you back to the easily recognized original, you may be fairly confident that the routine works.

Here is the recipe **TWOFFT**:

```
DECLARE SUB FOUR1 (DATQ!(), NN!, ISIGN!)

SUB TWOFFT (DATA1(), DATA2(), FFT1(), FFT2(), N)
        Given two real input arrays DATA1 and DATA2, each of length N, this routine calls FOUR1 and
        returns two complex output arrays, FFT1 and FFT2, each of complex length N (i.e. real length
        2*N), that contain the discrete Fourier transforms of the respective DATAs. N MUST be an
        integer power of 2.
C1R = .5
C1I = 0!
C2R = 0!
C2I = -.5
FOR J = 1 TO N
  FFT1(2 * J - 1) = DATA1(J)            Pack the two real arrays into one complex array.
  FFT1(2 * J) = DATA2(J)
NEXT J
CALL FOUR1(FFT1(), N, 1)               Transform the complex array.
FFT2(1) = FFT1(2)
FFT2(2) = 0!
FFT1(2) = 0!
N2 = 2 * (N + 2)
FOR J = 2 TO N / 2 + 1
  J2 = 2 * J
  CONJR = FFT1(N2 - J2 - 1)
  CONJI = -FFT1(N2 - J2)               Use symmetries to separate the two transforms.
  H1R = C1R * (FFT1(J2 - 1) + CONJR) - C1I * (FFT1(J2) + CONJI)
  H1I = C1I * (FFT1(J2 - 1) + CONJR) + C1R * (FFT1(J2) + CONJI)
  H2R = C2R * (FFT1(J2 - 1) - CONJR) - C2I * (FFT1(J2) - CONJI)
  H2I = C2I * (FFT1(J2 - 1) - CONJR) + C2R * (FFT1(J2) - CONJI)
```

```
   FFT1(J2 - 1) = H1R                    Ship them out in two complex arrays.
   FFT1(J2) = H1I
   FFT1(N2 - J2 - 1) = H1R
   FFT1(N2 - J2) = -H1I
   FFT2(J2 - 1) = H2R
   FFT2(J2) = H2I
   FFT2(N2 - J2 - 1) = H2R
   FFT2(N2 - J2) = -H2I
 NEXT J
 END SUB
```

A sample program using **TWOFFT** is the following:

```
DECLARE SUB PRNTFT (DATQ! (), N2!)
DECLARE SUB TWOFFT (DATA1! (), DATA2! (), FFT1! (), FFT2! (), N!)
DECLARE SUB FOUR1 (DATQ! (), NN!, ISIGN!)

'PROGRAM D12R2
'Driver for routine TWOFFT
CLS
N = 32
N2 = 2 * N
PER = 8!
PI = 3.14159
DIM DATA1(N), DATA2(N), FFT1(N2), FFT2(N2)
FOR I = 1 TO N
  X = 2! * PI * I / PER
  DATA1(I) = INT(COS(X) + .5)
  DATA2(I) = INT(SIN(X) + .5)
NEXT I
CALL TWOFFT(DATA1(), DATA2(), FFT1(), FFT2(), N)
PRINT "Fourier transform of first function:"
CALL PRNTFT(FFT1(), N2)
PRINT "Fourier transform of second function:"
CALL PRNTFT(FFT2(), N2)
'Invert transform
ISIGN = -1
CALL FOUR1(FFT1(), N, ISIGN)
PRINT "Inverted transform = first function:"
CALL PRNTFT(FFT1(), N2)
CALL FOUR1(FFT2(), N, ISIGN)
PRINT "Inverted transform = second function:"
CALL PRNTFT(FFT2(), N2)
END

SUB PRNTFT (DATQ(), N2)
PRINT "     n      Real(n)    Imag.(n)    Real(N-n)    Imag.(N-n)"
PRINT USING "######"; 0;
PRINT USING "#####.######"; DATQ(1); DATQ(2); DATQ(1); DATQ(2)
FOR I = 3 TO N2 / 2 + 1 STEP 2
  M = (I - 1) / 2
  NN2 = N2 + 2 - I
  PRINT USING "######"; M;
  PRINT USING "#####.######"; DATQ(I); DATQ(I + 1); DATQ(NN2); DATQ(NN2 + 1)
NEXT I
```

```
PRINT
PRINT " press RETURN to continue ... "
LINE INPUT DUM$
END SUB
```

REALFT performs the Fourier transform of a single real-valued data array. Sample routine D12R3 takes this function to be sinusoidal, and allows you to choose the period. After transforming, it simply plots the magnitude of each element of the transform. If the period you choose is a power of two, the transform will be nonzero in a single bin; otherwise there will be leakage to adjacent channels. D12R3 follows every transform by an inverse transform to make sure the original function is recovered.

Here is the recipe REALFT:

```
DECLARE SUB FOUR1 (DATQ!(), NN!, ISIGN!)

SUB REALFT (DATQ(), N, ISIGN)
        Calculates the Fourier transform of a set of 2N real-valued data points. Replaces this data
        (which is stored in array DATQ) by the positive frequency half of its complex Fourier transform.
        The real-valued first and last components of the complex transform are returned as elements
        DATQ(1) and DATQ(2) respectively. N must be a power of 2. This routine also calculates the
        inverse transform of a complex data array if it is the transform of real data. (Result in this
        case must be multiplied by 1/N.) Use double precision for the trigonometric recurrences.
THETA# = 3.141592653589793# / CDBL(N)        Initialize the recurrence.
C1 = .5
IF ISIGN = 1 THEN
   C2 = -.5
   CALL FOUR1(DATQ(), N, 1)                  The forward transform is here.
ELSE
   C2 = .5                                   Otherwise set up for an inverse transform.
   THETA# = -THETA#
END IF
WPR# = -2# * SIN(.5# * THETA#) ^ 2
WPI# = SIN(THETA#)
WR# = 1# + WPR#
WI# = WPI#
N2P3 = 2 * N + 3
FOR I = 2 TO INT(N / 2)                      Case I=1 done separately below.
   I1 = 2 * I - 1
   I2 = I1 + 1
   I3 = N2P3 - I2
   I4 = I3 + 1
   WRS# = CSNG(WR#)
   WIS# = CSNG(WI#)
   H1R = C1 * (DATQ(I1) + DATQ(I3))          The two separate transforms are separated
   H1I = C1 * (DATQ(I2) - DATQ(I4))          out of DATQ.
   H2R = -C2 * (DATQ(I2) + DATQ(I4))
   H2I = C2 * (DATQ(I1) - DATQ(I3))
   DATQ(I1) = H1R + WRS# * H2R - WIS# * H2I  Here they are recombined to form the true
   DATQ(I2) = H1I + WRS# * H2I + WIS# * H2R  transform of the original real data.
   DATQ(I3) = H1R - WRS# * H2R + WIS# * H2I
   DATQ(I4) = -H1I + WRS# * H2I + WIS# * H2R
   WTEMP# = WR#                    The recurrence.
   WR# = WR# * WPR# - WI# * WPI# + WR#
```

```
  WI# = WI# * WPR# + WTEMP# * WPI# + WI#
NEXT I
IF ISIGN = 1 THEN
  H1R = DATQ(1)
  DATQ(1) = H1R + DATQ(2)
  DATQ(2) = H1R - DATQ(2)
ELSE
  H1R = DATQ(1)
  DATQ(1) = C1 * (H1R + DATQ(2))
  DATQ(2) = C1 * (H1R - DATQ(2))
  CALL FOUR1(DATQ(), N, -1)
END IF
END SUB
```

> Squeeze the first and last data together to get them all within the original array.

> This is the inverse transform for the case ISIGN=-1.

A sample program using **REALFT** is the following:

```
DECLARE SUB REALFT (DATQ!(), N!, ISIGN!)

'PROGRAM D12R3
'Driver for routine REALFT
CLS
EPS = .001
NP = 32
WIDTQ = 50!
PI = 3.14159
DIM DATQ(NP), SIZE(NP)
N = NP / 2
DO
  PRINT "Period of sinusoid in channels (2-"; NP; "; OR 0 TO STOP)"
  INPUT PER
  IF PER <= 0! THEN EXIT DO
  FOR I = 1 TO NP
    DATQ(I) = COS(2! * PI * (I - 1) / PER)
  NEXT I
  CALL REALFT(DATQ(), N, 1)
  SIZE(1) = DATQ(1)
  BIG = SIZE(1)
  FOR I = 2 TO N
    SIZE(I) = SQR(DATQ(2 * I - 1) ^ 2 + DATQ(2 * I) ^ 2)
    IF I = 1 THEN SIZE(I) = DATQ(I)
    IF SIZE(I) > BIG THEN BIG = SIZE(I)
  NEXT I
  SCAL = WIDTQ / BIG
  FOR I = 1 TO N
    NLIM = INT(SCAL * SIZE(I) + EPS)
    PRINT USING "####"; I;
    PRINT " ";
    FOR J = 1 TO NLIM + 1
      PRINT "*";
    NEXT J
    PRINT
  NEXT I
  PRINT "press continue ..."
  LINE INPUT DUM$
  CALL REALFT(DATQ(), N, -1)
```

```
BIG = -1E+10
SMALL = 1E+10
FOR I = 1 TO NP
  IF DATQ(I) < SMALL THEN SMALL = DATQ(I)
  IF DATQ(I) > BIG THEN BIG = DATQ(I)
NEXT I
SCAL = WIDTQ / (BIG - SMALL)
FOR I = 1 TO NP
  NLIM = INT(SCAL * (DATQ(I) - SMALL) + EPS)
  PRINT USING "####"; I;
  PRINT " ";
  FOR J = 1 TO NLIM + 1
    PRINT "*";
  NEXT J
  PRINT
NEXT I
LOOP
END
```

SINFT performs a sine-transform of a real-valued array. The necessity for such a transform arises in solution methods for partial differential equations with certain kinds of boundary conditions (see Chapter 17). The sample program D12R4 works exactly as the previous program. Notice that in this program no distinction needs to be made between the transform and its inverse. They are identical.

Here is the recipe SINFT:

```
DECLARE SUB REALFT (DATQ!(), N!, ISIGN!)

SUB SINFT (Y(), N)
```
> Calculates the sine transform of a set of N real-valued data points stored in array Y. The number N must be a power of 2. On exit Y is replaced by its transform. This program, without changes, also calculates the inverse sine transform, but in this case the output array should be multiplied by 2/N. Use double precision in the trigonometric recurrences.

```
THETA# = 3.141592653589793# / CDBL(N)     Initialize the recurrence.
WR# = 1#
WI# = 0#
WPR# = -2# * SIN(.5# * THETA#) ^ 2
WPI# = SIN(THETA#)
Y(1) = 0!
M = N / 2
FOR J = 1 TO M
  WTEMP# = WR#
  WR# = WR# * WPR# - WI# * WPI# + WR#       Calculate the sine for the auxiliary array.
  WI# = WI# * WPR# + WTEMP# * WPI# + WI#    The cosine is needed to continue the recurrence.
  Y1 = WI# * (Y(J + 1) + Y(N - J + 1))      Construct the auxiliary array.
  Y2 = .5 * (Y(J + 1) - Y(N - J + 1))
  Y(J + 1) = Y1 + Y2                         Terms j and N − j are related
  Y(N - J + 1) = Y1 - Y2
NEXT J
CALL REALFT(Y(), M, 1)                       Transform the auxiliary array.
SUM = 0!
Y(1) = .5 * Y(1)                             Initialize the sum used for odd terms below.
Y(2) = 0!
```

```
FOR J = 1 TO N - 1 STEP 2
  SUM = SUM + Y(J)
  Y(J) = Y(J + 1)
  Y(J + 1) = SUM
NEXT J
END SUB
```

Even terms in the transform are determined directly.
Odd terms are determined by this running sum.

A sample program using **SINFT** is the following:

```
DECLARE SUB SINFT (Y!(), N!)

'PROGRAM D12R4
'Driver for routine SINFT
CLS
EPS = .001
NP = 16
WIDTQ = 30!
PI = 3.14159
DIM DATQ(NP), SIZE(NP)
DO
  PRINT "Period of sinusoid in channels (2-";
  PRINT USING "##"; NP;
  PRINT ")"
  INPUT PER
  IF PER <= 0! THEN EXIT DO
  FOR I = 1 TO NP
    DATQ(I) = SIN(2! * PI * (I - 1) / PER)
  NEXT I
  CALL SINFT(DATQ(), NP)
  BIG = -1E+10
  SMALL = 1E+10
  FOR I = 1 TO NP
    IF DATQ(I) < SMALL THEN SMALL = DATQ(I)
    IF DATQ(I) > BIG THEN BIG = DATQ(I)
  NEXT I
  SCAL = WIDTQ / (BIG - SMALL)
  FOR I = 1 TO NP
    NLIM = INT(SCAL * (DATQ(I) - SMALL) + EPS)
    PRINT USING "####"; I;
    PRINT " ";
    FOR J = 1 TO NLIM + 1
      PRINT "*";
    NEXT J
    PRINT
  NEXT I
  PRINT "press continue ..."
  LINE INPUT DUM$
  CALL SINFT(DATQ(), NP)
  BIG = -1E+10
  SMALL = 1E+10
  FOR I = 1 TO NP
    IF DATQ(I) < SMALL THEN SMALL = DATQ(I)
    IF DATQ(I) > BIG THEN BIG = DATQ(I)
  NEXT I
  SCAL = WIDTQ / (BIG - SMALL)
```

```
    FOR I = 1 TO NP
      NLIM = INT(SCAL * (DATQ(I) - SMALL) + EPS)
      PRINT USING "####"; I;
      PRINT " ";
      FOR J = 1 TO NLIM + 1
        PRINT "*";
      NEXT J
      PRINT
    NEXT I
LOOP
END
```

COSFT is a companion subroutine to SINFT that does the cosine transform. It also plays a role in partial differential equation solutions. Although program D12R5 is again the same as D12R3, you will notice some difference in solutions. The cosine transform of a cosine with a period that is a power of two does not give a transform that is nonzero in a single bin. It has some small values at other frequencies. This is due to our desire to cast the transform into something that calls REALFT, and therefore works on 2^N points rather than the more natural $2^N + 1$. The sample program will prove to you, however, that the transform expressed here is invertible. Notice that, unlike the sine transform, the cosine transform is not self-inverting.

Here is the recipe COSFT:

```
DECLARE SUB REALFT (DATQ!(), N!, ISIGN!)

SUB COSFT (Y(), N, ISIGN)
```
> Calculates the cosine transform of a set Y of N real-valued data points. The transformed data replace the original data in array Y. N must be a power of 2. Set ISIGN to +1 for a transform, and to −1 for an inverse transform. For an inverse transform, the output array should be multiplied by 2/N. Use double precision for the trigonometric recurrences.

```
THETA# = 3.141592653589793# / CDBL(N)       Initialize the recurrence.
WR# = 1#
WI# = 0#
WPR# = -2# * SIN(.5# * THETA#) ^ 2
WPI# = SIN(THETA#)
SUM = Y(1)
M = N / 2
FOR J = 1 TO M - 1                 J=M unnecessary since Y(N/2+1) unchanged
  WTEMP# = WR#
  WR# = WR# * WPR# - WI# * WPI# + WR#        Carry out the recurrence.
  WI# = WI# * WPR# + WTEMP# * WPI# + WI#
  Y1 = .5 * (Y(J + 1) + Y(N - J + 1))        Calculates the auxiliary function.
  Y2 = Y(J + 1) - Y(N - J + 1)
  Y(J + 1) = Y1 - WI# * Y2          The values for j and N − j are related.
  Y(N - J + 1) = Y1 + WI# * Y2
  SUM = SUM + WR# * Y2          Carry along this sum for later use in unfolding the transform.
NEXT J
CALL REALFT(Y(), M, 1)          Calculate the transform of the auxiliary function.
Y(2) = SUM                  SUM is the value in equation (12.3.19).
FOR J = 4 TO N STEP 2
  SUM = SUM + Y(J)             Equation (12.3.18).
  Y(J) = SUM
NEXT J
```

```
IF ISIGN = -1 THEN                          This code applies only to the inverse transform.
  EVEN = Y(1)                               Sum up the even and odd transform values as in equation
  ODD = Y(2)                                (12.3.22).
  FOR I = 3 TO N - 1 STEP 2
    EVEN = EVEN + Y(I)
    ODD = ODD + Y(I + 1)
  NEXT I
  ENFO = 2! * (EVEN - ODD)
  SUMO = Y(1) - ENFO                        Next, implement equation (12.3.24).
  SUME = 2! * ODD / CSNG(N) - SUMO
  Y(1) = .5 * ENFO
  Y(2) = Y(2) - SUME
  FOR I = 3 TO N - 1 STEP 2
    Y(I) = Y(I) - SUMO                      Finally, equation (12.3.21) gives us the true inverse cosine
    Y(I + 1) = Y(I + 1) - SUME              transform (excepting the factor 2/N).
  NEXT I
END IF
END SUB
```

A sample program using COSFT is the following:

```
DECLARE SUB COSFT (Y!(), N!, ISIGN!)

'PROGRAM D12R5
'Driver for routine COSFT
CLS
EPS = .001
NP = 16
WIDTQ = 30!
PI = 3.14159
DIM DATQ(NP), SIZE(NP)
DO
  PRINT "Period of cosine in channels (2-";
  PRINT USING "##"; NP;
  PRINT ")"
  INPUT PER
  IF PER <= 0! THEN EXIT DO
  FOR I = 1 TO NP
    DATQ(I) = COS(2! * PI * (I - 1) / PER)
  NEXT I
  CALL COSFT(DATQ(), NP, 1)
  BIG = -1E+10
  SMALL = 1E+10
  FOR I = 1 TO NP
    IF DATQ(I) < SMALL THEN SMALL = DATQ(I)
    IF DATQ(I) > BIG THEN BIG = DATQ(I)
  NEXT I
  SCAL = WIDTQ / (BIG - SMALL)
  FOR I = 1 TO NP
    NLIM = INT(SCAL * (DATQ(I) - SMALL) + EPS)
    PRINT USING "##"; I;
    PRINT USING "###.##"; DATQ(I);
    PRINT " ";
    FOR J = 1 TO NLIM + 1
      PRINT "*";
```

```
      NEXT J
      PRINT
    NEXT I
    PRINT "press continue ..."
    LINE INPUT DUM$
    CALL COSFT(DATQ(), NP, -1)
    BIG = -1E+10
    SMALL = 1E+10
    FOR I = 1 TO NP
      IF DATQ(I) < SMALL THEN SMALL = DATQ(I)
      IF DATQ(I) > BIG THEN BIG = DATQ(I)
    NEXT I
    SCAL = WIDTQ / (BIG - SMALL)
    FOR I = 1 TO NP
      NLIM = INT(SCAL * (DATQ(I) - SMALL) + EPS)
      PRINT USING "####"; I;
      PRINT " ";
      FOR J = 1 TO NLIM + 1
        PRINT "*";
      NEXT J
      PRINT
    NEXT I
LOOP
END
```

Subroutine **CONVLV** performs the convolution of a data set with a response function using an FFT. Sample program **D12R6** uses two functions that take on only the values 0.0 and 1.0. The data array **DATQ(I)** has sixteen values, and is zero everywhere except between **I=6** and **I=10** where it is 1.0. The response function **RESPNS(I)** has nine values and is zero except between **I=3** and **I=6** where it is 1.0. The expected value of the convolution is determined simply by flipping the response function end-to-end, moving it to the left by the desired shift, and counting how many nonzero channels of **RESPNS** fall on nonzero channels of **DATQ**. In this way, you should be able to verify the result from the program. The sample program, incidentally, does the calculation by this direct method for the purpose of comparison.

Here is the recipe **CONVLV**:

```
DECLARE SUB TWOFFT (DATA1!(), DATA2!(), FFT1!(), FFT2!(), N!)
DECLARE SUB REALFT (DATQ!(), N!, ISIGN!)

SUB CONVLV (DATQ(), N, RESPNS(), M, ISIGN, ANS())
      Convolves or deconvolves a real data set DATQ of length N (including any user-supplied zero
      padding) with a response function RESPNS, stored in wraparound order in a real array of length
      M ≤ N. (M should be an odd integer.) Wraparound order means that the first half of the array
      RESPNS contains the impulse response function at positive times, while the second half of
      the array contains the impulse response function at negative times, counting down from the
      highest element RESPNS(M). On input ISIGN is +1 for convolution, −1 for deconvolution.
      The answer is returned in the first N components of ANS. However, ANS must be supplied in
      the calling program with length at least 2*N, for consistency with TWOFFT. N MUST be an
      integer power of two.
DIM FFT(2 * N)
FOR I = 1 TO INT((M - 1) / 2)          Put RESPNS in array of length N.
  RESPNS(N + 1 - I) = RESPNS(M + 1 - I)
NEXT I
```

```
FOR I = INT((M + 3) / 2) TO N - INT((M - 1) / 2)      Pad with zeros.
  RESPNS(I) = 0!
NEXT I
CALL TWOFFT(DATQ(), RESPNS(), FFT(), ANS(), N)        FFT both at once.
NO2 = INT(N / 2)
FOR I = 1 TO NO2 + 1
  IF ISIGN = 1 THEN
    DUM = ANS(2 * I - 1)                    Multiply FFTs to convolve.
    ANS(2 * I - 1) = (FFT(2 * I - 1) * DUM - FFT(2 * I) * ANS(2 * I)) / NO2
    ANS(2 * I) = (FFT(2 * I - 1) * ANS(2 * I) + FFT(2 * I) * DUM) / NO2
  ELSEIF ISIGN = -1 THEN
    IF DUM = 0! AND ANS(2 * I) = 0! THEN
      PRINT "deconvolving at a response zero"
      EXIT SUB
    END IF
    ANS = FFT(2 * I - 1) * DUM + FFT(2 * I) * ANS(2 * I)  Divide to deconvolve.
    ANS(2 * I - 1) = ANS / (DUM * DUM + ANS(2 * I) * ANS(2 * I)) / NO2
    ANS = FFT(2 * I) * DUM - FFT(2 * I - 1) * ANS(2 * I)
    ANS(2 * I) = ANS / (DUM * DUM + ANS(2 * I) * ANS(2 * I)) / NO2
  ELSE
    PRINT "no meaning for ISIGN"
    EXIT SUB
  END IF
NEXT I
ANS(2) = ANS(2 * NO2 + 1)                Pack last element with first for REALFT.
CALL REALFT(ANS(), NO2, -1)              Inverse transform back to time domain.
ERASE FFT
END SUB
```

A sample program using **CONVLV** is the following:

```
DECLARE SUB CONVLV (DATQ!(), N!, RESPNS!(), M!, ISIGN!, ANS!())

'PROGRAM D12R6
'Driver for routine CONVLV
CLS
N = 16
N2 = 32
M = 9
PI = 3.14159265#
DIM DATQ(N), RESPNS(M), RESP(N), ANS(N2)
FOR I = 1 TO N
  DATQ(I) = 0!
  IF I >= (N / 2 - N / 8) AND I <= (N / 2 + N / 8) THEN DATQ(I) = 1!
NEXT I
FOR I = 1 TO M
  RESPNS(I) = 0!
  IF I > 2 AND I < 7 THEN RESPNS(I) = 1!
  RESP(I) = RESPNS(I)
NEXT I
ISIGN = 1
CALL CONVLV(DATQ(), N, RESP(), M, ISIGN, ANS())
'Compare with a direct convolution
PRINT "  I         CONVLV         Expected"
FOR I = 1 TO N
```

```
    CMP = 0!
    FOR J = 1 TO M / 2
      CMP = CMP + DATQ(((I - J - 1 + N) MOD N) + 1) * RESPNS(J + 1)
      CMP = CMP + DATQ(((I + J - 1) MOD N) + 1) * RESPNS(M - J + 1)
    NEXT J
    CMP = CMP + DATQ(I) * RESPNS(1)
    PRINT USING "###"; I;
    PRINT USING "########.######"; ANS(I); CMP
  NEXT I
  END
```

CORREL calculates the correlation function of two data sets. Sample program D12R7 defines DATA1(I) as an array of 64 values which are all zero except from I=25 to I=39, where they are one. DATA2(I) is defined in the same way. Therefore, the correlation being performed is an autocorrelation. The sample routine compares the result of the calculation as performed by CORREL with that found by a direct calculation. In this case the calculation may be done manually simply by successively shifting DATA2 with respect to DATA1 and counting the number of nonzero channels of the two that overlap.

Here is the recipe CORREL:

```
DECLARE SUB TWOFFT (DATA1!(), DATA2!(), FFT1!(), FFT2!(), N!)
DECLARE SUB REALFT (DATQ!(), N!, ISIGN!)

SUB CORREL (DATA1(), DATA2(), N, ANS())
    Computes the correlation of two real data sets DATA1 and DATA2, each of length N (including
    any user-supplied zero padding). N MUST be an integer power of two. The answer is returned
    as the first N points in ANS stored in wraparound order, i.e. correlations at increasingly negative
    lags are in ANS(N) on down to ANS(N/2+1), while correlations at increasingly positive lags
    are in ANS(1) (zero lag) on up to ANS(N/2). Note that ANS must be supplied in the calling
    program with length at least 2*N, since it is also used as working space. Sign convention of
    this routine: if DATA1 lags DATA2, i.e. is shifted to the right of it, then ANS will show a peak
    at positive lags.
DIM FFT(2 * N)
CALL TWOFFT(DATA1(), DATA2(), FFT(), ANS(), N)         Transform both DATA vectors at once.
NO2 = INT(N / 2)                       Normalization for inverse FFT.
FOR I = 1 TO NO2 + 1
    DUM = ANS(2 * I - 1)               Multiply to find FFT of their correlation.
    ANS(2 * I - 1) = (FFT(2 * I - 1) * DUM + FFT(2 * I) * ANS(2 * I)) / CSNG(NO
    ANS(2 * I) = (FFT(2 * I) * DUM - FFT(2 * I - 1) * ANS(2 * I)) / CSNG(NO2)
NEXT I
ANS(2) = ANS(N + 1)                    Pack first and last into one element.
CALL REALFT(ANS(), NO2, -1)            Inverse transform gives correlation.
ERASE FFT
END SUB
```

A sample program using CORREL is the following:

```
DECLARE SUB CORREL (DATA1!(), DATA2!(), N!, ANS!())

'PROGRAM D12R7
'Driver for routine CORREL
CLS
N = 64
N2 = 128
PI = 3.1415927#
DIM DATA1(N), DATA2(N), ANS(N2)
FOR I = 1 TO N
  DATA1(I) = 0!
  IF I > (N / 2 - N / 8) AND I < (N / 2 + N / 8) THEN DATA1(I) = 1!
  DATA2(I) = DATA1(I)
NEXT I
CALL CORREL(DATA1(), DATA2(), N, ANS())
'Calculate directly
PRINT "  n          CORREL       Direct Calc."
PRINT
FOR I = 0 TO 16
  CMP = 0!
  FOR J = 1 TO N
    CMP = CMP + DATA1(((I + J - 1) MOD N) + 1) * DATA2(J)
  NEXT J
  PRINT USING "###"; I;
  PRINT USING "########.######"; ANS(I + 1); CMP
NEXT I
END
```

SPCTRM does a spectral estimate of a data set by reading it in as segments, windowing, Fourier transforming, and accumulating the power spectrum. Data segments may or may not be overlapped at the decision of the user. In sample program D12R8 the spectral data are read in from a file called SPCTRL.DAT containing 1200 numbers and included on the *Numerical Recipes Examples Diskette*. It is analyzed first with overlap and then without. The results are tabulated side by side for comparison.

Here is the recipe SPCTRM:

```
DECLARE FUNCTION WINDOQ! (J!, FACM!, FACP!)
DECLARE SUB FOUR1 (DATQ!(), NN!, ISIGN!)

SUB SPCTRM (P(), M, K, OVRLAP%, W1(), W2()) STATIC
            Reads data from SPCTRL.DAT and returns as P(J) the data's power (mean square amplitude)
            at frequency (J-1)/(2*M) cycles per gridpoint, for J=1,2,...,M, based on (2*K+1)*M
            data points (if OVRLAP% is -1) or 4*K*M data points (if OVRLAP% is 0.). The number of
            segments of the data is 2*K in both cases: the routine calls FOUR1 K-times, each call with 2
            partitions each of 2*M real data points. W1 and W2 are user-supplied workspaces of length 4*M
            and M respectively. Windowing is performed by a routine of the fixed name WINDOQ.
MM = M + M                        Useful factors.
M4 = MM + MM
M44 = M4 + 4
M43 = M4 + 3
DEN = 0!
FACM = M - .5                     Factors used by the window statement function.
FACP = 1! / (M + .5)
```

```
SUMW = 0!                               Accumulate the squared sum of the weights.
FOR J = 1 TO MM
  SUMW = SUMW + WINDOQ(J, FACM, FACP) ^ 2
NEXT J
FOR J = 1 TO M                          Initialize the spectrum to zero.
  P(J) = 0!
NEXT J
IF OVRLAP% THEN
  FOR J = 1 TO M                        Initialize the "save" half-buffer.
    W2(J) = VAL(INPUT$(12, #1))
    IF J MOD 4 = 0 THEN DUM$ = INPUT$(2, #1)
  NEXT J
END IF
JN& = 0
FOR KK = 1 TO K                         Loop over data set segments in groups of two.
  FOR JOFF = -1 TO 0                    Get two complete segments into workspace.
    IF OVRLAP% THEN
      FOR J = 1 TO M
        W1(JOFF + J + J) = W2(J)
      NEXT J
      FOR J = 1 TO M
        W2(J) = VAL(INPUT$(12, #1))
        IF J MOD 4 = 0 THEN DUM$ = INPUT$(2, #1)
      NEXT J
      JOFFN = JOFF + MM
      FOR J = 1 TO M
        W1(JOFFN + J + J) = W2(J)
      NEXT J
    ELSE
      FOR J = JOFF + 2 TO M4 STEP 2
        W1(J) = VAL(INPUT$(12, #1))
        JN& = JN& + 1
        IF JN& MOD 4 = 0 THEN DUM$ = INPUT$(2, #1)
      NEXT J
    END IF
  NEXT JOFF
  FOR J = 1 TO MM                       Apply the window to the data.
    J2 = J + J
    W = WINDOQ(J, FACM, FACP)
    W1(J2) = W1(J2) * W
    W1(J2 - 1) = W1(J2 - 1) * W
  NEXT J
  CALL FOUR1(W1(), MM, 1)               Fourier transform the windowed data.
  P(1) = P(1) + W1(1) ^ 2 + W1(2) ^ 2   Sum results into previous segments.
  FOR J = 2 TO M
    J2 = J + J
    P(J) = P(J) + W1(J2) ^ 2 + W1(J2 - 1) ^ 2 + W1(M44 - J2) ^ 2
    P(J) = P(J) + W1(M43 - J2) ^ 2
  NEXT J
  DEN = DEN + SUMW
NEXT KK
DEN = M4 * DEN                          Correct normalization.
FOR J = 1 TO M
  P(J) = P(J) / DEN                     Normalize the output.
NEXT J
```

```
END SUB

FUNCTION WINDOQ (J, FACM, FACP)
WINDOQ = 1! - ABS(((J - 1) - FACM) * FACP)
'WINDOQ = 1!
'WINDOQ = 1! - (((J - 1) * FACM) * FACP) ^ 2
END FUNCTION
```

Defines Parzen window.
Alternative for square window.
Alternative for Welch window.

A sample program using **SPCTRM** is the following:

```
DECLARE SUB SPCTRM (P!(), M!, K!, OVRLAP%, W1!(), W2!())

'PROGRAM D12R8
'Driver for routine SPCTRM
CLS
M = 16
M4 = 4 * M
DIM P(M), Q(M), W1(M4), W2(M)
OPEN "SPCTRL.DAT" FOR INPUT AS #1
K = 8
OVRLAP% = -1
CALL SPCTRM(P(), M, K, OVRLAP%, W1(), W2())
CLOSE #1
OPEN "SPCTRL.DAT" FOR INPUT AS #1
K = 16
OVRLAP% = 0
CALL SPCTRM(Q(), M, K, OVRLAP%, W1(), W2())
CLOSE #1
PRINT "Spectrum of data in file SPCTRL.DAT"
PRINT "          Overlapped      Non-Overlapped"
FOR J = 1 TO M
  PRINT USING "####"; J;
  PRINT USING "##########.######"; P(J); Q(J)
NEXT J
END
```

MEMCOF and **EVLMEM** are used to perform spectral analysis by the maximum entropy method. **MEMCOF** finds the coefficients for a model spectrum, the magnitude squared of the inverse of a polynomial series. Sample program **D12R9** determines the coefficients for 1000 numbers from the file **SPCTRL.DAT** and simply prints the results for comparison to the following table:

```
Coefficients for spectral estimation of SPCTRL.DAT
  a[ 1] =     1.261539
  a[ 2] =    -0.007695
  a[ 3] =    -0.646778
  a[ 4] =    -0.280603
  a[ 5] =     0.163693
  a[ 6] =     0.347674
  a[ 7] =     0.111247
  a[ 8] =    -0.337141
  a[ 9] =    -0.358043
  a[10] =     0.378774
     a0 =     0.003511
```

Here is the recipe **MEMCOF**:

```
SUB MEMCOF (DATQ(), N, M, PM, COF(), WK1(), WK2(), WKM())
```
Given a real vector of DATQ of length N, and given M, this routine returns a vector COF of length M with $COF(J) = a_j$, and a scalar $PM = a_0$, which are the coefficients for Maximum Entropy Method spectral estimation. The user must provide workspace vectors WK1, WK2, and WKM of lengths N, N, and M, respectively.

```
P = 0!
FOR J = 1 TO N
  P = P + DATQ(J) ^ 2
NEXT J
PM = P / N
WK1(1) = DATQ(1)
WK2(N - 1) = DATQ(N)
FOR J = 2 TO N - 1
  WK1(J) = DATQ(J)
  WK2(J - 1) = DATQ(J)
NEXT J
FOR K = 1 TO M
  PNEUM = 0!
  DENOM = 0!
  FOR J = 1 TO N - K
    PNEUM = PNEUM + WK1(J) * WK2(J)
    DENOM = DENOM + WK1(J) ^ 2 + WK2(J) ^ 2
  NEXT J
  COF(K) = 2! * PNEUM / DENOM
  PM = PM * (1! - COF(K) ^ 2)
  FOR I = 1 TO K - 1
    COF(I) = WKM(I) - COF(K) * WKM(K - I)
  NEXT I
```
The algorithm is recursive, building up the answer for larger and larger values of M until the desired value is reached. At this point in the algorithm, one could return the vector COF and scalar PM for an MEM spectral estimate of K (rather than M) terms.

```
  IF K = M THEN EXIT SUB
  FOR I = 1 TO K
    WKM(I) = COF(I)
  NEXT I
  FOR J = 1 TO N - K - 1
    WK1(J) = WK1(J) - WKM(K) * WK2(J)
    WK2(J) = WK2(J + 1) - WKM(K) * WK1(J + 1)
  NEXT J
NEXT K
PRINT "never get here"
END SUB
```

A sample program using MEMCOF is the following:

```
DECLARE SUB MEMCOF (DATQ!(), N!, M!, PM!, COF!(), WK1!(), WK2!(), WKM!())

'PROGRAM D12R9
'Driver for routine MEMCOF
CLS
N = 1000
M = 10
DIM DATQ(N), COF(M), WK1(N), WK2(N), WKM(M)
OPEN "SPCTRL.DAT" FOR INPUT AS #1
```

```
FOR I = 1 TO N
  DATQ(I) = VAL(INPUT$(12, #1))
  IF I MOD 4 = 0 THEN DUM$ = INPUT$(2, #1)
NEXT I
CLOSE #1
CALL MEMCOF(DATQ(), N, M, PM, COF(), WK1(), WK2(), WKM())
PRINT "Coeff. for spectral estim. of SPCTRL.DAT"
PRINT
FOR I = 1 TO M
  PRINT "a[";
  PRINT USING "##"; I;
  PRINT "] =";
  PRINT USING "#####.######"; COF(I)
NEXT I
PRINT
PRINT "a0 =";
PRINT USING "#####.######"; PM
END
```

EVLMEM uses coefficients from MEMCOF to generate a spectral estimate. The example D12R10 uses the same data from SPCTRL.DAT and prints the spectral estimate. You may compare the result to:

```
Power spectrum estimate of data in SPCTRL.DAT
     f*delta        power
    0.000000      0.026023
    0.031250      0.029266
    0.062500      0.193087
    0.093750      0.139241
    0.125000     29.915518
    0.156250      0.003878
    0.187500      0.000633
    0.218750      0.000334
    0.250000      0.000437
    0.281250      0.001331
    0.312500      0.000780
    0.343750      0.000451
    0.375000      0.000784
    0.406250      0.001381
    0.437500      0.000649
    0.468750      0.000775
    0.500000      0.001716
```

Here is the recipe EVLMEM:

```
DECLARE FUNCTION EVLMEM! (FDT!, COF!(), M!, PM!)

FUNCTION EVLMEM (FDT, COF(), M, PM)
```

Given COF, M, PM as returned by MEMCOF, this function returns the power spectrum estimate $P(f)$ as a function of FDT $= f\Delta$.

```
THETA# = 6.28318530717959# * FDT    Do trigonometric recurrences in double precision.
WPR# = COS(THETA#)                  Set up for recurrence relations.
WPI# = SIN(THETA#)
WR# = 1#
WI# = 0#
SUMR = 1!                           These will accumulate the denominator of (12.8.4).
SUMI = 0!
```

```
FOR I = 1 TO M                          Loop over the terms in the sum.
  WTEMP# = WR#
  WR# = WR# * WPR# - WI# * WPI#
  WI# = WI# * WPR# + WTEMP# * WPI#
  SUMR = SUMR - COF(I) * CSNG(WR#)
  SUMI = SUMI - COF(I) * CSNG(WI#)
NEXT I
EVLMEM = PM / (SUMR ^ 2 + SUMI ^ 2)        Equation (12.8.4).
END FUNCTION
```

A sample program using **EVLMEM** is the following:

```
DECLARE SUB MEMCOF (DATQ!(), N!, M!, PM!, COF!(), WK1!(), WK2!(), WKM!())
DECLARE FUNCTION EVLMEM! (FDT!, COF!(), M!, PM!)

'PROGRAM D12R10
'Driver for routine EVLMEM
CLS
N = 1000
M = 10
NFDT = 16
DIM DATQ(N), COF(M), WK1(N), WK2(N), WKM(M)
OPEN "SPCTRL.DAT" FOR INPUT AS #1
FOR I = 1 TO N
  DATQ(I) = VAL(INPUT$(12, #1))
  IF I MOD 4 = 0 THEN DUM$ = INPUT$(2, #1)
NEXT I
CLOSE #1
CALL MEMCOF(DATQ(), N, M, PM, COF(), WK1(), WK2(), WKM())
PRINT "Power spectrum estimate of data in SPCTRL.DAT"
PRINT "     f*delta        power"
FOR I = 0 TO NFDT
  FDT = .5 * I / NFDT
  PRINT USING "#####.######"; FDT; EVLMEM(FDT, COF(), M, PM)
NEXT I
END
```

Notice that once **MEMCOF** has determined coefficients, we may evaluate the estimate at any intervals we wish. Notice also that we have built a spectral peak into the noisy data in **SPCTRL.DAT**.

Linear prediction is carried out by routines **PREDIC**, **MEMCOF**, and **FIXRTS**. **MEMCOF** produces the linear prediction coefficients from the data set. **FIXRTS** massages the coefficients so that all roots of the characteristic polynomial fall inside the unit circle of the complex domain, thus insuring stability of the prediction algorithm. Finally, **PREDIC** predicts future data points based on the modified coefficients. Sample program **D12R11** demonstrates the operation of **FIXRTS**. The coefficients provided in the **DATA** statement for **D(I)** are those appropriate to the polynomial $(z - 1)^6 = 1$. This equation has six roots on a circle of radius one, centered at $(1.0, 0.0)$ in the complex plane. Some of these lie within the unit circle and some outside. The ones outside are moved by **FIXRTS** according to $z_i \rightarrow 1/z_i^*$. You can easily figure these out by hand and check the results. Also, the sample routine calculates $(z - 1)^6$ for each of the adjusted roots, and thereby shows which have been changed and which have not.

Here is the recipe **FIXRTS**:

```
DECLARE SUB ZROOTS (A!(), M!, ROOTS!(), POLISH%)

SUB FIXRTS (D(), NPOLES)
```
Given the LP coefficients D(J), J=1...NPOLES, this routine finds all roots of the characteristic polynomial (12.10.3), reflects any roots that are outside the unit circle back inside, and
then returns a modified set of D(J)'s. The routine ZROOTS of numbers is referenced.
```
DIM A(2, NPOLES + 1), ROOTS(2, NPOLES + 1) 'Complex numbers
A(1, NPOLES + 1) = 1!
A(2, NPOLES + 1) = 0!
FOR J = NPOLES TO 1 STEP -1          Set up complex coefficients for polynomial root finder.
  A(1, J) = -D(NPOLES + 1 - J)
  A(2, J) = 0!
NEXT J
POLISH% = -1
CALL ZROOTS(A(), NPOLES, ROOTS(), POLISH%)     Find all the roots.
FOR J = 1 TO NPOLES                            Look for a...
  IF SQR(ROOTS(1, J) ^ 2 + ROOTS(2, J) ^ 2) > 1! THEN  root outside the unit circle,
    DUM = ROOTS(1, J)                                    and reflect it back inside.
    ROOTS(1, J) = DUM / (DUM ^ 2 + ROOTS(2, J) ^ 2)
    ROOTS(2, J) = ROOTS(2, J) / (DUM ^ 2 + ROOTS(2, J) ^ 2)
  END IF
NEXT J
A(1, 1) = -ROOTS(1, 1)               Now reconstruct the polynomial coefficients,
A(2, 1) = -ROOTS(2, 1)
A(1, 2) = 1!
A(2, 2) = 0!
FOR J = 2 TO NPOLES                  by looping over the roots
  A(1, J + 1) = 1!
  A(2, J + 1) = 0!
  FOR I = J TO 2 STEP -1             and synthetically multiplying.
    DUM = A(1, I)
    A(1, I) = A(1, I - 1) - (ROOTS(1, J) * DUM - ROOTS(2, J) * A(2, I))
    A(2, I) = A(2, I - 1) - (ROOTS(2, J) * DUM + ROOTS(1, J) * A(2, I))
  NEXT I
  DUM = A(1, 1)
  A(1, 1) = -(ROOTS(1, J) * DUM - ROOTS(2, J) * A(2, 1))
  A(2, 1) = -(ROOTS(2, J) * DUM + ROOTS(1, J) * A(2, 1))
NEXT J
FOR J = 1 TO NPOLES               The polynomial coefficients are guaranteed to be real,
  D(NPOLES + 1 - J) = -A(1, J)    so we need only return the real part as new LP coefficients.
NEXT J
ERASE ROOTS, A
END SUB
```

A sample program using **FIXRTS** is the following:

```
DECLARE SUB FIXRTS (D!(), NPOLES!)
DECLARE SUB ZROOTS (A!(), M!, ROOTS!(), POLISH%)


'PROGRAM D12R11
'Driver for routine FIXRTS
CLS
NPOLES = 6
NPOL = NPOLES + 1
DIM D(NPOLES)
DIM ZCOEF(2, NPOL), ZEROS(2, NPOLES), Z(2)
FOR I = 1 TO NPOLES
  READ D(I)
NEXT I
DATA 6.0,-15.0,20.0,-15.0,6.0,0.0
'Finding roots of (z-1.0)^6=1.0
'First print roots
ZCOEF(1, NPOLES + 1) = 1!
ZCOEF(2, NPOLES + 1) = 0!
FOR I = NPOLES TO 1 STEP -1
  ZCOEF(1, I) = -D(NPOLES + 1 - I)
  ZCOEF(2, I) = 0!
NEXT I
POLISH% = -1
CALL ZROOTS(ZCOEF(), NPOLES, ZEROS(), POLISH%)
PRINT "Roots of (z-1.0)^6 = 1.0"
PRINT "                    Root                    (z-1.0)^6"
FOR I = 1 TO NPOLES
  Z(1) = 1!
  Z(2) = 0!
  FOR J = 1 TO 6
    DUM = Z(1)
    Z(1) = DUM * (ZEROS(1, I) - 1!) - Z(2) * ZEROS(2, I)
    Z(2) = DUM * ZEROS(2, I) + Z(2) * (ZEROS(1, I) - 1!)
  NEXT J
  PRINT USING "######"; I;
  PRINT USING "#####.######"; ZEROS(1, I); ZEROS(2, I); Z(1); Z(2)
NEXT I
PRINT
'Now fix them to lie within unit circle
CALL FIXRTS(D(), NPOLES)
'Check results
ZCOEF(1, NPOLES + 1) = 1!
ZCOEF(2, NPOLES + 1) = 0!
FOR I = NPOLES TO 1 STEP -1
  ZCOEF(1, I) = -D(NPOLES + 1 - I)
  ZCOEF(2, I) = 0!
NEXT I
CALL ZROOTS(ZCOEF(), NPOLES, ZEROS(), POLISH%)
PRINT "Roots reflected in unit circle"
PRINT "                    Root                    (z-1.0)^6"
FOR I = 1 TO NPOLES
  Z(1) = 1!
  Z(2) = 0!
  FOR J = 1 TO 6
    DUM = Z(1)
```

```
    Z(1) = DUM * (ZEROS(1, I) - 1!) - Z(2) * ZEROS(2, I)
    Z(2) = DUM * ZEROS(2, I) + Z(2) * (ZEROS(1, I) - 1!)
  NEXT J
  PRINT USING "######"; I;
  PRINT USING "#####.######"; ZEROS(1, I); ZEROS(2, I); Z(1); Z(2)
NEXT I
END
```

PREDIC carries out the job of performing the prediction. The function chosen for investigation in sample program D12R12 is

$$\text{FNF}(N) = \exp(-N/\text{NPTS})\sin(2\pi N/50) + \exp(-2N/\text{NPTS})\sin(2.2\pi N/50)$$

the sum of two sine waves of similar period and exponentially decaying amplitudes. On the basis of 500 data points, and working with coefficients representing ten poles, the routine predicts 20 future points. The quality of this prediction may be judged by comparing these 20 points with the evaluations of FNF(N) that are provided.

Here is the recipe PREDIC:

```
SUB PREDIC (DATQ(), NDATA, D(), NPOLES, FUTURE(), NFUT)
       Given DATQ(J), J=1...NDATA, and given the data's LP coefficients D(I), I=1...NPOLES,
       this routine applies equation (12.10.1) to predict the next NFUT data points, which it returns
       in the array FUTURE. Note that the routine references only the last NPOLES values of DATQ,
       as initial values for the prediction.
DIM REQ(NPOLES)
FOR J = 1 TO NPOLES
  REQ(J) = DATQ(NDATA + 1 - J)
NEXT J
FOR J = 1 TO NFUT
  DISCRP = 0!                        This is where you would put in a known discrepancy if you
  SUM = DISCRP                       were reconstructing a function by linear predictive coding
  FOR K = 1 TO NPOLES                rather than extrapolating a function by linear prediction. See
    SUM = SUM + D(K) * REQ(K)        text.
  NEXT K
  FOR K = NPOLES TO 2 STEP -1        [If you know how to implement circular arrays, you can avoid
    REQ(K) = REQ(K - 1)              this shifting of coefficients!]
  NEXT K
  REQ(1) = SUM
  FUTURE(J) = SUM
NEXT J
ERASE REQ
END SUB
```

A sample program using PREDIC is the following:

```
DECLARE SUB PREDIC (DATQ!(), NDATA!, D!(), NPOLES!, FUTURE!(), NFUT!)
DECLARE SUB MEMCOF (DATQ!(), N!, M!, PM!, COF!(), WK1!(), WK2!(), WKM!())
DECLARE SUB FIXRTS (D!(), NPOLES!)

'PROGRAM D12R12
'Driver for routine PREDIC
CLS
NPTS = 500
NPOLES = 10
NFUT = 20
PI = 3.1415926#
DIM DATQ(NPTS), D(NPOLES), WK1(NPTS), WK2(NPTS), WKM(NPOLES), FUTURE(NFUT)
DEF FNF (N)
  DUM = EXP(-N / NPTS)
  FNF = DUM * SIN(PI * N / 25!) + DUM * DUM * SIN(2.2 * PI * N / 50!)
END DEF
FOR I = 1 TO NPTS
  DATQ(I) = FNF(I)
NEXT I
CALL MEMCOF(DATQ(), NPTS, NPOLES, DUM, D(), WK1(), WK2(), WKM())
CALL FIXRTS(D(), NPOLES)
CALL PREDIC(DATQ(), NPTS, D(), NPOLES, FUTURE(), NFUT)
PRINT "    I    Actual     PREDIC"
FOR I = 1 TO NFUT
  PRINT USING "######"; I;
  PRINT USING "#####.#####"; FNF(I + NPTS); FUTURE(I)
NEXT I
END
```

FOURN is a routine for performing N-dimensional Fourier transforms. We have used it in sample program **D12R13** to transform a 3-dimensional complex data array of dimensions $4 \times 8 \times 16$. The function analyzed is not that easy to visualize, but it is very easy to calculate. The test conducted here is to perform a 3-dimensional transform and inverse transform in succession, and to compare the result with the original array. Ratios are provided for convenience.

Here is the recipe **FOURN**:

```
SUB FOURN (DATQ(), NN(), NDIM, ISIGN)
```
> Replaces **DATQ** by its **NDIM**-dimensional discrete Fourier transform, if **ISIGN** is input as 1. **NN** is an integer array of length **NDIM**, containing the lengths of each dimension (number of complex values), which MUST all be powers of 2. **DATQ** is a real array of length twice the product of these lengths. If **ISIGN** is input as −1, **DATQ** is replaced by its inverse transform times the product of the lengths of all dimensions.
```
NTOT = 1
FOR IDIM = 1 TO NDIM                Compute total number of complex values.
  NTOT = NTOT * NN(IDIM)
NEXT IDIM
NPREV = 1
FOR IDIM = 1 TO NDIM                Main loop over the dimensions.
  N = NN(IDIM)
  NREM = NTOT / (N * NPREV)
  IP1 = 2 * NPREV
  IP2 = IP1 * N
```

```
    IP3 = IP2 * NREM
    I2REV = 1
    FOR I2 = 1 TO IP2 STEP IP1        This is the bit reversal section of the routine.
      IF I2 < I2REV THEN
        FOR I1 = I2 TO I2 + IP1 - 2 STEP 2
          FOR I3 = I1 TO IP3 STEP IP2
            I3REV = I2REV + I3 - I2
            TEMPR = DATQ(I3)
            TEMPI = DATQ(I3 + 1)
            DATQ(I3) = DATQ(I3REV)
            DATQ(I3 + 1) = DATQ(I3REV + 1)
            DATQ(I3REV) = TEMPR
            DATQ(I3REV + 1) = TEMPI
          NEXT I3
        NEXT I1
      END IF
      IBIT = IP2 / 2
      WHILE IBIT >= IP1 AND I2REV > IBIT
        I2REV = I2REV - IBIT
        IBIT = IBIT / 2
      WEND
      I2REV = I2REV + IBIT
    NEXT I2
    IFP1 = IP1                        Here begins the Danielson-Lanczos section of the routine.
    WHILE IFP1 < IP2
      IFP2 = 2 * IFP1
      THETA# = ISIGN * 6.28318530717959# / (IFP2 / IP1)    Initialize for the trig. re-
      WPR# = -2# * SIN(.5# * THETA#) ^ 2                   currence.
      WPI# = SIN(THETA#)
      WR# = 1#
      WI# = 0#
      FOR I3 = 1 TO IFP1 STEP IP1
        FOR I1 = I3 TO I3 + IP1 - 2 STEP 2
          FOR I2 = I1 TO IP3 STEP IFP2
            K1 = I2                   Danielson-Lanczos formula:
            K2 = K1 + IFP1
            TEMPR = CSNG(WR#) * DATQ(K2) - CSNG(WI#) * DATQ(K2 + 1)
            TEMPI = CSNG(WR#) * DATQ(K2 + 1) + CSNG(WI#) * DATQ(K2)
            DATQ(K2) = DATQ(K1) - TEMPR
            DATQ(K2 + 1) = DATQ(K1 + 1) - TEMPI
            DATQ(K1) = DATQ(K1) + TEMPR
            DATQ(K1 + 1) = DATQ(K1 + 1) + TEMPI
          NEXT I2
        NEXT I1
        WTEMP# = WR#                  Trigonometric recurrence.
        WR# = WR# * WPR# - WI# * WPI# + WR#
        WI# = WI# * WPR# + WTEMP# * WPI# + WI#
      NEXT I3
      IFP1 = IFP2
    WEND
    NPREV = N * NPREV
  NEXT IDIM
END SUB
```

A sample program using **FOURN** is the following:

```
DECLARE SUB FOURN (DATQ!(), NN!(), NDIM!, ISIGN!)

'PROGRAM D12R13
'Driver for routine FOURN
CLS
NDIM = 3
NDAT = 1024
DIM NN(NDIM), DATQ(NDAT)
FOR I = 1 TO NDIM
  NN(I) = 2 * (2 ^ I)
NEXT I
FOR I = 1 TO NN(3)
  FOR J = 1 TO NN(2)
    FOR K = 1 TO NN(1)
      L = K + (J - 1) * NN(1) + (I - 1) * NN(2) * NN(1)
      LL = 2 * L - 1
      DATQ(LL) = CSNG(LL)
      DATQ(LL + 1) = CSNG(LL + 1)
    NEXT K
  NEXT J
NEXT I
ISIGN = 1
CALL FOURN(DATQ(), NN(), NDIM, ISIGN)
ISIGN = -1
PRINT "Double 3-dimensional Transform"
PRINT
PRINT "        Double Transf.        Original Data        Ratio"
PRINT "    Real        Imag.        Real        Imag.        Real        ";
PRINT "Imag."
PRINT
CALL FOURN(DATQ(), NN(), NDIM, ISIGN)
FOR I = 1 TO 4
  J = 2 * I
  K = 2 * J
  L = K + (J - 1) * NN(1) + (I - 1) .* NN(2) * NN(1)
  LL = 2 * L - 1
  PRINT USING "#########.##"; DATQ(LL); DATQ(LL + 1); CSNG(LL); CSNG(LL + 1);
  PRINT USING "#########.##"; DATQ(LL) / LL; DATQ(LL + 1) / (LL + 1)
NEXT I
PRINT
PRINT "The product of transform lengths is:"; NN(1) * NN(2) * NN(3)
END
```

Chapter 13: Statistical Description of Data

Chapter 13 of *Numerical Recipes* covers the subject of descriptive statistics, the representation of data in terms of its statistical properties, and the use of such properties to compare data sets. There are three subroutines that characterize data sets. MOMENT returns the average, average deviation, standard deviation, variance, skewness, and kurtosis of a data array. MDIAN1 and MDIAN2 both find the median of an array. The former also sorts the array.

Most of the remaining subroutines compare data sets. TTEST compares the means of two data sets having the same variance; TUTEST does the same for two sets having different variance; and TPTEST does it for paired samples, correcting for covariance. FTEST is a test of whether two data arrays have significantly different variance. The question of whether two distributions are different is treated by four subroutines (pertaining to whether the data are binned or continuous, and whether data are compared to a model distribution or to other data). Specifically,

1. CHSONE compares binned data to a model distribution.
2. CHSTWO compares two binned data sets.
3. KSONE compares the cumulative distribution function of an unbinned data set to a given function.
4. KSTWO compares the cumulative distribution functions of two unbinned data sets.

The next set of subroutines tests for associations between nominal variables. CNTAB1 and CNTAB2 both check for associations in a two-dimensional contingency table, the first calculating on the basis of χ^2, and the second by evaluating entropies. Linear correlation is represented by Pearson's r, or the linear correlation coefficient, which is calculated with routine PEARSN. Alternatively, the data can be investigated with a nonparametric or rank correlation, using SPEAR to find Spearman's rank correlation r_s. Kendall's τ uses rank ordering of ordinal data to test for monotonic correlations. KENDL1 does this for two data arrays of the same size, while KENDL2 applies it to contingency tables.

One final routine SMOOFT makes no attempt to describe or compare data statistically. It seeks, instead, to smooth out the statistical fluctuations, usually for the purpose of visual presentation.

★   ★   ★   ★

Subroutine MOMENT calculates successive moments of a given distribution of data. The example program D13R1 creates an unusual distribution, one that has a sinusoidal distribution of values (over a half-period of the sine, so the distribution is a symmetrical peak). We have worked out the moments of such a distribution theoretically and recorded them in the program for comparison. The data are discrete and will only approximate these values.

Here is the recipe MOMENT:

```
SUB MOMENT (DATQ(), N, AVE, ADEV, SDEV, VAR, SKEW, CURT)
```
        Given an array of DATQ of length N, this routine returns its mean AVE, average deviation ADEV,
        standard deviation SDEV, variance VAR, skewness SKEW, and kurtosis CURT.
```
IF N <= 1 THEN PRINT "N must be at least 2": EXIT SUB
S = 0!                              First pass to get the mean.
FOR J = 1 TO N
  S = S + DATQ(J)
NEXT J
AVE = S / N
ADEV = 0!                           Second pass to get the first (absolute), second, third, and
VAR = 0!                            fourth moments of the deviation from the mean.
SKEW = 0!
CURT = 0!
FOR J = 1 TO N
  S = DATQ(J) - AVE
  ADEV = ADEV + ABS(S)
  P = S * S
  VAR = VAR + P
  P = P * S
  SKEW = SKEW + P
  P = P * S
  CURT = CURT + P
NEXT J
ADEV = ADEV / N                     Put the pieces together according to the conventional defini-
VAR = VAR / (N - 1)                 tions.
SDEV = SQR(VAR)
IF VAR <> 0! THEN
  SKEW = SKEW / (N * SDEV ^ 3)
  CURT = CURT / (N * VAR ^ 2) - 3!
ELSE
  PRINT "no skew or kurtosis when zero variance"
END IF
END SUB
```

A sample program using MOMENT is the following:

```
DECLARE SUB MOMENT (DATQ!(), N!, AVE!, ADEV!, SDEV!, VAR!, SKEW!, CURT!)

'PROGRAM D13R1
'Driver for routine MOMENT
CLS
PI = 3.14159265#
NPTS = 10000
NBIN = 100
NDAT = NPTS + NBIN
DIM DATQ(NDAT)
I = 1
FOR J = 1 TO NBIN
  X = PI * J / NBIN
  NLIM = CINT(SIN(X) * PI / 2! * NPTS / NBIN)
  FOR K = 1 TO NLIM
    DATQ(I) = X
    I = I + 1
  NEXT K
```

```
NEXT J
PRINT "Moments of a sinusoidal distribution"
PRINT
CALL MOMENT(DATQ(), I - 1, AVE, ADEV, SDEV, VAR, SKEW, CURT)
PRINT "                            Calculated     Expected"
PRINT
PRINT "Mean :                      ";
PRINT USING "#######.####"; AVE; PI / 2!
PRINT "Average Deviation :      ";
PRINT USING "#######.####"; ADEV; .570796
PRINT "Standard Deviation :      ";
PRINT USING "#######.####"; SDEV; .683667
PRINT "Variance :                  ";
PRINT USING "#######.####"; VAR; .467401
PRINT "Skewness :                  ";
PRINT USING "#######.####"; SKEW; 0!
PRINT "Kurtosis :                  ";
PRINT USING "#######.####"; CURT; -.806249
END
```

MDIAN1 and MDIAN2 both find the median of a distribution. In programs D13R2 and D13R3 we allow this distribution to be Gaussian, as produced by routine GASDEV. This distribution should have a mean of zero and variance of one. MDIAN1 also sorts the data, so D13R2 prints the sorted data to show that it is done properly. Example D13R3 has nothing to show from MDIAN2 but the median itself, and it is checked by comparing to the result from MDIAN1.

Here is the recipe MDIAN1:

```
DECLARE SUB SORT (N!, RA!())

SUB MDIAN1 (X(), N, XMED)
     Given an array X of N numbers, returns their median value XMED. The array X is modified and
     returned sorted into ascending order.
CALL SORT(N, X())                    This routine is in §8.2.
N2 = INT(N / 2)
IF 2 * N2 = N THEN
  XMED = .5 * (X(N2) + X(N2 + 1))
ELSE
  XMED = X(N2 + 1)
END IF
END SUB
```

A sample program using MDIAN1 is the following:

```
DECLARE SUB MDIAN1 (X!(), N!, XMED!)
DECLARE FUNCTION GASDEV! (IDUM&)

'PROGRAM D13R2
'Driver for routine MDIAN1
CLS
NPTS = 50
DIM DATQ(NPTS)
IDUM& = -5
FOR I = 1 TO NPTS
```

```
  DATQ(I) = GASDEV(IDUM&)
NEXT I
CALL MDIAN1(DATQ(), NPTS, XMED)
PRINT "Gaussian distrib., zero mean, unit variance"
PRINT
PRINT "Median of data set is";
PRINT USING "#####.######"; XMED
PRINT
PRINT "Sorted data"
FOR I = 1 TO 50
  PRINT USING "#####.######"; DATQ(I);
  IF I MOD 5 = 0 THEN PRINT
NEXT I
END
```

Here is the recipe `MDIAN2`:

```
SUB MDIAN2 (X(), N, XMED)
```
> Given an array X of N numbers, returns their median value XMED. The array X is not modified, and is accessed sequentially in each consecutive pass.

```
BIG = 1E+30
AFAC = 1.5
AMP = 1.5
```
> Here, AMP is an overconvergence factor: on each iteration, we move the guess by this factor more than (13.2.4) would naively indicate. AFAC is a factor used to optimize the size of the "smoothing constant" EPS at each iteration.

`A = .5 * (X(1) + X(N))`	This can be any first guess for the median.
`EPS = ABS(X(N) - X(1))`	This can be any first guess for the characteristic spacing of
`AP = BIG`	the data points near the median.
`AM = -BIG`	AP and AM are upper and lower bounds on the median.
`1 SUM = 0!`	Here we start one pass through the data.
`SUMX = 0!`	
`NP = 0`	Number of points above the current guess,
`NM = 0`	and below it.
`XP = BIG`	Value of the point above and closest to the guess,
`XM = -BIG`	and below and closest.
`FOR J = 1 TO N`	Go through the points,
`  XX = X(J)`	
`  IF XX <> A THEN`	omit a zero denominator in the sums,
`    IF XX > A THEN`	update the diagnostics,
`      NP = NP + 1`	
`      IF XX < XP THEN XP = XX`	
`    ELSEIF XX < A THEN`	
`      NM = NM + 1`	
`      IF XX > XM THEN XM = XX`	
`    END IF`	
`    DUM = 1! / (EPS + ABS(XX - A))`	The smoothing constant is used here.
`    SUM = SUM + DUM`	Accumulate the sums.
`    SUMX = SUMX + XX * DUM`	
`  END IF`	
`NEXT J`	
`IF NP - NM >= 2 THEN`	Guess is too low; make another pass,
`  AM = A`	with a new lower bound,

```
    DUM = 0!
    IF SUMX / SUM - A > DUM THEN DUM = SUMX / SUM - A
    AA = XP + DUM * AMP                    a new best guess
    IF AA > AP THEN AA = .5 * (A + AP)     (but no larger than the upper bound)
    EPS = AFAC * ABS(AA - A)               and a new smoothing factor.
    A = AA
    GOTO 1
  ELSEIF NM - NP >= 2 THEN                 Guess is too high; make another pass,
    AP = A                                 with a new upper bound,
    DUM = 0!
    IF SUMX / SUM - A < DUM THEN DUM = SUMX / SUM - A
    AA = XM + DUM * AMP                    a new best guess
    IF AA < AM THEN AA = .5 * (A + AM)     (but no smaller than the lower bound)
    EPS = AFAC * ABS(AA - A)               and a new smoothing factor.
    A = AA
    GOTO 1
  ELSE                                     Got it!
    IF N MOD 2 = 0 THEN                    For even N median is always an average.
      IF NP = NM THEN
        XMED = .5 * (XP + XM)
      ELSEIF NP > NM THEN
        XMED = .5 * (A + XP)
      ELSE
        XMED = .5 * (XM + A)
      END IF
    ELSE                                   For odd N median is always one point.
      IF NP = NM THEN
        XMED = A
      ELSEIF NP > NM THEN
        XMED = XP
      ELSE
        XMED = XM
      END IF
    END IF
  END IF
END IF
END SUB
```

A sample program using **MDIAN2** is the following:

```
DECLARE SUB MDIAN2 (X!(), N!, XMED!)
DECLARE SUB MDIAN1 (X!(), N!, XMED!)
DECLARE FUNCTION GASDEV! (IDUM&)

'PROGRAM D13R3
'Driver for routine MDIAN2
CLS
NPTS = 50
DIM DATQ(NPTS)
IDUM& = -5
FOR I = 1 TO NPTS
  DATQ(I) = GASDEV(IDUM&)
NEXT I
CALL MDIAN2(DATQ(), NPTS, XMED)
PRINT "Gaussian distrib., zero mean, unit variance"
PRINT
```

```
PRINT "Median according to MDIAN2 is";
PRINT USING "#####.######"; XMED
CALL MDIAN1(DATQ(), NPTS, XMED)
PRINT "Median according to MDIAN1 is";
PRINT USING "#####.######"; XMED
END
```

Student's *t*-test is a test of two data sets for significantly different means. It is applied by D13R4 to two Gaussian data sets DATA1 and DATA2 that are generated by GASDEV. DATA2 is originally given an artificial shift of its mean to the right of that of DATA1, by NSHFT/2 units of EPS. Then DATA1 is successively shifted NSHFT times to the right by EPS and compared to DATA2 by TTEST. At about step NSHFT/2, the two distributions should superpose and indicate populations with the same mean. Notice that the two populations have the same variance (i.e. 1.0), as required by TTEST.

Here is the recipe TTEST:

```
DECLARE SUB AVEVAR (DATQ!(), N!, AVE!, VAR!)
DECLARE FUNCTION BETAI! (A!, B!, X!)
```

```
SUB TTEST (DATA1(), N1, DATA2(), N2, T, PROB)
```
> Given the arrays DATA1 of length N1 and DATA2 of length N2, this routine returns Student's *t* as T, and its significance as PROB, small values of PROB indicating that the arrays have significantly different means. The data arrays are assumed to be drawn from populations with the same true variance.

```
CALL AVEVAR(DATA1(), N1, AVE1, VAR1)
CALL AVEVAR(DATA2(), N2, AVE2, VAR2)
DF = N1 + N2 - 2                                     Degrees of freedom.
VAR = ((N1 - 1) * VAR1 + (N2 - 1) * VAR2) / DF       Pooled variance.
T = (AVE1 - AVE2) / SQR(VAR * (1! / N1 + 1! / N2))
PROB = BETAI(.5 * DF, .5, DF / (DF + T ^ 2))         See equation (6.3.9).
END SUB
```

A sample program using TTEST is the following:

```
DECLARE SUB TTEST (DATA1!(), N1!, DATA2!(), N2!, T!, PROB!)
DECLARE FUNCTION GASDEV! (IDUM&)

'PROGRAM D13R4
'Driver for routine TTEST
CLS
NPTS = 1024
MPTS = 512
EPS = .02
NSHFT = 10
DIM DATA1(NPTS), DATA2(MPTS)
'Generate Gaussian distributed data
IDUM& = -5
FOR I = 1 TO NPTS
  DATA1(I) = GASDEV(IDUM&)
NEXT I
IDUM& = -11
FOR I = 1 TO MPTS
  DATA2(I) = INT(NSHFT / 2!) * EPS + GASDEV(IDUM&)
```

```
NEXT I
PRINT "        Shift        T        Probability"
FOR I = 1 TO NSHFT + 1
  CALL TTEST(DATA1(), NPTS, DATA2(), MPTS, T, PROB)
  SHIFT = (I - 1) * EPS
  PRINT USING "#########.##"; SHIFT; T; PROB
  FOR J = 1 TO NPTS
    DATA1(J) = DATA1(J) + EPS
  NEXT J
NEXT I
END
```

AVEVAR is an auxiliary routine for **TTEST**. It finds the average and variance of a data set. Sample program **D13R5** generates a series of Gaussian distributions for I=1,..,11, and gives each a shift of $(I-1)$**EPS** and a variance of I^2. This progression allows you easily to check the operation of **AVEVAR** "by eye".

Here is the recipe **AVEVAR**:

```
SUB AVEVAR (DATQ(), N, AVE, VAR)
    Given array DATQ of length N, returns its mean as AVE and its variance as VAR.
AVE = 0!
VAR = 0!
FOR J = 1 TO N
  AVE = AVE + DATQ(J)
NEXT J
AVE = AVE / N
FOR J = 1 TO N
  S = DATQ(J) - AVE
  VAR = VAR + S * S
NEXT J
VAR = VAR / (N - 1)
END SUB
```

A sample program using **AVEVAR** is the following:

```
DECLARE SUB AVEVAR (DATQ!(), N!, AVE!, VAR!)
DECLARE FUNCTION GASDEV! (IDUM&)

'PROGRAM D13R5
'Driver for routine AVEVAR
CLS
NPTS = 1000
EPS = .1
DIM DATQ(NPTS)
'Generate Gaussian distributed data
IDUM& = -5
PRINT "        Shift        Average        Variance"
FOR I = 1 TO 11
  SHIFT = (I - 1) * EPS
  FOR J = 1 TO NPTS
    DATQ(J) = SHIFT + I * GASDEV(IDUM&)
  NEXT J
  CALL AVEVAR(DATQ(), NPTS, AVE, VAR)
```

```
    PRINT USING "#########.##"; SHIFT; AVE; VAR
NEXT I
END
```

TUTEST also does Student's *t*-test, but applies to the comparison of means of two distributions with different variance. The example D13R6A employs the comparison used on TTEST but gives the two distributions DATA1 and DATA2 variances of 1.0 and 4.0 respectively.

Here is the recipe TUTEST:

```
DECLARE SUB AVEVAR (DATQ!(), N!, AVE!, VAR!)
DECLARE FUNCTION BETAI! (A!, B!, X!)

SUB TUTEST (DATA1(), N1, DATA2(), N2, T, PROB)
          Given the arrays DATA1 of length N1 and DATA2 of length N2, this routine returns Student's
          t as T, and its significance as PROB, small values of PROB indicating that the arrays have
          significantly different means. The data arrays are allowed to be drawn from populations with
          unequal variances.
CALL AVEVAR(DATA1(), N1, AVE1, VAR1)
CALL AVEVAR(DATA2(), N2, AVE2, VAR2)
T = (AVE1 - AVE2) / SQR(VAR1 / N1 + VAR2 / N2)
DUM = (VAR1 / N1 + VAR2 / N2) ^ 2
DF = DUM / ((VAR1 / N1) ^ 2 / (N1 - 1) + (VAR2 / N2) ^ 2 / (N2 - 1))
PROB = BETAI(.5 * DF, .5, DF / (DF + T ^ 2))
END SUB
```

A sample program using TUTEST is the following:

```
DECLARE SUB TUTEST (DATA1!(), N1!, DATA2!(), N2!, T!, PROB!)
DECLARE FUNCTION GASDEV! (IDUM&)

'PROGRAM D13R6A
'Driver for routine TUTEST
CLS
NPTS = 5000
MPTS = 1000
EPS = .02
VAR1 = 1!
VAR2 = 4!
NSHFT = 10
DIM DATA1(NPTS), DATA2(MPTS)
'Generate two Gaussian distributions of different variance
IDUM& = -51773
FCTR1 = SQR(VAR1)
FOR I = 1 TO NPTS
  DATA1(I) = FCTR1 * GASDEV(IDUM&)
NEXT I
FCTR2 = SQR(VAR2)
FOR I = 1 TO MPTS
  DATA2(I) = (NSHFT / 2!) * EPS + FCTR2 * GASDEV(IDUM&)
NEXT I
PRINT "Distribution #1 : variance = ";
PRINT USING "###.##"; VAR1
PRINT "Distribution #2 : variance = ";
```

```
PRINT USING "###.##"; VAR2
PRINT
PRINT "        Shift        T        Probability"
FOR I = 1 TO NSHFT + 1
  CALL TUTEST(DATA1(), NPTS, DATA2(), MPTS, T, PROB)
  SHIFT = (I - 1) * EPS
  PRINT USING "#########.##"; SHIFT; T; PROB
  FOR J = 1 TO NPTS
    DATA1(J) = DATA1(J) + EPS
  NEXT J
NEXT I
END
```

TPTEST goes a step further, and compares two distributions not only having different variances, but also perhaps having point by point correlations. The example D13R6B creates two situations, one with correlated and one with uncorrelated distributions. It does this by way of three data sets. DATA1 is a simple Gaussian distribution of zero mean and unit variance. DATA2 is DATA1 plus some additional Gaussian fluctuations of smaller amplitude. DATA3 is similar to DATA2 but generated with independent calls to GASDEV so that its fluctuations ought not to have any correlation with those of DATA1. DATA1 is then given an offset with respect to the others and they are successively shifted as in previous routines. At each step of the shift TPTEST was applied. Our results are given below:

Shift	Correlated: T	Probability	Uncorrelated: T	Probability
.01	2.9264	0.0036	0.6028	0.5469
.02	2.1948	0.0286	0.4521	0.6514
.03	1.4632	0.1440	0.3014	0.7632
.04	0.7316	0.4647	0.1507	0.8802
.05	0.0000	1.0000	0.0000	1.0000
.06	-0.7316	0.4647	-0.1507	0.8802
.07	-1.4632	0.1440	-0.3014	0.7632
.08	-2.1948	0.0286	-0.4521	0.6514
.09	-2.9264	0.0036	-0.6028	0.5469
.10	-3.6580	0.0003	-0.7536	0.4514
.11	-4.3896	0.0000	-0.9043	0.3663

Here is the recipe TPTEST:

```
DECLARE SUB AVEVAR (DATQ!(), N!, AVE!, VAR!)
DECLARE FUNCTION BETAI! (A!, B!, X!)

SUB TPTEST (DATA1(), DATA2(), N, T, PROB)
    Given the paired arrays DATA1 and DATA2, both of length N, this routine returns Student's t
    for paired data as T, and its significance as PROB, small values of PROB indicating a significant
    difference of means.
CALL AVEVAR(DATA1(), N, AVE1, VAR1)
CALL AVEVAR(DATA2(), N, AVE2, VAR2)
COV = 0!
FOR J = 1 TO N
  COV = COV + (DATA1(J) - AVE1) * (DATA2(J) - AVE2)
NEXT J
DF = N - 1
COV = COV / DF
SD = SQR((VAR1 + VAR2 - 2! * COV) / CSNG(N))
```

```
T = (AVE1 - AVE2) / SD
PROB = BETAI(.5 * DF, .5, DF / (DF + T ^ 2))
END SUB
```

A sample program using **TPTEST** is the following:

```
DECLARE SUB TPTEST (DATA1!(), DATA2!(), N!, T!, PROB!)
DECLARE SUB AVEVAR (DATQ!(), N!, AVE!, VAR!)
DECLARE FUNCTION GASDEV! (IDUM&)

'PROGRAM D13R6B
'Driver for routine TPTEST
'Compare two correlated distributions vs. two
'uncorrelated distributions
CLS
NPTS = 500
EPS = .01
NSHFT = 10
ANOISE = .3
DIM DATA1(NPTS), DATA2(NPTS), DATA3(NPTS)
IDUM& = -5
PRINT "                Correlated:               Uncorrelated:"
PRINT " Shift      T       Probability      T       Probability"
OFFSET = (NSHFT / 2) * EPS
FOR J = 1 TO NPTS
  GAUSS = GASDEV(IDUM&)
  DATA1(J) = GAUSS
  DATA2(J) = GAUSS + ANOISE * GASDEV(IDUM&)
  DATA3(J) = GASDEV(IDUM&) + ANOISE * GASDEV(IDUM&)
NEXT J
CALL AVEVAR(DATA1(), NPTS, AVE1, VAR1)
CALL AVEVAR(DATA2(), NPTS, AVE2, VAR2)
CALL AVEVAR(DATA3(), NPTS, AVE3, VAR3)
FOR J = 1 TO NPTS
  DATA1(J) = DATA1(J) - AVE1 + OFFSET
  DATA2(J) = DATA2(J) - AVE2
  DATA3(J) = DATA3(J) - AVE3
NEXT J
FOR I = 1 TO NSHFT + 1
  SHIFT = I * EPS
  FOR J = 1 TO NPTS
    DATA2(J) = DATA2(J) + EPS
    DATA3(J) = DATA3(J) + EPS
  NEXT J
  CALL TPTEST(DATA1(), DATA2(), NPTS, T1, PROB1)
  CALL TPTEST(DATA1(), DATA3(), NPTS, T2, PROB2)
  PRINT USING "###.##"; SHIFT;
  PRINT USING "#######.####"; T1; PROB1;
  PRINT "      ";
  PRINT USING "#######.####"; T2; PROB2
NEXT I
END
```

The *F*-test (subroutine FTEST) is a test for differing variances between two distributions. For demonstration purposes, sample program D13R7 generates two distributions DATA1 and DATA2 having Gaussian distributions of unit variance. The values of a third array DATA3 are then set by multiplying DATA2 by a series of values FACTR which takes its variance from 1.0 to 1.1 in ten equal steps. The effect of this on the *F*-test can be evaluated from the probabilities PROB.

Here is the recipe FTEST:

```
DECLARE SUB AVEVAR (DATQ!(), N!, AVE!, VAR!)
DECLARE FUNCTION BETAI! (A!, B!, X!)

SUB FTEST (DATA1(), N1, DATA2(), N2, F, PROB)
        Given the arrays DATA1 of length N1 and DATA2 of length N2, this routine returns the value
        of F, and its significance as PROB. Small values of PROB indicate that the two arrays have
        significantly different variances.
CALL AVEVAR(DATA1(), N1, AVE1, VAR1)
CALL AVEVAR(DATA2(), N2, AVE2, VAR2)
IF VAR1 > VAR2 THEN              Make F the ratio of the larger variance to the smaller one.
  F = VAR1 / VAR2
  DF1 = N1 - 1
  DF2 = N2 - 1
ELSE
  F = VAR2 / VAR1
  DF1 = N2 - 1
  DF2 = N1 - 1
END IF
PROB = 2! * BETAI(.5 * DF2, .5 * DF1, DF2 / (DF2 + DF1 * F))
IF PROB > 1! THEN PROB = 2! - PROB
END SUB
```

A sample program using FTEST is the following:

```
DECLARE SUB FTEST (DATA1!(), N1!, DATA2!(), N2!, F!, PROB!)
DECLARE FUNCTION GASDEV! (IDUM&)

'PROGRAM D13R7
'Driver for routine FTEST
CLS
NPTS = 1000
MPTS = 500
EPS = .01
NVAL = 10
DIM DATA1(NPTS), DATA2(MPTS), DATA3(MPTS)
'Generate two Gaussian distributions with
'different variances
IDUM& = -13
FOR J = 1 TO NPTS
  DATA1(J) = GASDEV(IDUM&)
NEXT J
FOR J = 1 TO MPTS
  DATA2(J) = GASDEV(IDUM&)
NEXT J
PRINT "   Variance 1 =   1.00"
PRINT "   Variance 2      Ratio       Probability"
```

```
FOR I = 1 TO NVAL + 1
  VAR = 1! + (I - 1) * EPS
  FACTOR = SQR(VAR)
  FOR J = 1 TO MPTS
    DATA3(J) = FACTOR * DATA2(J)
  NEXT J
  CALL FTEST(DATA1(), NPTS, DATA3(), MPTS, F, PROB)
  PRINT USING "######.####"; VAR;
  PRINT USING "#########.####"; F; PROB
NEXT I
END
```

CHSONE and CHSTWO compare two distributions on the basis of a χ^2 test to see if they are different. CHSONE, specifically, compares a data distribution to an expected distribution. Sample program D13R8 generates an exponential distribution BINS(I) of data using routine EXPDEV. It then creates an array EBINS(I) which is the expected result (a smooth exponential decay in the absence of statistical fluctuations). EBINS and BINS are compared by CHSONE to give χ^2 and a probability that they represent the same distribution.

Here is the recipe CHSONE:

```
DECLARE FUNCTION GAMMQ! (A!, X!)

SUB CHSONE (BINS(), EBINS(), NBINS, KNSTRN, DF, CHSQ, PROB)
```
> Given the array BINS of length NBINS, containing the observed numbers of events, and an array EBINS of length NBINS containing the expected numbers of events, and given the number of constraints KNSTRN (normally zero), this routine returns (trivially) the number of degrees of freedom DF, and (nontrivially) the chi-square CHSQ and the significance PROB. A small value of PROB indicates a significant difference between the distributions BINS and EBINS. Note that BINS and EBINS are both real arrays, although BINS will normally contain integer values.
```
DF = NBINS - 1 - KNSTRN
CHSQ = 0!
FOR J = 1 TO NBINS
  IF EBINS(J) <= 0! THEN PRINT "bad expected number": EXIT SUB
  CHSQ = CHSQ + (BINS(J) - EBINS(J)) ^ 2 / EBINS(J)
NEXT J
PROB = GAMMQ(.5 * DF, .5 * CHSQ)          Chi-square probability function. See §6.2.
END SUB
```

A sample program using CHSONE is the following:

```
DECLARE SUB CHSONE (BINS!(), EBINS!(), NBINS!, KNSTRN!, DF!, CHSQ!, PROB!)
DECLARE FUNCTION EXPDEV! (IDUM&)

'PROGRAM D13R8
'Driver for routine CHSONE
CLS
NBINS = 10
NPTS = 2000
DIM BINS(NBINS), EBINS(NBINS)
IDUM& = -15
FOR J = 1 TO NBINS
  BINS(J) = 0!
NEXT J
```

```
FOR I = 1 TO NPTS
  X = EXPDEV(IDUM&)
  IBIN = INT(X * NBINS / 3!) + 1
  IF IBIN <= NBINS THEN BINS(IBIN) = BINS(IBIN) + 1!
NEXT I
FOR I = 1 TO NBINS
  EBINS(I) = 3! * NPTS / NBINS * EXP(-3! * (I - .5) / NBINS)
NEXT I
CALL CHSONE(BINS(), EBINS(), NBINS, -1, DF, CHSQ, PROB)
PRINT "         Expected         Observed"
FOR I = 1 TO NBINS
  PRINT USING "############.##"; EBINS(I); BINS(I)
NEXT I
PRINT
PRINT "        Chi-squared:    ";
PRINT USING ".####^^^^"; CHSQ
PRINT "        Probability:    ";
PRINT USING ".####^^^^"; PROB
END
```

CHSTWO compares two binned distributions BINS1 and BINS2, again using a χ^2 test. Sample program D13R9 prepares these distributions both in the same way. Each is composed of 2000 random numbers, drawn from an exponential deviate, and placed into 10 bins. The two data sets are then analyzed by CHSTWO to calculate χ^2 and probability PROB.

Here is the recipe CHSTWO:

```
DECLARE FUNCTION GAMMQ! (A!, X!)

SUB CHSTWO (BINS1(), BINS2(), NBINS, KNSTRN, DF, CHSQ, PROB)
        Given the arrays BINS1 and BINS2, both of length NBINS, containing two sets of binned
        data, and given the number of additional constraints KNSTRN (normally 0 or −1), this routine
        returns the number of degrees of freedom DF, the chi-square CHSQ and the significance PROB.
        A small value of PROB indicates a significant difference between the distributions BINS1 and
        BINS2. Note that BINS1 and BINS2 are both real arrays, although they will normally contain
        integer values.
DF = NBINS - 1 - KNSTRN
CHSQ = 0!
FOR J = 1 TO NBINS
  IF BINS1(J) = 0! AND BINS2(J) = 0! THEN
    DF = DF - 1!                         No data means one less degree of freedom.
  ELSE
    CHSQ = CHSQ + (BINS1(J) - BINS2(J)) ^ 2 / (BINS1(J) + BINS2(J))
  END IF
NEXT J
PROB = GAMMQ(.5 * DF, .5 * CHSQ)   Chi-square probability function. See §6.2.
END SUB
```

A sample program using **CHSTWO** is the following:

```
DECLARE SUB CHSTWO (BINS1!(), BINS2!(), NBINS!, KNSTRN!, DF!, CHSQ!, PROB!)
DECLARE FUNCTION EXPDEV! (IDUM&)

'PROGRAM D13R9
'Driver for routine CHSTWO
CLS
NBINS = 10
NPTS = 2000
DIM BINS1(NBINS), BINS2(NBINS)
IDUM& = -17
FOR J = 1 TO NBINS
  BINS1(J) = 0!
  BINS2(J) = 0!
NEXT J
FOR I = 1 TO NPTS
  X = EXPDEV(IDUM&)
  IBIN = INT(X * NBINS / 3!) + 1
  IF IBIN <= NBINS THEN BINS1(IBIN) = BINS1(IBIN) + 1!
  X = EXPDEV(IDUM&)
  IBIN = INT(X * NBINS / 3!) + 1
  IF IBIN <= NBINS THEN BINS2(IBIN) = BINS2(IBIN) + 1!
NEXT I
CALL CHSTWO(BINS1(), BINS2(), NBINS, -1, DF, CHSQ, PROB)
PRINT "        Dataset 1      Dataset 2"
FOR I = 1 TO NBINS
  PRINT USING "###########.##"; BINS1(I); BINS2(I)
NEXT I
PRINT
PRINT "       Chi-squared:    ";
PRINT USING ".####^^^^"; CHSQ
PRINT "       Probability:    ";
PRINT USING ".####^^^^"; PROB
END
```

The Kolmogorov-Smirnov test used in **KSONE** and **KSTWO** applies to unbinned distributions with a single independent variable. **KSONE** uses the K-S criterion to compare a single data set to an expected distribution, and **KSTWO** uses it to compare two data sets. Sample program **D13R10** creates data sets with Gaussian distributions and with stepwise increasing variance, and compares their cumulative distribution function to the expected result for a Gaussian distribution of unit variance. This result is the error function and is generated by routine **ERF**. Increasing variance in the test destribution should reduce the likelihood that it was drawn from the same distribution represented by the comparison function.

Here is the recipe **KSONE**:

```
DECLARE SUB SORT (N!, RA!())
DECLARE FUNCTION ERF! (Y!)
DECLARE FUNCTION PROBKS! (ALAM!)
DECLARE FUNCTION FUNC! (X!)
```

```
SUB KSONE (DATQ(), N, DUM, D, PROB)
```
Given an array of N values, DATQ, and given a user-supplied function of a single variable FUNC which is a cumulative distribution function ranging from 0 (for smallest values of its argument) to 1 (for largest values of its argument), this routine returns the K–S statistic D, and the significance level PROB. Small values of PROB show that the cumulative distribution function of DATQ is significantly different from FUNC. The array DATQ is modified by being sorted into ascending order.

```
CALL SORT(N, DATQ())                      If the data are already sorted into ascending order, then this
EN = N                                    call can be omitted.
D = 0!
FO = 0!                                   Data's c.d.f. before the next step.
FOR J = 1 TO N                            Loop over the sorted data points.
  FQ = J / EN                             Data's c.d.f. after this step.
  FF = FUNC(DATQ(J))                      Compare to the user-supplied function.
  DT = ABS(FO - FF)                       Maximum distance.
  IF ABS(FQ - FF) > DT THEN DT = ABS(FQ - FF)
  IF DT > D THEN D = DT
  FO = FQ
NEXT J
PROB = PROBKS(SQR(EN) * D)                Compute significance.
END SUB
```

A sample program using KSONE is the following:

```
DECLARE FUNCTION FUNC! (X!)
DECLARE FUNCTION ERF! (X!)
DECLARE SUB KSONE (DATQ!(), N!, DUM!, D!, PROB!)
DECLARE FUNCTION GASDEV! (IDUM&)

'PROGRAM D13R10
'Driver for routine KSONE
CLS
NPTS = 1000
EPS = .1
DIM DATQ(NPTS)
IDUM& = -5
PRINT "   Variance Ratio      K-S Statistic        Probability"
PRINT
FOR I = 1 TO 11
  VAR = 1! + (I - 1) * EPS
  FACTR = SQR(VAR)
  FOR J = 1 TO NPTS
    DATQ(J) = FACTR * ABS(GASDEV(IDUM&))
  NEXT J
  CALL KSONE(DATQ(), NPTS, DUM, D, PROB)
  PRINT USING "#######.######"; VAR;
  PRINT USING "###########.######"; D;
  PRINT "           ";
  PRINT USING ".####^^^^"; PROB
NEXT I
```

```
END

FUNCTION FUNC (X)
Y = X / SQR(2!)
FUNC = ERF(Y)
END FUNCTION
```

KSTWO compares the cumulative distribution functions of two unbinned data sets, DATA1 and DATA2. In sample program D13R11, they are both Gaussian distributions, but DATA2 is given a stepwise increase of variance. In other respects, D13R11 is like D13R10.

Here is the recipe KSTWO:


```
DECLARE SUB SORT (N!, RA!())
DECLARE FUNCTION PROBKS! (ALAM!)

SUB KSTWO (DATA1(), N1, DATA2(), N2, D, PROB)
```
              Given an array DATA1 of N1 values, and an array DATA2 of N2 values, this routine returns the
              K–S statistic D, and the significance level PROB for the null hypothesis that the data sets are
              drawn from the same distribution. Small values of PROB show that the cumulative distribution
              function of DATA1 is significantly different from that of DATA2. The arrays DATA1 and DATA2
              are modified by being sorted into ascending order.
```
CALL SORT(N1, DATA1())
CALL SORT(N2, DATA2())
EN1 = N1
EN2 = N2
J1 = 1                            Next value of DATA1 to be processed.
J2 = 1                            Ditto, DATA2.
FQ1 = 0!
FQ2 = 0!
D = 0!
WHILE J1 <= N1 AND J2 <= N2       If we are not done...
  D1 = DATA1(J1)
  D2 = DATA2(J2)
  IF D1 <= D2 THEN                Next step is in DATA1.
    FQ1 = J1 / EN1
    J1 = J1 + 1
  END IF
  IF D2 <= D1 THEN                Next step is in DATA2.
    FQ2 = J2 / EN2
    J2 = J2 + 1
  END IF
  DT = ABS(FQ2 - FQ1)
  IF DT > D THEN D = DT
WEND
PROB = PROBKS(SQR(EN1 * EN2 / (EN1 + EN2)) * D)        Compute significance.
END SUB
```

A sample program using KSTWO is the following:

```
DECLARE SUB KSTWO (DATA1! (), N1!, DATA2! (), N2!, D!, PROB!)
DECLARE FUNCTION GASDEV! (IDUM&)

'PROGRAM D13R11
'Driver for routine KSTWO
CLS
N1 = 2000
N2 = 1000
EPS = .1
DIM DATA1(N1), DATA2(N2)
IDUM& = -1357
FOR J = 1 TO N1
  DATA1(J) = GASDEV(IDUM&)
NEXT J
PRINT "     Variance Ratio      K-S Statistic      Probability"
PRINT
IDUM& = -2468
FOR I = 1 TO 11
  VAR = 1! + (I - 1) * EPS
  FACTR = SQR(VAR)
  FOR J = 1 TO N2
    DATA2(J) = FACTR * GASDEV(IDUM&)
  NEXT J
  CALL KSTWO(DATA1(), N1, DATA2(), N2, D, PROB)
  PRINT USING "########.######"; VAR;
  PRINT USING "############.######"; D;
  PRINT "          ";
  PRINT USING ".####^^^^"; PROB
NEXT I
END
```

PROBKS is an auxiliary routine for KSONE and KSTWO which calculates the function $Q_{ks}(\lambda)$ used to evaluate the probability that the two distributions being compared are the same. There is no independent means of producing this function, so in sample program D13R12 we have chosen simply to graph it. Our output is reproduced below.

Here is the recipe PROBKS:

```
DECLARE FUNCTION PROBKS! (ALAM!)

FUNCTION PROBKS (ALAM)
EPS1 = .001
EPS2 = 1E-08
A2 = -2! * ALAM * ALAM
FAC = 2!
DUM = 0!
TERMBF = 0!                    Previous term in sum.
FOR J = 1 TO 100
  TERM = FAC * EXP(A2 * J * J)
  DUM = DUM + TERM
  IF ABS(TERM) <= EPS1 * TERMBF OR ABS(TERM) <= EPS2 * DUM THEN
    PROBKS = DUM
```

```
   EXIT FUNCTION
   END IF
   FAC = -FAC                          Alternating signs in sum.
   TERMBF = ABS(TERM)
 NEXT J
 PROBKS = 1!                           Get here only by failing to converge.
 END FUNCTION
```

A sample program using PROBKS is the following:

```
DECLARE FUNCTION PROBKS! (ALAM!)

'PROGRAM D13R12
'Driver for routine PROBKS
CLS
DIM TEXT$(50)
PRINT "Probability func. for Kolmogorov-Smirnov statistic"
PRINT
PRINT " Lambda:      Value:       Graph:"
NPTS = 20
EPS = .1
SCALE = 40!
FOR I = 1 TO NPTS
  ALAM = I * EPS
  VALUE = PROBKS(ALAM)
  TEXT$(1) = "*"
  FOR J = 1 TO 50
    IF J <= CINT(SCALE * VALUE) THEN
      TEXT$(J) = "*"
    ELSE
      TEXT$(J) = " "
    END IF
  NEXT J
  PRINT USING "##.######"; ALAM;
  PRINT USING "#####.######"; VALUE;
  PRINT "    ";
  FOR J = 1 TO 50
    PRINT TEXT$(J);
  NEXT J
  PRINT
NEXT I
END
```

```
Probability func. for Kolmogorov-Smirnov statistic
Lambda:    Value:      Graph:
0.100000   1.000000    **************************************************
0.200000   1.000000    **************************************************
0.300000   0.999991    ************************************************
0.400000   0.997192    ********************************************
0.500000   0.963945    *****************************************
0.600000   0.864283    ***********************************
0.700000   0.711235    ****************************
0.800000   0.544142    *********************
0.900000   0.392731    ***************
1.000000   0.270000    ***********
```

```
1.100000     0.177718     *******
1.200000     0.112250     ****
1.300000     0.068092     ***
1.400000     0.039682     **
1.500000     0.022218     *
1.600000     0.011952
1.700000     0.006177
1.800000     0.003068
1.900000     0.001464
2.000000     0.000671
```

Subroutine **CNTAB1** analyzes a two-dimensional contingency table and returns several parameters describing any association between its nominal variables. Sample program **D13R3** supplies a table from a file **TABLE.DAT** which is listed in the Appendix to this chapter. The table shows the rate of certain accidents, tabulated on a monthly basis. These data are listed, as well as their statistical properties, by D13R3. We found the results to be:

Chi-squared	5026.30
Degrees of Freedom	88.00
Probability	.0000
Cramer-V	.0772
Contingency Coeff.	.2134

Here is the recipe CNTAB1:

```
DECLARE FUNCTION GAMMQ! (A!, X!)

SUB CNTAB1 (NN(), NI, NJ, CHISQ, DF, PROB, CRAMRV, CCC)
        Given a two-dimensional contingency table in the form of an integer array NN(I,J), where
        I ranges from 1 to NI, J ranges from 1 to NJ, this routine returns the chi-square CHISQ,
        the number of degrees of freedom DF, the significance level PROB (small values indicating
        a significant association), and two measures of association, Cramer's V (CRAMRV) and the
        contingency coefficient C (CCC).
TINY = 1E-30                          A small number.
DIM SUMI(NI), SUMJ(NJ)
SUM = 0                               Will be total number of events.
NNI = NI                              Number of rows
NNJ = NJ                              and columns.
FOR I = 1 TO NI                       Get the row totals.
  SUMI(I) = 0!
  FOR J = 1 TO NJ
    SUMI(I) = SUMI(I) + NN(I, J)
    SUM = SUM + NN(I, J)
  NEXT J
  IF SUMI(I) = 0! THEN NNI = NNI - 1      Eliminate any zero rows by reducing the number.
NEXT I
FOR J = 1 TO NJ                       Get the column totals.
  SUMJ(J) = 0!
  FOR I = 1 TO NI
    SUMJ(J) = SUMJ(J) + NN(I, J)
  NEXT I
  IF SUMJ(J) = 0! THEN NNJ = NNJ - 1      Eliminate any zero columns.
NEXT J
DF = INT(NNI * NNJ - NNI - NNJ + 1)      Corrected number of degrees of freedom.
CHISQ = 0!
FOR I = 1 TO NI                       Do the chi-square sum.
  FOR J = 1 TO NJ
```

```
    EXPCTD = SUMJ(J) * SUMI(I) / SUM
    CHISQ = CHISQ + (NN(I, J) - EXPCTD) ^ 2 / (EXPCTD + TINY)   Here TINY guar-
    NEXT J                                                      antees that any eliminated row or column will not contribute
NEXT I                                                          to the sum.
PROB = GAMMQ(.5 * DF, .5 * CHISQ)          Chi-square probability function.
NDUM = NNI - 1
IF NNJ - 1 < NDUM THEN NDUM = NNJ - 1
CRAMRV = SQR(CHISQ / (SUM * NDUM))
CCC = SQR(CHISQ / (CHISQ + SUM))
ERASE SUMJ, SUMI
END SUB
```

A sample program using **CNTAB1** is the following:

```
DECLARE SUB CNTAB1 (NN!(), NI!, NJ!, CHISQ!, DF!, PROB!, CRAMRV!, CCC!)

'PROGRAM D13R13
'Driver for routine CNTAB1
'Contingency table in file TABLE.DAT
CLS
NDAT = 9
NMON = 12
DIM NMBR(NDAT, NMON)
DIM FATE$(NDAT), MON$(NMON)
OPEN "TABLE.DAT" FOR INPUT AS #1
LINE INPUT #1, DUM$
LINE INPUT #1, TITLE$
LINE INPUT #1, DUM$
TEXT$ = LEFT$(DUM$, 15)
FOR I = 1 TO 12
  MON$(I) = MID$(DUM$, 5 * I + 11, 5)
NEXT I
LINE INPUT #1, DUM$
FOR I = 1 TO NDAT
  LINE INPUT #1, DUM$
  FATE$(I) = LEFT$(DUM$, 15)
  FOR J = 1 TO 12
    NMBR(I, J) = VAL(MID$(DUM$, 5 * J + 11, 5))
  NEXT J
NEXT I
CLOSE #1
PRINT TITLE$
PRINT
PRINT "                ";
FOR I = 1 TO 12
  PRINT MON$(I);
NEXT I
FOR I = 1 TO NDAT
  PRINT FATE$(I);
  FOR J = 1 TO 12
    PRINT USING "#####"; NMBR(I, J);
  NEXT J
  PRINT
NEXT I
PRINT
```

```
CALL CNTAB1(NMBR(), NDAT, NMON, CHISQ, DF, PROB, CRAMRV, CCC)
PRINT "Chi-squared          ";
PRINT USING "###########.##"; CHISQ
PRINT "Degrees of Freedom    ";
PRINT USING "###########.##"; DF
PRINT "Probability          ";
PRINT USING "#########.####"; PROB
PRINT "Cramer-V             ";
PRINT USING "#########.####"; CRAMRV
PRINT "Contingency Coeff.    ";
PRINT USING "#########.####"; CCC
END
```

The test looks for any association between accidents and the months in which they occur. TABLE.DAT clearly shows some. Drownings, for example, happen mostly in the summer. CNTAB2 carries out a similar analysis on TABLE.DAT but measures associations on the basis of entropy. Sample program D13R14 prints out the following entropies for the table:

Entropy of Table	4.0368
Entropy of x-distribution	1.5781
Entropy of y-distribution	2.4820
Entropy of y given x	2.4588
Entropy of x given y	1.5548
Dependency of y on x	.0094
Dependency of x on y	.0147
Symmetrical dependency	.0114

Here is the recipe CNTAB2:

```
SUB CNTAB2 (NN(), NI, NJ, H, HX, HY, HYGX, HXGY, UYGX, UXGY, UXY)
     Given a two-dimensional contingency table in the form of an integer array NN(I,J), where I
     labels the x variable and ranges from 1 to NI, J labels the y variable and ranges from 1 to
     NJ, this routine returns the entropy H of the whole table, the entropy HX of the x distribution,
     the entropy HY of the y distribution, the entropy HYGX of y given x, the entropy HXGY of x
     given y, the dependency UYGX of y on x (eq. 13.6.15), the dependency UXGY of x on y (eq.
     13.6.16), and the symmetrical dependency UXY (eq. 13.6.17).
TINY = 1E-30                         A small number.
DIM SUMI(NI), SUMJ(NJ)
SUM = 0
FOR I = 1 TO NI                      Get the row totals.
  SUMI(I) = 0!
  FOR J = 1 TO NJ
    SUMI(I) = SUMI(I) + NN(I, J)
    SUM = SUM + NN(I, J)
  NEXT J
NEXT I
FOR J = 1 TO NJ                      Get the column totals.
  SUMJ(J) = 0!
  FOR I = 1 TO NI
    SUMJ(J) = SUMJ(J) + NN(I, J)
  NEXT I
NEXT J
HX = 0!
FOR I = 1 TO NI                      Entropy of the x distribution,
  IF SUMI(I) <> 0! THEN
```

```
      P = SUMI(I) / SUM
      HX = HX - P * LOG(P)
    END IF
  NEXT I
  HY = O!                              and of the y distribution.
  FOR J = 1 TO NJ
    IF SUMJ(J) <> O! THEN
      P = SUMJ(J) / SUM
      HY = HY - P * LOG(P)
    END IF
  NEXT J
  H = O!
  FOR I = 1 TO NI                      Total entropy: loop over both x
    FOR J = 1 TO NJ                    and y.
      IF NN(I, J) <> O THEN
        P = NN(I, J) / SUM
        H = H - P * LOG(P)
      END IF
    NEXT J
  NEXT I
  HYGX = H - HX                        Uses equation (13.6.18),
  HXGY = H - HY                        as does this.
  UYGX = (HY - HYGX) / (HY + TINY)     Equation (13.6.15).
  UXGY = (HX - HXGY) / (HX + TINY)     Equation (13.6.16).
  UXY = 2! * (HX + HY - H) / (HX + HY + TINY)      Equation (13.6.17).
  ERASE SUMJ, SUMI
END SUB
```

A sample program using **CNTAB2** is the following:

```
DECLARE SUB CNTAB2 (NN!(), NI!, NJ!, H!, HX!, HY!, HYGX!, HXGY!, UYGX!, UXGY!, UX

'PROGRAM D13R14
'Driver for routine CNTAB2
'Contingency table in file TABLE.DAT
CLS
NI = 9
NMON = 12
DIM NMBR(NI, NMON)
DIM FATE$(NI), MON$(NMON)
OPEN "TABLE.DAT" FOR INPUT AS #1
LINE INPUT #1, DUM$
LINE INPUT #1, TITLE$
LINE INPUT #1, DUM$
TEXT$ = LEFT$(DUM$, 15)
FOR I = 1 TO 12
  MON$(I) = MID$(DUM$, 5 * I + 11, 5)
NEXT I
LINE INPUT #1, DUM$
FOR I = 1 TO NI
  LINE INPUT #1, DUM$
  FATE$(I) = LEFT$(DUM$, 15)
  FOR J = 1 TO 12
    NMBR(I, J) = VAL(MID$(DUM$, 5 * J + 11, 5))
  NEXT J
```

```
NEXT I
CLOSE #1
PRINT TITLE$
PRINT
PRINT "                    ";
FOR I = 1 TO 12
  PRINT MON$(I);
NEXT I
FOR I = 1 TO NI
  PRINT FATE$(I);
  FOR J = 1 TO 12
    PRINT USING "#####"; NMBR(I, J);
  NEXT J
  PRINT
NEXT I
PRINT
CALL CNTAB2(NMBR(), NI, NMON, H, HX, HY, HYGX, HXGY, UYGX, UXGY, UXY)
PRINT "Entropy of Table         ";
PRINT USING "#######.####"; H
PRINT "Entropy of x-distribution ";
PRINT USING "#######.####"; HX
PRINT "Entropy of y-distribution ";
PRINT USING "#######.####"; HY
PRINT "Entropy of y given x      ";
PRINT USING "#######.####"; HYGX
PRINT "Entropy of x given y      ";
PRINT USING "#######.####"; HXGY
PRINT "Dependency of y on x      ";
PRINT USING "#######.####"; UYGX
PRINT "Dependency of x on y      ";
PRINT USING "#######.####"; UXGY
PRINT "Symmetrical dependency    ";
PRINT USING "#######.####"; UXY
END
```

The dependencies of x on y and y on x indicate the degree to which the type of accident can be predicted by knowing the month, or vice-versa.

PEARSN makes an examination of two ordinal or continuous variables to find linear correlations. It returns a linear correlation coefficient R, a probability of correlation PROB, and Fisher's z. Sample program D13R15 sets up data pairs in arrays DOSE and SPORE which show hypothetical data for the spore count from plants exposed to various levels of γ-rays. The results of applying PEARSN to this data set are compared with the correct results by the program.

Here is the recipe PEARSN:

DECLARE FUNCTION BETAI! (A!, B!, X!)

SUB PEARSN (X(), Y(), N, R, PROB, Z)

> Given two arrays X and Y of length N, this routine computes their correlation coefficient r (returned as R), the significance level at which the null hypothesis of zero correlation is disproved

(PROB whose small value indicates a significant correlation), and Fisher's z (returned as Z), whose value can be used in further statistical tests as described above.

```
TINY = 1E-20                          Will regularize the unusual case of complete correlation.
AX = 0!
AY = 0!
FOR J = 1 TO N                        Find the means.
  AX = AX + X(J)
  AY = AY + Y(J)
NEXT J
AX = AX / N
AY = AY / N
SXX = 0!
SYY = 0!
SXY = 0!
FOR J = 1 TO N                        Compute the correlation coefficient.
  XT = X(J) - AX
  YT = Y(J) - AY
  SXX = SXX + XT ^ 2
  SYY = SYY + YT ^ 2
  SXY = SXY + XT * YT
NEXT J
R = SXY / SQR(SXX * SYY)
Z = .5 * LOG(((1! + R) + TINY) / ((1! - R) + TINY))    Fisher's z transformation.
DF = N - 2
T = R * SQR(DF / (((1! - R) + TINY) * ((1! + R) + TINY)))    Equation (13.7.5).
PROB = BETAI(.5 * DF, .5, DF / (DF + T ^ 2))    Student's t probability.
'PROB = ERFCC(ABS(Z * SQR(N - 1!)) / 1.4142136#)    For large N, this easier computation of
END SUB                                             PROB, using the short routine ERFCC,
                                                    would give approximately the same value.
```

A sample program using PEARSN is the following:

```
DECLARE SUB PEARSN (X!(), Y!(), N!, R!, PROB!, Z!)

'PROGRAM D13R15
'Driver for routine PEARSN
CLS
DIM DOSE(10), SPORE(10)
FOR I = 1 TO 10
  READ DOSE(I)
NEXT I
DATA 56.1,64.1,70.0,66.6,82.,91.3,90.,99.7,115.3,110.
FOR I = 1 TO 10
  READ SPORE(I)
NEXT I
DATA 0.11,0.4,0.37,0.48,0.75,0.66,0.71,1.2,1.01,0.95
PRINT "Effect of Gamma Rays on Man-in-the-Moon Marigolds"
PRINT "Count Rate (cpm)          Pollen Index"
FOR I = 1 TO 10
  PRINT USING "#######.##"; DOSE(I);
  PRINT "                  ";
  PRINT USING "#######.##"; SPORE(I)
NEXT I
PRINT
CALL PEARSN(DOSE(), SPORE(), 10, R, PROB, Z)
```

```
PRINT "                    PEARSN          Expected"
PRINT "Corr. Coeff.        ";
PRINT USING ".######^^^^"; R;
PRINT "    ";
PRINT USING ".######^^^^"; .906959
PRINT "Probability         ";
PRINT USING ".######^^^^"; PROB;
PRINT "    ";
PRINT USING ".######^^^^"; 2.9265E-04
PRINT "Fisher's Z          ";
PRINT USING ".######^^^^"; Z;
PRINT "    ";
PRINT USING ".######^^^^"; 1.51011
END
```

Rank order correlation may be done with **SPEAR** to compare two distributions **DATA1** and **DATA2** for correlation. Correlations are reported both in terms of **D**, the sum-squared difference in ranks, and **RS**, Spearman's rank correlation parameter. Sample program **D13R16** applies the calculation to the data in table **TABLE2.DAT** (see Appendix) which shows the solar flux incident on various cities during different months of the year. It then checks for correlations between columns of the table, considering each column as a separate data set. In this fashion it looks for correlations between the July solar flux and that of other months. The probability of such correlations are shown by **PROBD** and **PROBRS**. Our results are:

```
Correlation of sampled U.S. solar radiation (July with other months)
Month      D       St. Dev.    PROBD      Spearman R   PROBRS
 jul      .00     -4.358899   .000013     .993965     .000000
 aug    122.00    -3.958458   .000075     .901959     .000000
 sep    218.00    -3.643896   .000269     .832831     .000005
 oct    384.00    -3.098495   .001945     .704372     .000526
 nov    390.50    -3.077642   .002086     .701205     .000572
 dec    622.00    -2.318075   .020445     .526751     .017022
 jan    644.50    -2.244251   .024816     .509796     .021662
 feb    483.50    -2.772503   .005563     .631122     .002844
 mar    497.00    -2.728208   .006368     .620949     .003480
 apr    405.50    -3.027925   .002462     .688158     .000796
 may    264.00    -3.492371   .000479     .794870     .000028
 jun    121.50    -3.960099   .000075     .902336     .000000
```

Here is the recipe **SPEAR**:

```
DECLARE SUB CRANK (N!, W!(), S!)
DECLARE SUB SORT2 (N!, RA!(), RB!())
DECLARE FUNCTION ERFCC! (X!)
DECLARE FUNCTION BETAI! (A!, B!, X!)
```

```
SUB SPEAR (DATA1(), DATA2(), N, WKSP1(), WKSP2(), D, ZD, PROBD, RS, PROBRS)
```
Given two data arrays, DATA1 and DATA2, each of length N, and given two work spaces of equal length, this routine returns their sum-squared difference of ranks as D, the number of standard deviations by which D deviates from its null-hypothesis expected value as ZD, the two-sided significance level of this deviation as PROBD, Spearman's rank correlation r_s as RS, and the two-sided significance level of its deviation from zero as PROBRS. The work spaces can be identical to the data arrays, but in that case the data arrays are destroyed. The external

routines CRANK (below) and SORT2 (§8.2) are used. A small value of either PROBD or PROBRS indicates a significant correlation (RS positive) or anticorrelation (RS negative).

```
FOR J = 1 TO N
  WKSP1(J) = DATA1(J)
  WKSP2(J) = DATA2(J)
NEXT J
CALL SORT2(N, WKSP1(), WKSP2())          Sort each of the data arrays, and convert the entries to
CALL CRANK(N, WKSP1(), SF)               ranks. The values SF and SG return the sums ∑(f_k^3 − f_k)
CALL SORT2(N, WKSP2(), WKSP1())          and ∑(g_m^3 − g_m) respectively.
CALL CRANK(N, WKSP2(), SG)
D = 0!
FOR J = 1 TO N                           Sum the squared difference of ranks.
  D = D + (WKSP1(J) - WKSP2(J)) ^ 2
NEXT J
EN = INT(N)
EN3N = EN ^ 3 - EN
AVED = EN3N / 6! - (SF + SG) / 12!                      Expectation value of D,
FAC = (1! - SF / EN3N) * (1! - SG / EN3N)
VARD = ((EN - 1!) * EN ^ 2 * (EN + 1!) ^ 2 / 36!) * FAC   and variance of D give
ZD = (D - AVED) / SQR(VARD)                             number of standard deviations,
PROBD = ERFCC(ABS(ZD) / 1.4142136#)                    and significance.
RS = (1! - (6! / EN3N) * (D + (SF + SG) / 12!)) / SQR(FAC)   Rank correlation coef-
FAC = (1! + RS) * (1! - RS)                             ficient,
IF FAC > 0! THEN
  T = RS * SQR((EN - 2!) / FAC)                         and its t value,
  DF = EN - 2!
  PROBRS = BETAI(.5 * DF, .5, DF / (DF + T ^ 2))        give its significance.
ELSE
  PROBRS = 0!
END IF
END SUB
```

A sample program using SPEAR is the following:

```
DECLARE SUB SPEAR (DATA1!(), DATA2!(), N!, WKSP1!(), WKSP2!(), D!, ZD!, PROBD!,
RS!, PROBRS!)


'PROGRAM D13R16
'Driver for routine SPEAR
CLS
NDAT = 20
NMON = 12
DIM DATA1(NDAT), DATA2(NDAT), RAYS(NDAT, NMON)
DIM WKSP1(NDAT), WKSP2(NDAT), AVE(NDAT), ZLAT(NDAT)
DIM CITY$(NDAT), MON$(NMON)
OPEN "TABLE2.DAT" FOR INPUT AS #1
LINE INPUT #1, DUM$
LINE INPUT #1, TITLE$
LINE INPUT #1, DUM$
TEXT$ = LEFT$(DUM$, 15)
FOR I = 1 TO 12
  MON$(I) = MID$(DUM$, 4 * I + 12, 4)
NEXT I
LINE INPUT #1, DUM$
```

```
FOR I = 1 TO NDAT
  LINE INPUT #1, DUM$
  CITY$(I) = LEFT$(DUM$, 15)
  FOR J = 1 TO 12
    RAYS(I, J) = VAL(MID$(DUM$, 4 * J + 12, 4))
  NEXT J
  AVE(I) = VAL(MID$(DUM$, 66, 5))
  ZLAT(I) = VAL(MID$(DUM$, 71, 5))
NEXT I
CLOSE #1
PRINT TITLE$
PRINT "                   ";
FOR I = 1 TO 12
  PRINT MON$(I);
NEXT I
PRINT
FOR I = 1 TO NDAT
  PRINT CITY$(I);
  FOR J = 1 TO 12
    PRINT USING "####"; CINT(RAYS(I, J));
  NEXT J
  PRINT
NEXT I
PRINT
'Check temperature correlations between different months
PRINT "Are sunny summer places also sunny winter places?"
PRINT "Check correlation of sampled U.S. solar radiation ";
PRINT "(july with other months)"
PRINT
PRINT "Month          D     St. Dev.      PROBD      Spearman R     PROBRS"
PRINT
FOR I = 1 TO NDAT
  DATA1(I) = RAYS(I, 1)
NEXT I
FOR J = 1 TO 12
  FOR I = 1 TO NDAT
    DATA2(I) = RAYS(I, J)
  NEXT I
  CALL SPEAR(DATA1(), DATA2(), NDAT, WKSP1(), WKSP2(), D, ZD, PROBD, RS, PROBRS)
  PRINT MON$(J);
  PRINT USING "##########.##"; D;
  PRINT "    ";
  PRINT USING "##.######"; ZD;
  PRINT USING "#####.#####"; PROBD;
  PRINT USING "########.#####"; RS;
  PRINT USING "#####.#####"; PROBRS
NEXT J
END
```

CRANK is an auxiliary routine for SPEAR and is used in conjunction with SORT2. The latter sorts an array, and CRANK then assigns ranks to each data entry, including the midranking of ties. Sample program D13R17 uses the solar flux data of TABLE2.DAT (see Appendix) to illustrate. Each column of the solar flux table is replaced by the rank order of its entries. You can check the rank order chart against the chart of original values to verify the ordering.

Here is the recipe CRANK:

```
SUB CRANK (N, W(), S)
        Given a sorted array W of N elements, replaces the elements by their rank, including midranking
        of ties, and returns as S the sum of f³ − f, where f is the number of elements in each tie.
S = 0!
J = 1
WHILE J < N                          The next rank to be assigned.
  IF W(J + 1) <> W(J) THEN           "DO WHILE" structure.
    W(J) = J                         Not a tie.
    J = J + 1
  ELSE                               A tie:
    FOR JT = J + 1 TO N              How far does it go?
      IF W(JT) <> W(J) THEN GOTO 2
    NEXT JT
    JT = N + 1                       If here, it goes all the way to the last element.
2   RANK = .5 * (J + JT - 1)         This is the mean rank of the tie,
    FOR JI = J TO JT - 1             so enter it into all the tied entries,
      W(JI) = RANK
    NEXT JI
    T = JT - J
    S = S + T ^ 3 - T                and update S.
    J = JT
  END IF
WEND
IF J = N THEN W(N) = N               If the last element was not tied, this is its rank.
END SUB
```

A sample program using CRANK is the following:

```
DECLARE SUB SORT2 (NDAT!, DATQ!(), ORDER!())
DECLARE SUB CRANK (NDAT!, DATQ!(), S!)

'PROGRAM D13R17
'Driver for routine CRANK
CLS
NDAT = 20
NMON = 12
DIM DATQ(NDAT), RAYS(NDAT, NMON)
DIM ORDER(NDAT), AVE(NDAT), ZLAT(NDAT)
DIM CITY$(NDAT), MON$(NMON)
OPEN "TABLE2.DAT" FOR INPUT AS #1
LINE INPUT #1, DUM$
LINE INPUT #1, TITLE$
LINE INPUT #1, DUM$
TEXT$ = LEFT$(DUM$, 15)
FOR I = 1 TO 12
```

```
  MON$(I) = MID$(DUM$, 4 * I + 12, 4)
NEXT I
LINE INPUT #1, DUM$
FOR I = 1 TO NDAT
  LINE INPUT #1, DUM$
  CITY$(I) = LEFT$(DUM$, 15)
  FOR J = 1 TO 12
    RAYS(I, J) = VAL(MID$(DUM$, 4 * J + 12, 4))
  NEXT J
  AVE(I) = VAL(MID$(DUM$, 66, 5))
  ZLAT(I) = VAL(MID$(DUM$, 71, 5))
NEXT I
CLOSE #1
PRINT TITLE$
PRINT "                    ";
FOR I = 1 TO 12
  PRINT MON$(I);
NEXT I
PRINT
FOR I = 1 TO NDAT
  PRINT CITY$(I);
  FOR J = 1 TO 12
    PRINT USING "####"; CINT(RAYS(I, J));
  NEXT J
  PRINT
NEXT I
'Replace solar flux in each column by rank order
DO
  INPUT "Number of month (1-12)"; MONTH
  IF MONTH = 0 THEN EXIT DO
  MONTH = (MONTH + 5) MOD 12 + 1
  FOR I = 1 TO NDAT
    DATQ(I) = RAYS(I, MONTH)
    ORDER(I) = I
  NEXT I
  CALL SORT2(NDAT, DATQ(), ORDER())
  CALL CRANK(NDAT, DATQ(), S)
  PRINT "Month of"; MON$(MONTH)
  PRINT "City                  Rank      Solar Flux    Latitude"
  FOR I = 1 TO NDAT
    NN = CINT(ORDER(I))
    PRINT CITY$(NN);
    PRINT USING "##########.#"; DATQ(I); RAYS(NN, MONTH); ZLAT(NN)
  NEXT I
LOOP
END
```

KENDL1 and KENDL2 test for monotonic correlations of ordinal data. They differ in that KENDL1 compares two data sets of the same rank, while KENDL2 operates on a contingency table. Sample program D13R18, for example, uses KENDL1 to look for pair correlations in our five random number routines. That is to say, it tests for randomness by seeing if two consecutive numbers from the generator have a monotonic correlation. It uses the random number generators RAN0, ..., RAN4, one at a time, to generate 200 pairs of random numbers each. Then KENDL1 tests for correlation of the pairs, and a chart is made showing Kendall's τ, the standard deviation from the null hypotheses, and the probability. For a better test of the generators, you may wish to increase the number of pairs NDAT. It would also be a good idea to see how your result depends on the value of the seed IDUM&.

Here is the recipe KENDL1:

```
DECLARE FUNCTION ERFCC! (X!)

SUB KENDL1 (DATA1(), DATA2(), N, TAU, Z, PROB)
```
> Given data arrays DATA1 and DATA2, each of length N, this program returns Kendall's τ as TAU, its number of standard deviations from zero as Z, and its two-sided significance level as PROB. Small values of PROB indicate a significant correlation (TAU positive) or anticorrelation (TAU negative).

```
N1 = 0                                    This will be the argument of one square root in (13.8.8),
N2 = 0                                    and this the other.
IQ = 0                                    This will be the numerator in (13.8.8).
FOR J = 1 TO N - 1                        Loop over first member of pair,
  FOR K = J + 1 TO N                      and second member.
    A1 = DATA1(J) - DATA1(K)
    A2 = DATA2(J) - DATA2(K)
    AA = A1 * A2
    IF AA <> 0! THEN                      Neither array has a tie.
      N1 = N1 + 1
      N2 = N2 + 1
      IF AA > 0! THEN
        IQ = IQ + 1
      ELSE
        IQ = IQ - 1
      END IF
    ELSE                                  One or both arrays have ties.
      IF A1 <> 0! THEN N1 = N1 + 1        An "extra x" event.
      IF A2 <> 0! THEN N2 = N2 + 1        An "extra y" event.
    END IF
  NEXT K
NEXT J
TAU = CSNG(IQ) / SQR(CSNG(N1) * CSNG(N2))  Equation (13.8.8).
VAR = (4! * N + 10!) / (9! * N * (N - 1!))  Equation (13.8.9).
Z = TAU / SQR(VAR)
PROB = ERFCC(ABS(Z) / 1.4142136#)          Significance.
END SUB
```

A sample program using KENDL1 is the following:

```
DECLARE SUB KENDL1 (DATA1!(), DATA2!(), N!, TAU!, Z!, PROB!)
DECLARE FUNCTION RAN0! (IDUM&)
DECLARE FUNCTION RAN1! (IDUM&)
DECLARE FUNCTION RAN2! (IDUM&)
DECLARE FUNCTION RAN3! (IDUM&)
DECLARE FUNCTION RAN4! (IDUM&)

'PROGRAM D13R18
'Driver for routine KENDL1
'Look for correlations in RAN0 ... RAN4
CLS
```

```
NDAT = 200
DIM DATA1(NDAT), DATA2(NDAT), TEXT$(5)
FOR I = 1 TO 5
  READ TEXT$(I)
NEXT I
DATA "RAN0","RAN1","RAN2","RAN3","RAN4"
PRINT "Pair correlations of RAN0 ... RAN4"
PRINT
PRINT " Program        Kendall Tau     Std. Dev.      Probability"
PRINT
FOR I = 1 TO 5
  IDUM& = -1357
  FOR J = 1 TO NDAT
    IF I = 1 THEN
      DATA1(J) = RAN0(IDUM&)
      DATA2(J) = RAN0(IDUM&)
    ELSEIF I = 2 THEN
      DATA1(J) = RAN1(IDUM&)
      DATA2(J) = RAN1(IDUM&)
    ELSEIF I = 3 THEN
      DATA1(J) = RAN2(IDUM&)
      DATA2(J) = RAN2(IDUM&)
    ELSEIF I = 4 THEN
      DATA1(J) = RAN3(IDUM&)
      DATA2(J) = RAN3(IDUM&)
    ELSEIF I = 5 THEN
      DATA1(J) = RAN4(IDUM&)
      DATA2(J) = RAN4(IDUM&)
    END IF
  NEXT J
  CALL KENDL1(DATA1(), DATA2(), NDAT, TAU, Z, PROB)
  PRINT "   "; TEXT$(I); "   ";
  PRINT USING "########.######"; TAU; Z; PROB
NEXT I
END
```

Sample program D13R19, for subroutine KENDL2, prepares a contingency table based on the routines IRBIT1 and IRBIT2. You may recall that these routines generate random binary sequences. The program checks the sequences by breaking them into groups of three bits. Each group is treated as a three-bit binary number. Two consecutive groups then act as indices into an 8×8 contingency table that records how many times each possible sequence of six bits (two groups) occurs. For each random bit generator, NDAT=1000 samples are taken. Then the contingency table TAQ(K,L) is analyzed by KENDL2 to find Kendall's τ, the standard deviation, and the probability. Notice that Kendall's τ can only be applied when both variables are ordinal (here, the numbers 0 to 7), and that the test is specifically for monotonic correlations. In this case we are actually testing whether the larger 3-bit binary numbers tend to be followed by others of their own kind. Within the program, we have expressed this roughly as a test of whether ones or zeros tend to come in groups more than they should.

Here is the recipe KENDL2:

```
DECLARE FUNCTION ERFCC! (X!)

SUB KENDL2 (TAQ(), I, J, IP, JP, TAU, Z, PROB)
```
Given a two-dimensional table TAQ of physical dimension (IP,JP) and logical dimension (I,J), such that TAQ(K,L) contains the number of events falling in bin K of one variable and bin L of another, this program returns Kendall's τ as TAU, its number of standard deviations from zero as Z, and its two-sided significance level as PROB. Small values of PROB indicate a significant correlation (TAU positive) or anticorrelation (TAU negative) between the two variables. Although TAQ is a floating point array, it will normally contain integral values.

```
EN1 = 0!                              See KENDL1 above.
EN2 = 0!
S = 0!
NN = I * J                            Total number of entries in contingency table.
POINTS = TAQ(I, J)
FOR K = 0 TO NN - 2                   Loop over entries in table,
  KI = INT(K / J)                     decoding a row
  KJ = K - J * KI                     and a column.
  POINTS = POINTS + TAQ(KI + 1, KJ + 1)         Increment the total count of events.
  FOR L = K + 1 TO NN - 1             Loop over other member of the pair,
    LI = INT(L / J)                   decoding its row
    LJ = L - J * LI                   and column.
    M1 = LI - KI
    M2 = LJ - KJ
    MM = M1 * M2
    PAIRS = TAQ(KI + 1, KJ + 1) * TAQ(LI + 1, LJ + 1)
    IF MM <> 0 THEN                   Not a tie.
      EN1 = EN1 + PAIRS
      EN2 = EN2 + PAIRS
      IF MM > 0 THEN                  Concordant, or
        S = S + PAIRS
      ELSE                            discordant.
        S = S - PAIRS
      END IF
    ELSE
      IF M1 <> 0 THEN EN1 = EN1 + PAIRS
      IF M2 <> 0 THEN EN2 = EN2 + PAIRS
    END IF
  NEXT L
NEXT K
TAU = S / SQR(EN1 * EN2)
VAR = (4! * POINTS + 10!) / (9! * POINTS * (POINTS - 1!))
Z = TAU / SQR(VAR)
PROB = ERFCC(ABS(Z) / 1.4142136#)
END SUB
```

A sample program using KENDL2 is the following:

```
DECLARE SUB KENDL2 (TAQ!(), I!, J!, IP!, JP!, TAU!, Z!, PROB!)
DECLARE FUNCTION IRBIT1! (X!)
DECLARE FUNCTION IRBIT2! (X!)

'PROGRAM D13R19
'Driver for routine KENDL2
'Look for "ones-after-zeros" in IRBIT1 and IRBIT2 sequences
CLS
```

```
NDAT = 1000
IP = 8
JP = 8
DIM TAQ(IP, JP), TEXT$(8)
FOR I = 1 TO 8
  READ TEXT$(I)
NEXT I
DATA "000","001","010","011","100","101","110","111"
PRINT "Are ones followed by zeros and vice-versa?"
PRINT
I = IP
J = JP
FOR IFUNC = 1 TO 2
  ISEED = 2468
  PRINT "Test of IRBIT";
  PRINT USING "#"; IFUNC
  PRINT
  FOR K = 1 TO I
    FOR L = 1 TO J
      TAQ(K, L) = 0!
    NEXT L
  NEXT K
  FOR M = 1 TO NDAT
    K = 1
    FOR N = 0 TO 2
      IF IFUNC = 1 THEN
        K = K + IRBIT1(ISEED) * 2 ^ N
      ELSE
        K = K + IRBIT2(ISEED) * 2 ^ N
      END IF
    NEXT N
    L = 1
    FOR N = 0 TO 2
      IF IFUNC = 1 THEN
        L = L + IRBIT1(ISEED) * 2 ^ N
      ELSE
        L = L + IRBIT2(ISEED) * 2 ^ N
      END IF
    NEXT N
    TAQ(K, L) = TAQ(K, L) + 1!
  NEXT M
  CALL KENDL2(TAQ(), I, J, IP, JP, TAU, Z, PROB)
  PRINT "    ";
  FOR N = 1 TO 8
    PRINT "   "; TEXT$(N);
  NEXT N
  PRINT
  PRINT
  FOR N = 1 TO 8
    PRINT TEXT$(N);
    FOR M = 1 TO 8
      PRINT USING "######"; CINT(TAQ(N, M));
    NEXT M
    PRINT
  NEXT N
```

```
  PRINT
  PRINT "       Kendall Tau      Std. Dev.      Probability"
  PRINT USING "########.######"; TAU; Z; PROB
  PRINT
  PRINT "Press RETURN to continue ..."
  LINE INPUT DUM$
  PRINT
NEXT IFUNC
END
```

SMOOFT is a subroutine for smoothing data. This is not a mathematically valuable procedure since it always reduces the information content of the data. However, it is a satisfactory tool for data presentation, as it may help to make evident important features of the data. Sample program D13R20 prepares an artificial data set Y(I) with a broad maximum. It then adds noise from a Gaussian deviate. Subsequently the data are plotted three times; first the original data, then following each of two consecutive applications of SMOOFT. You will notice that the second use of SMOOFT is almost entirely ineffectual, but the first makes a significant change in the presentational quality of the graph.

Here is the recipe SMOOFT:

```
DECLARE SUB REALFT (DATQ!(), N!, ISIGN!)

SUB SMOOFT (Y(), N, PTS)
        Smooths an array Y of length N, with a window whose full width is of order PTS neighboring
        points, a user supplied value. Y is modified.
M = 2
NMIN = N + INT(2! * PTS)        Minimum size of padded array, including buffer against wraparound.
WHILE M < NMIN                  Find the next larger power of 2.
  M = 2 * M
WEND
IF M > 1024 THEN PRINT "M too small": EXIT SUB
CONSQ = (PTS / M) ^ 2           Useful constants below.
Y1 = Y(1)
YN = Y(N)
RN1 = 1! / (N - 1!)
FOR J = 1 TO N                  Remove the linear trend and transfer data.
  Y(J) = Y(J) - RN1 * (Y1 * (N - J) + YN * (J - 1))
NEXT J
FOR J = N + 1 TO M              Zero pad.
  Y(J) = 0!
NEXT J
MO2 = INT(M / 2)
CALL REALFT(Y(), MO2, 1)        Fourier transform.
Y(1) = Y(1) / MO2
FAC = 1!                        Window function.
FOR J = 1 TO MO2 - 1           Multiply the data by the window function.
  K = 2 * J + 1
  IF FAC <> 0! THEN
    FAC = (1! - CONSQ * J ^ 2) / MO2
    IF FAC < 0! THEN FAC = 0!
    Y(K) = FAC * Y(K)
    Y(K + 1) = FAC * Y(K + 1)
```

```
      ELSE                          Don't do unnecessary multiplies after window function is zero.
        Y(K) = 0!
        Y(K + 1) = 0!
      END IF
    NEXT J
    FAC = (1! - .25 * PTS ^ 2) / MO2    Last point.
    IF FAC < 0! THEN FAC = 0!
    Y(2) = FAC * Y(2)
    CALL REALFT(Y(), MO2, -1)          Inverse Fourier transform.
    FOR J = 1 TO N                     Restore the linear trend.
      Y(J) = RN1 * (Y1 * (N - J) + YN * (J - 1)) + Y(J)
    NEXT J
    END SUB
```

A sample program using SMOOFT is the following:

```
DECLARE SUB SMOOFT (Y!(), N!, PTS!)
DECLARE FUNCTION GASDEV! (IDUM&)

'PROGRAM D13R20
'Driver for routine SMOOFT
CLS
MMAX = 1024
N = 100
HASH = .05
SCALE = 100!
PTS = 10!
DIM Y(MMAX), TEXT$(64)
IDUM& = -7
FOR I = 1 TO N
  Y(I) = 3! * I / N * EXP(-3! * I / N)
  Y(I) = Y(I) + HASH * GASDEV(IDUM&)
NEXT I
FOR K = 1 TO 3
  NSTP = INT(N / 20)
  PRINT " Data:        Graph:"
  FOR I = 1 TO N STEP NSTP
    FOR J = 1 TO 64
      TEXT$(J) = " "
    NEXT J
    IBAR = SCALE * Y(I)
    FOR J = 1 TO 64
      IF J <= IBAR THEN TEXT$(J) = "*"
    NEXT J
    PRINT USING "##.######"; Y(I);
    PRINT "   ";
    FOR J = 1 TO 64
      PRINT TEXT$(J);
    NEXT J
    PRINT
  NEXT I
  PRINT
  PRINT "press RETURN to smooth ...";
  LINE INPUT DUM$
  PRINT
```

```
PRINT
CALL SMOOFT(Y(), N, PTS)
NEXT K
END
```

Appendix

File TABLE.DAT:

Accidental Deaths by Month and Type (1979)

Month:	jan	feb	mar	apr	may	jun	jul	aug	sep	oct	nov	dec
Motor Vehicle	3298	3304	4241	4291	4594	4710	4914	4942	4861	4914	4563	4892
Falls	1150	1034	1089	1126	1142	1100	1112	1099	1114	1079	999	1181
Drowning	180	190	370	530	800	1130	1320	990	580	320	250	212
Fires	874	768	630	516	385	324	277	272	271	381	533	760
Choking	299	264	258	247	273	269	251	269	271	279	297	266
Fire-arms	168	142	122	140	153	142	147	160	162	172	266	230
Poisons	298	277	346	263	253	239	268	228	240	260	252	241
Gas-poison	267	193	144	127	70	63	55	53	60	118	150	172
Other	1264	1234	1172	1220	1547	1339	1419	1453	1359	1308	1264	1246

File TABLE2.DAT:

Average solar radiation (watts/square meter) for selected cities

Month:	jul	aug	sep	oct	nov	dec	jan	feb	mar	apr	may	jun	ave	lat
Atlanta, GA	257	246	201	166	30	102	106	140	184	236	258	271	192	34.0
Barrow, AK	208	123	56	20	0	0	0	18	87	184	248	256	100	71.0
Bismark, ND	296	251	185	132	78	60	76	121	170	217	267	284	178	47.0
Boise, ID	324	275	221	152	88	60	69	113	164	235	284	309	191	43.5
Boston, MA	240	206	165	115	70	58	67	96	142	176	228	242	150	42.5
Caribou, ME	246	218	161	102	53	51	66	111	178	194	229	232	153	47.0
Cleveland, OH	267	239	182	127	68	56	60	87	151	182	253	271	162	41.5
Dodge City, KS	311	287	239	184	138	113	123	153	202	256	275	315	216	38.0
El Paso, TX	324	309	278	224	178	151	160	209	266	317	346	353	260	32.0
Fresno, CA	323	293	243	182	117	77	90	143	212	264	308	337	216	37.0
Greensboro, NC	263	235	197	156	118	95	97	134	171	227	257	273	185	36.0
Honolulu, HI	305	293	271	245	208	176	175	200	234	262	300	297	247	21.0
Little Rock, AR	270	250	214	167	118	91	96	127	173	220	256	272	188	35.0
Miami, FL	260	246	216	188	171	154	166	201	238	263	267	257	219	26.0
New York, NY	251	238	175	127	77	62	71	102	151	183	220	255	159	41.0
Omaha, NE	275	252	192	142	96	80	99	134	172	224	248	272	182	21.0
Rapid City, SD	288	262	208	152	99	76	90	135	193	235	259	287	190	44.0
Seattle, WA	242	209	150	84	44	29	34	60	118	174	216	228	132	47.5
Tucson, AZ	304	286	281	216	172	144	151	195	264	322	358	343	253	41.0
Washington, DC	267	190	196	145	75	64	101	124	153	182	215	247	163	39.0

Chapter 14: Modeling of Data

Chapter 14 of *Numerical Recipes* deals with the fitting of a model function to a set of data, in order to summarize the data in terms of a few model parameters. Both traditional least-squares fitting and robust fitting are considered. Fits to a straight line are carried out by routine FIT. More general linear least-squares fits are handled by LFIT and COVSRT. (Remember that the term "linear" here refers not to a linear dependence of the fitting function on its argument, but rather to a linear dependence of the function on its fitting parameters.) In cases where LFIT fails, owing probably to near degeneracy of some basis functions, the answer may still be found using SVDFIT and SVDVAR. In fact, these are generally recommended in preference to LFIT because they never (?) fail. For nonlinear least-squares fits, the Levenberg-Marquardt method is discussed, and is implemented in MRQMIN, which makes use also of COVSRT and MRQCOF.

Robust estimation is discussed in several forms, and illustrated by routine MEDFIT which fits a straight line to data points based on the criterion of least absolute deviations rather than least-squares deviations. ROFUNC is an auxiliary function for MEDFIT.

$$\star \quad \star \quad \star \quad \star$$

Routine FIT fits a set of N data points $(X(I),Y(I))$, with standard deviations $SIG(I)$, to the linear model $y = A + Bx$. It uses χ^2 as the criterion for goodness-of-fit. To demonstrate FIT, we generate some noisy data in sample program D14R1. For NPT values of I we take $x = 0.1I$ and $y = 1 - 2x$ plus some values drawn from a Gaussian distribution to represent noise. Then we make two calls to FIT, first performing the fit without allowance for standard deviations $SIG(I)$, and then with such allowance. Since $SIG(I)$ has been set to the constant value SPREAD, it should not affect the resulting parameter values. The values output from this routine are:

Ignoring standard deviation:

```
A =  0.936991      Uncertainty:  0.099560
B = -1.979427      Uncertainty:  0.017116
Chi-squared:      23.922630
Goodness-of-fit:   1.000000
```

Including standard deviation:

```
A =  0.936991      Uncertainty:  0.100755
B = -1.979427      Uncertainty:  0.017321
Chi-squared:      95.690506
Goodness-of-fit:   0.547179
```

Here is the recipe FIT:

```
DECLARE FUNCTION GAMMQ! (A!, X!)

SUB FIT (X(), Y(), NDATA, SIG(), MWT, A, B, SIGA, SIGB, CHI2, Q)
```

Given a set of NDATA points X(I),Y(I) with standard deviations SIG(I), fit them to a straight line $y = a + bx$ by minimizing χ^2. Returned are A,B and their respective probable uncertainties SIGA and SIGB, the chi-square CHI2, and the goodness-of-fit probability Q (that the fit would have χ^2 this large or larger). If MWT=0 on input, then the standard deviations are assumed to be unavailable: Q is returned as 1.0 and the normalization of CHI2 is to unit standard deviation on all points.

```
SX = 0!                              Initialize sums to zero.
SY = 0!
ST2 = 0!
B = 0!
IF MWT <> 0 THEN                     Accumulate sums ...
  SS = 0!
  FOR I = 1 TO NDATA                 ...with weights
    WT = 1! / SIG(I) ^ 2
    SS = SS + WT
    SX = SX + X(I) * WT
    SY = SY + Y(I) * WT
  NEXT I
ELSE
  FOR I = 1 TO NDATA                 ...or without weights.
    SX = SX + X(I)
    SY = SY + Y(I)
  NEXT I
  SS = CSNG(NDATA)
END IF
SXOSS = SX / SS
IF MWT <> 0 THEN
  FOR I = 1 TO NDATA
    T = (X(I) - SXOSS) / SIG(I)
    ST2 = ST2 + T * T
    B = B + T * Y(I) / SIG(I)
  NEXT I
ELSE
  FOR I = 1 TO NDATA
    T = X(I) - SXOSS
    ST2 = ST2 + T * T
    B = B + T * Y(I)
  NEXT I
END IF
B = B / ST2                          Solve for A, B, σₐ and σᵦ.
A = (SY - SX * B) / SS
SIGA = SQR((1! + SX * SX / (SS * ST2)) / SS)
SIGB = SQR(1! / ST2)
CHI2 = 0!                            Calculate χ².
IF MWT = 0 THEN
  FOR I = 1 TO NDATA
    CHI2 = CHI2 + (Y(I) - A - B * X(I)) ^ 2
  NEXT I
  Q = 1!
  SIGDAT = SQR(CHI2 / (NDATA - 2))    For unweighted data evaluate typical SIG using CHI2,
  SIGA = SIGA * SIGDAT                and adjust the standard deviations.
  SIGB = SIGB * SIGDAT
```

```
ELSE
  FOR I = 1 TO NDATA
    CHI2 = CHI2 + ((Y(I) - A - B * X(I)) / SIG(I)) ^ 2
  NEXT I
  Q = GAMMQ(.5 * (NDATA - 2), .5 * CHI2)        §6.2
END IF
END SUB
```

A sample program using **FIT** is the following:

```
DECLARE SUB FIT (X!(), Y!(), NDATA!, SIG!(), MWT!, A!, B!, SIGA!, SIGB!, CHI2!,
Q!)
DECLARE FUNCTION GASDEV! (IDUM&)

'PROGRAM D14R1
'Driver for routine FIT
CLS
NPT = 100
SPREAD = .5
DIM X(NPT), Y(NPT), SIG(NPT)
IDUM& = -117
FOR I = 1 TO NPT
  X(I) = .1 * I
  Y(I) = -2! * X(I) + 1! + SPREAD * GASDEV(IDUM&)
  SIG(I) = SPREAD
NEXT I
FOR MWT = 0 TO 1
  CALL FIT(X(), Y(), NPT, SIG(), MWT, A, B, SIGA, SIGB, CHI2, Q)
  IF MWT = 0 THEN
    PRINT "Ignoring standard deviation"
  ELSE
    PRINT "Including standard deviation"
  END IF
  PRINT "   A = ";
  PRINT USING "##.######"; A;
  PRINT "      Uncertainty:    ";
  PRINT USING "##.######"; SIGA
  PRINT "   B = ";
  PRINT USING "##.######"; B;
  PRINT "      Uncertainty:    ";
  PRINT USING "##.######"; SIGB
  PRINT "   Chi-squared: ";
  PRINT USING "#######.#####"; CHI2
  PRINT "   Goodness-of-fit: ";
  PRINT USING "###.######"; Q
  PRINT
  PRINT
NEXT MWT
END
```

LFIT carries out the same sort of fit but this time does a linear least-squares fit to a more general function. In sample program D14R2 the chosen function is a linear sum of powers of x, generated by subroutine FUNCS. For convenience in checking the result we have generated data according to $y = 1 + 2x + 3x^2 + \cdots$. This series is truncated depending on the choice of NTERM, and some Gaussian noise is added to simulate realistic data. The SIG(I) are taken as constant errors. LFIT is called three times to fit the same data. The first time LISTA(I) is set to I, so that the fitted parameters should be returned in the order $A(1) \approx 1.0$, $A(2) \approx 2.0$, $A(3) \approx 3.0$. Then, as a test of the LISTA feature, which determines which parameters are to be fit and in which order, the array LISTA(I) is reversed. Finally, the fit is restricted to odd-numbered parameters, while even-numbered parameters are fixed. In this case the elements of the covariance matrix associated with fixed parameters should be zero. In D14R2, we have set NTERM=3 to fit a quadratic. You may wish to try something larger.

Here is the recipe LFIT:

```
DECLARE SUB FUNCS (X!, AFUNC!(), MA!)
DECLARE SUB GAUSSJ (A!(), N!, NP!, B!(), M!, MP!)
DECLARE SUB COVSRT (COVAR!(), NCVM!, MA!, LISTA!(), MFIT!)

SUB LFIT (X(), Y(), SIG(), NDATA, A(), MA, LISTA(), MFIT, COVAR(), NCVM, CHISQ,
DUM)
```
> Given a set of NDATA points X(I),Y(I) with individual standard deviations SIG(I), use χ^2 minimization to determine MFIT of MA coefficients A of a function that depends linearly on A, $y = \sum_i A_i \times \text{AFUNC}_i(x)$. The array LISTA renumbers the parameters so that the first MFIT elements correspond to the parameters actually being determined; the remaining MA–MFIT elements are held fixed at their input values. The program returns values for the MA fit parameters A, $\chi^2 = \text{CHISQ}$, and the covariance matrix COVAR(I,J). NCVM is the physical dimension of COVAR(NCVM,NCVM) in the calling routine. The user supplies a subroutine FUNCS(X,AFUNC,MA) that returns the MA basis functions evaluated at x =X in the array AFUNC.

```
DIM BETA(MA, 1), AFUNC(MA)
KK = MFIT + 1                          Check to see that LISTA contains a proper permutation of
FOR J = 1 TO MA                        the coefficients and fill in any missing members.
  IHIT = 0
  FOR K = 1 TO MFIT
    IF LISTA(K) = J THEN IHIT = IHIT + 1
  NEXT K
  IF IHIT = 0 THEN
    LISTA(KK) = J
    KK = KK + 1
  ELSEIF IHIT > 1 THEN
    PRINT "Improper set in LISTA"
    EXIT SUB
  END IF
NEXT J
IF KK <> (MA + 1) THEN PRINT "Improper set in LISTA": EXIT SUB
FOR J = 1 TO MFIT                      Initialize the (symmetric) matrix.
  FOR K = 1 TO MFIT
    COVAR(J, K) = 0!
  NEXT K
  BETA(J, 1) = 0!
NEXT J
FOR I = 1 TO NDATA                     Loop over data to accumulate coefficients of the normal equa-
                                       tions.
```

```
CALL FUNCS(X(I), AFUNC(), MA)
YM = Y(I)
IF MFIT < MA THEN                    Subtract off dependences on known pieces of the fitting func-
   FOR J = MFIT + 1 TO MA            tion.
      YM = YM - A(LISTA(J)) * AFUNC(LISTA(J))
   NEXT J
END IF
SIG2I = 1! / SIG(I) ^ 2
FOR J = 1 TO MFIT
   WT = AFUNC(LISTA(J)) * SIG2I
   FOR K = 1 TO J
      COVAR(J, K) = COVAR(J, K) + WT * AFUNC(LISTA(K))
   NEXT K
   BETA(J, 1) = BETA(J, 1) + YM * WT
NEXT J
NEXT I
IF MFIT > 1 THEN
   FOR J = 2 TO MFIT                 Fill in above the diagonal from symmetry.
      FOR K = 1 TO J - 1
         COVAR(K, J) = COVAR(J, K)
      NEXT K
   NEXT J
END IF
CALL GAUSSJ(COVAR(), MFIT, NCVM, BETA(), 1, 1)        Matrix solution.
FOR J = 1 TO MFIT
   A(LISTA(J)) = BETA(J, 1)         Partition solution to appropriate coefficients A.
NEXT J
CHISQ = 0!                          Evaluate χ² of the fit.
FOR I = 1 TO NDATA
   CALL FUNCS(X(I), AFUNC(), MA)
   SUM = 0!
   FOR J = 1 TO MA
      SUM = SUM + A(J) * AFUNC(J)
   NEXT J
   CHISQ = CHISQ + ((Y(I) - SUM) / SIG(I)) ^ 2
NEXT I
CALL COVSRT(COVAR(), NCVM, MA, LISTA(), MFIT)        Sort covariance matrix to true or-
ERASE AFUNC, BETA                                    der of fitting coefficients.
END SUB
```

A sample program using `LFIT` is the following:

```
DECLARE SUB LFIT (X!(), Y!(), SIG!(), NDATA!, A!(), MA!, LISTA!(), MFIT!, COVAR!(),
NCVM!, CHISQ!, DUM!)
DECLARE FUNCTION GASDEV! (IDUM&)

'PROGRAM D14R2
'Driver for routine LFIT
CLS
NPT = 100
SPREAD = .1
NTERM = 3
DIM X(NPT), Y(NPT), SIG(NPT), A(NTERM), COVAR(NTERM, NTERM), LISTA(NTERM)
IDUM& = -911
FOR I = 1 TO NPT
```

```
    X(I) = .1 * I
    Y(I) = CSNG(NTERM)
    FOR J = NTERM - 1 TO 1 STEP -1
      Y(I) = J + Y(I) * X(I)
    NEXT J
    Y(I) = Y(I) + SPREAD * GASDEV(IDUM&)
    SIG(I) = SPREAD
  NEXT I
  MFIT = NTERM
  FOR I = 1 TO MFIT
    LISTA(I) = I
  NEXT I
  CALL LFIT(X(), Y(), SIG(), NPT, A(), NTERM, LISTA(), MFIT, COVAR(), NTERM, CHISQ,
  DUM)
  PRINT " Parameter          Uncertainty"
  FOR I = 1 TO NTERM
    PRINT "    A(";
    PRINT USING "#"; I;
    PRINT ") = ";
    PRINT USING "#.######"; A(I);
    PRINT USING "####.######"; SQR(COVAR(I, I))
  NEXT I
  PRINT
  PRINT "  Chi-squared = ";
  PRINT USING "#.######^^^^"; CHISQ
  PRINT
  PRINT "  Full covariance matrix"
  FOR I = 1 TO NTERM
    FOR J = 1 TO NTERM
      PRINT "    ";
      PRINT USING "#.##^^^^"; COVAR(I, J);
    NEXT J
    PRINT
  NEXT I
  PRINT
  PRINT "press RETURN to continue..."
  LINE INPUT DUM$
  PRINT
  'Now test the LISTA feature
  FOR I = 1 TO NTERM
    LISTA(I) = NTERM + 1 - I
  NEXT I
  CALL LFIT(X(), Y(), SIG(), NPT, A(), NTERM, LISTA(), MFIT, COVAR(), NTERM, CHISQ,
  DUM)
  PRINT " Parameter          Uncertainty"
  FOR I = 1 TO NTERM
    PRINT "    A(";
    PRINT USING "#"; I;
    PRINT ") = ";
    PRINT USING "#.######"; A(I);
    PRINT USING "####.######"; SQR(COVAR(I, I))
  NEXT I
  PRINT
  PRINT "  Chi-squared = ";
  PRINT USING "#.######^^^^"; CHISQ
```

```
PRINT
PRINT "  Full covariance matrix"
FOR I = 1 TO NTERM
  FOR J = 1 TO NTERM
    PRINT "    ";
    PRINT USING "#.##^^^^"; COVAR(I, J);
  NEXT J
  PRINT
NEXT I
PRINT
PRINT "press RETURN to continue..."
LINE INPUT DUM$
PRINT
'Now check results of restricting fit parameters
II = 1
FOR I = 1 TO NTERM
  IF (I MOD 2) = 1 THEN
    LISTA(II) = I
    II = II + 1
  END IF
NEXT I
MFIT = II - 1
CALL LFIT(X(), Y(), SIG(), NPT, A(), NTERM, LISTA(), MFIT, COVAR(), NTERM, CHISQ,
DUM)
PRINT "  Parameter        Uncertainty"
FOR I = 1 TO NTERM
  PRINT "   A(";
  PRINT USING "#"; I;
  PRINT ") = ";
  PRINT USING "#.######"; A(I);
  PRINT USING "####.######"; SQR(COVAR(I, I))
NEXT I
PRINT
PRINT "  Chi-squared = ";
PRINT USING "#.######^^^^"; CHISQ
PRINT
PRINT "  Full covariance matrix"
FOR I = 1 TO NTERM
  FOR J = 1 TO NTERM
    PRINT "    ";
    PRINT USING "#.##^^^^"; COVAR(I, J);
  NEXT J
  PRINT
NEXT I
END

SUB FUNCS (X, AFUNC(), MA)
AFUNC(1) = 1!
FOR I = 2 TO MA
  AFUNC(I) = X * AFUNC(I - 1)
NEXT I
END SUB
```

COVSRT is used in conjunction with LFIT (and later with the routine SVDFIT) to redistribute the covariance matrix COVAR so that it represents the true order of coefficients, rather than the order in which they were fit. In sample routine D14R3 an artificial 10×10 covariance matrix COVAR(I,J) is created, which is all zeros except for the upper left 5×5 section, for which the elements are COVAR(I,J)=I+J-1. Then three tests are performed.

1. By setting LISTA(I) = 2I for $I = 1,\ldots,5$ and MFIT=5, we spread the elements so that alternate elements are zero.
2. By taking LISTA(I) $= \text{MFIT} + 1 - I$ for $I = 1,\ldots,5$ we put the elements in reverse order, but leave them in an upper left-hand block.
3. With LISTA(I) $= 12 - 2I$ for $I = 1,\ldots,5$ we both spread and reverse the elements.

Here is the recipe COVSRT:

```
SUB COVSRT (COVAR(), NCVM, MA, LISTA(), MFIT)
        Given the covariance matrix COVAR of a fit for MFIT of MA total parameters, and their order-
        ing LISTA(I), repack the covariance matrix to the true order of the parameters. Elements
        associated with fixed parameters will be zero. NCVM is the physical dimension of COVAR.
FOR J = 1 TO MA - 1                  Zero all elements below diagonal.
  FOR I = J + 1 TO MA
    COVAR(I, J) = 0!
  NEXT I
NEXT J
FOR I = 1 TO MFIT - 1                 Repack off-diagonal elements of fit into correct locations be-
  FOR J = I + 1 TO MFIT              low diagonal.
    IF LISTA(J) > LISTA(I) THEN
      COVAR(LISTA(J), LISTA(I)) = COVAR(I, J)
    ELSE
      COVAR(LISTA(I), LISTA(J)) = COVAR(I, J)
    END IF
  NEXT J
NEXT I
SWAQ = COVAR(1, 1)                    Temporarily store original diagonal elements in top row, and
FOR J = 1 TO MA                       zero the diagonal.
  COVAR(1, J) = COVAR(J, J)
  COVAR(J, J) = 0!
NEXT J
COVAR(LISTA(1), LISTA(1)) = SWAQ
FOR J = 2 TO MFIT                     Now sort elements into proper order on diagonal.
  COVAR(LISTA(J), LISTA(J)) = COVAR(1, J)
NEXT J
FOR J = 2 TO MA                       Finally, fill in above diagonal by symmetry.
  FOR I = 1 TO J - 1
    COVAR(I, J) = COVAR(J, I)
  NEXT I
NEXT J
END SUB
```

A sample program using **COVSRT** is the following:

```
DECLARE SUB COVSRT (COVAR!(), NCVM!, MA!, LISTA!(), MFIT!)

'PROGRAM D14R3
'Driver for routine COVSRT
CLS
MA = 10
MFIT = 5
DIM COVAR(MA, MA), LISTA(MFIT)
FOR I = 1 TO MA
  FOR J = 1 TO MA
    COVAR(I, J) = 0!
    IF I <= 5 AND J <= 5 THEN
      COVAR(I, J) = I + J - 1
    END IF
  NEXT J
NEXT I
PRINT " Original matrix"
FOR I = 1 TO MA
  FOR J = 1 TO MA
    PRINT USING "##.#"; COVAR(I, J);
  NEXT J
  PRINT
NEXT I
PRINT " press RETURN to continue..."
LINE INPUT DUM$
PRINT
'Test 1 - spread by 2
PRINT " Test #1 - Spread by two"
FOR I = 1 TO MFIT
  LISTA(I) = 2 * I
NEXT I
CALL COVSRT(COVAR(), MA, MA, LISTA(), MFIT)
FOR I = 1 TO MA
  FOR J = 1 TO MA
    PRINT USING "##.#"; COVAR(I, J);
  NEXT J
  PRINT
NEXT I
PRINT " press RETURN to continue..."
LINE INPUT DUM$
PRINT
'Test 2 - reverse
PRINT " Test #2 - Reverse"
FOR I = 1 TO MA
  FOR J = 1 TO MA
    COVAR(I, J) = 0!
    IF I <= 5 AND J <= 5 THEN
      COVAR(I, J) = I + J - 1
    END IF
  NEXT J
NEXT I
FOR I = 1 TO MFIT
  LISTA(I) = MFIT + 1 - I
```

```
NEXT I
CALL COVSRT(COVAR(), MA, MA, LISTA(), MFIT)
FOR I = 1 TO MA
  FOR J = 1 TO MA
    PRINT USING "##.#"; COVAR(I, J);
  NEXT J
  PRINT
NEXT I
PRINT " press RETURN to continue..."
LINE INPUT DUM$
PRINT
'Test 3 - spread and reverse
PRINT " Test #3 - Spread and reverse"
FOR I = 1 TO MA
  FOR J = 1 TO MA
    COVAR(I, J) = 0!
    IF I <= 5 AND J <= 5 THEN
      COVAR(I, J) = I + J - 1
    END IF
  NEXT J
NEXT I
FOR I = 1 TO MFIT
  LISTA(I) = MA + 2 - 2 * I
NEXT I
CALL COVSRT(COVAR(), MA, MA, LISTA(), MFIT)
FOR I = 1 TO MA
  FOR J = 1 TO MA
    PRINT USING "##.#"; COVAR(I, J);
  NEXT J
  PRINT
NEXT I
END
```

Routine SVDFIT is recommended in preference to LFIT for performing linear least-squares fits. The sample program D14R4 puts SVDFIT to work on the data generated according to

$$F(x) = 1 + 2x + 3x^2 + 4x^3 + 5x^4 + \text{Gaussian noise.}$$

This data is fit first to a five-term polynomial sum, and then to a five-term Legendre polynomial sum. In each case SIG(I), the measurement fluctuation in y, is taken to be constant. For the polynomial fit, the resulting coefficients should clearly have the values A(I) $\approx$ I. For Legendre polynomials the expected results are:

$$A(1) \approx 3.0$$

$$A(2) \approx 4.4$$

$$A(3) \approx 4.9$$

$$A(4) \approx 1.6$$

$$A(5) \approx 1.1$$

Here is the recipe SVDFIT:

```
DECLARE SUB FPOLY (X!, P!(), NP!)
DECLARE SUB FLEG (X!, PL!(), NL!)
DECLARE SUB SVDCMP (A!(), M!, N!, MP!, NP!, W!(), V!())
DECLARE SUB SVBKSB (U!(), W!(), V!(), M!, N!, MP!, NP!, B!(), X!())

SUB SVDFIT (X(), Y(), SIG(), NDATA, A(), MA, U(), V(), W(), MP, NP, CHISQ, FUNCS$)
```

Given a set of NDATA points $X(I), Y(I)$ with individual standard deviations $SIG(I)$, use χ^2 minimization to determine the MA coefficients A of the fitting function $y = \sum_i A_i \times AFUNC_i(x)$. Here we solve the fitting equations using singular value decomposition of the NDATA by MA matrix, as in §2.9. Arrays U, V, W provide work space on input, on output they define the singular value decomposition, and can be used to obtain the covariance matrix. MP, NP are the physical dimensions of the matrices U, V, W, as indicated below. It is necessary that $MP \geq NDATA$, $NP \geq MA$. The program returns values for the MA fit parameters A, and χ^2, $CHISQ$. The user supplies a subroutine $FUNCS(X, AFUNC, MA)$ that returns the MA basis functions evaluated at $x = X$ in the array $AFUNC$. Here $FUNCS\$$ is input as either the string "FPOLY" or "FLEG".

```
TOL = .00001
DIM B(NDATA), AFUNC(MA)
FOR I = 1 TO NDATA                  Accumulate coefficients of the fitting matrix.
  IF FUNCS$ = "FPOLY" THEN CALL FPOLY(X(I), AFUNC(), MA)
  IF FUNCS$ = "FLEG" THEN CALL FLEG(X(I), AFUNC(), MA)
  TMP = 1! / SIG(I)
  FOR J = 1 TO MA
    U(I, J) = AFUNC(J) * TMP
  NEXT J
  B(I) = Y(I) * TMP
NEXT I
CALL SVDCMP(U(), NDATA, MA, MP, NP, W(), V())      Singular value decomposition.
WMAX = 0!                           Edit the singular values, given TOL from the parameter state-
FOR J = 1 TO MA                     ment, between here ...
  IF W(J) > WMAX THEN WMAX = W(J)
NEXT J
THRESH = TOL * WMAX
FOR J = 1 TO MA
  IF W(J) < THRESH THEN W(J) = 0!
NEXT J                              ...and here.
CALL SVBKSB(U(), W(), V(), NDATA, MA, MP, NP, B(), A())
CHISQ = 0!                          Evaluate chi-square.
FOR I = 1 TO NDATA
  IF FUNCS$ = "FPOLY" THEN CALL FPOLY(X(I), AFUNC(), MA)
  IF FUNCS$ = "FLEG" THEN CALL FLEG(X(I), AFUNC(), MA)
  SUM = 0!
  FOR J = 1 TO MA
    SUM = SUM + A(J) * AFUNC(J)
  NEXT J
  CHISQ = CHISQ + ((Y(I) - SUM) / SIG(I)) ^ 2
NEXT I
ERASE AFUNC, B
END SUB
```

A sample program using SVDFIT is the following:

```
DECLARE SUB SVDVAR (V!(), MA!, NP!, W!(), CVM!(), NCVM!)
DECLARE SUB SVDFIT (X!(), Y!(), SIG!(), NDATA!, A!(), MA!, U!(), V!(), W!(), MP!,
NP!, CHISQ!, FUNCS$)
DECLARE FUNCTION GASDEV! (IDUM&)

'PROGRAM D14R4
'Driver for routine SVDFIT
CLS
NPT = 100
SPREAD = .02
NPOL = 5
DIM X(NPT), Y(NPT), SIG(NPT), A(NPOL), CVM(NPOL, NPOL)
DIM U(NPT, NPOL), V(NPOL, NPOL), W(NPOL)
'Polynomial fit
IDUM& = -911
MP = NPT
NP = NPOL
FOR I = 1 TO NPT
  X(I) = .02 * I
  Y(I) = 1! + X(I) * (2! + X(I) * (3! + X(I) * (4! + X(I) * 5!)))
  Y(I) = Y(I) * (1! + SPREAD * GASDEV(IDUM&))
  SIG(I) = Y(I) * SPREAD
NEXT I
CALL SVDFIT(X(), Y(), SIG(), NPT, A(), NPOL, U(), V(), W(), MP, NP, CHISQ, "FPOLY
CALL SVDVAR(V(), NPOL, NP, W(), CVM(), NPOL)
PRINT "Polynomial fit:"
FOR I = 1 TO NPOL
  PRINT USING "#####.######"; A(I);
  PRINT "   +-";
  PRINT USING "###.######"; SQR(CVM(I, I))
NEXT I
PRINT "Chi-squared";
PRINT USING "#####.######"; CHISQ
PRINT
CALL SVDFIT(X(), Y(), SIG(), NPT, A(), NPOL, U(), V(), W(), MP, NP, CHISQ, "FLEG"
CALL SVDVAR(V(), NPOL, NP, W(), CVM(), NPOL)
PRINT "Legendre polynomial fit"
FOR I = 1 TO NPOL
  PRINT USING "#####.######"; A(I);
  PRINT "   +-";
  PRINT USING "###.######"; SQR(CVM(I, I))
NEXT I
PRINT "Chi-squared";
PRINT USING "#####.######"; CHISQ
END
```

SVDVAR is used with SVDFIT to evaluate the covariance matrix CVM of a fit with MA parameters. In program D14R5, we provide input vector W and array V for this routine via two data statements, and calculate the covariance matrix CVM determined from them. We have also done the calculation by hand and recorded the correct results in array TRU for comparison.

Here is the recipe SVDVAR:

```
SUB SVDVAR (V(), MA, NP, W(), CVM(), NCVM)
```
> To evaluate the covariance matrix CVM of the fit for MA parameters obtained by SVDFIT, call this routine with matrices V, W as returned from SVDFIT. NP, NCVM give the physical dimensions of V, W, CVM as indicated below.

```
DIM WTI(MA)
FOR I = 1 TO MA
  WTI(I) = 0!
  IF W(I) <> 0! THEN WTI(I) = 1! / (W(I) * W(I))
NEXT I
FOR I = 1 TO MA                    Sum contributions to covariance matrix (14.3.20).
  FOR J = 1 TO I
    SUM = 0!
    FOR K = 1 TO MA
      SUM = SUM + V(I, K) * V(J, K) * WTI(K)
    NEXT K
    CVM(I, J) = SUM
    CVM(J, I) = SUM
  NEXT J
NEXT I
ERASE WTI
END SUB
```

A sample program using SVDVAR is the following:

```
DECLARE SUB SVDVAR (V!(), MA!, NP!, W!(), CVM!(), NCVM!)

'PROGRAM D14R5
'Driver for routine SVDVAR
CLS
MP = 6
MA = 3
NCVM = MA
DIM V(MP, MP), W(MP), CVM(NCVM, NCVM), TRU(MA, MA)
FOR I = 1 TO MP
  READ W(I)
NEXT I
DATA 0.0,1.0,2.0,3.0,4.0,5.0
FOR J = 1 TO MP
  FOR I = 1 TO MP
    READ V(I, J)
  NEXT I
NEXT J
DATA 1.0,2.0,3.0,4.0,5.0,6.0,1.0,2.0,3.0,4.0,5.0,6.0,1.0,2.0,3.0,4.0,5.0,6.0
DATA 1.0,2.0,3.0,4.0,5.0,6.0,1.0,2.0,3.0,4.0,5.0,6.0,1.0,2.0,3.0,4.0,5.0,6.0
FOR J = 1 TO MA
  FOR I = 1 TO MA
    READ TRU(I, J)
  NEXT I
NEXT J
DATA 1.25,2.5,3.75,2.5,5.0,7.5,3.75,7.5,11.25
PRINT "Matrix V"
FOR I = 1 TO MP
  FOR J = 1 TO MP
```

```
    PRINT USING "#####.######"; V(I, J);
  NEXT J
  PRINT
NEXT I
PRINT
PRINT "Vector W"
FOR I = 1 TO MP
  PRINT USING "#####.######"; W(I);
NEXT I
PRINT
CALL SVDVAR(V(), MA, MP, W(), CVM(), NCVM)
PRINT
PRINT "Covariance matrix from SVDVAR"
FOR I = 1 TO MA
  FOR J = 1 TO MA
    PRINT USING "#####.######"; CVM(I, J);
  NEXT J
  PRINT
NEXT I
PRINT
PRINT "Expected covariance matrix"
FOR I = 1 TO MA
  FOR J = 1 TO MA
    PRINT USING "#####.######"; TRU(I, J);
  NEXT J
  PRINT
NEXT I
END
```

Routines **FPOLY** and **FLEG** are used with sample program **D14R4** to generate the powers of x and the Legendre polynomials, respectively. In the case of **FPOLY**, sample program **D14R6** is used to list the powers of x generated by **FPOLY** so that they may be checked "by eye". For **FLEG**, the generated polynomials in program **D14R7** are compared to values from routine **PLGNDR**.

Here is the recipe **FPOLY**:

```
SUB FPOLY (X, P(), NP)
    Fitting routine for a polynomial of degree NP-1, with NP coefficients.
P(1) = 1!
FOR J = 2 TO NP
  P(J) = P(J - 1) * X
NEXT J
END SUB
```

A sample program using **FPOLY** is the following:

```
DECLARE SUB FPOLY (X!, P!(), NP!)

'PROGRAM D14R6
'Driver for FPOLY
CLS
NVAL = 15
DX = .1
NPOLY = 5
DIM AFUNC(NPOLY)
PRINT "                              Powers of X"
PRINT
PRINT "        X        X^0       X^1       X^2       X^3       X^4"
FOR I = 1 TO NVAL
  X = I * DX
  CALL FPOLY(X, AFUNC(), NPOLY)
  PRINT USING "#####.####"; X;
  FOR J = 1 TO NPOLY
    PRINT USING "#####.####"; AFUNC(J);
  NEXT J
  PRINT
NEXT I
END
```

Here is the recipe FLEG:

```
SUB FLEG (X, PL(), NL)
```
Fitting routine for an expansion with NL Legendre polynomials PL, evaluated using the recurrence relation as in §4.5.
```
PL(1) = 1!
PL(2) = X
IF NL > 2 THEN
  TWOX = 2! * X
  F2 = X
  D = 1!
  FOR J = 3 TO NL
    F1 = D
    F2 = F2 + TWOX
    D = D + 1!
    PL(J) = (F2 * PL(J - 1) - F1 * PL(J - 2)) / D
  NEXT J
END IF
END SUB
```

A sample program using FLEG is the following:

```
DECLARE SUB FLEG (X!, PL!(), NL!)
DECLARE FUNCTION PLGNDR! (L!, M!, X!)

'PROGRAM D14R7
'Driver for routine FLEG
CLS
NVAL = 5
DX = .2
NPOLY = 5
```

```
DIM AFUNC(NPOLY)
PRINT "                          Legendre Polynomials"
PRINT
PRINT "        N=1        N=2        N=3        N=4        N=5"
FOR I = 1 TO NVAL
  X = I * DX
  CALL FLEG(X, AFUNC(), NPOLY)
  PRINT "X =";
  PRINT USING "###.##"; X
  FOR J = 1 TO NPOLY
    PRINT USING "#####.####"; AFUNC(J);
  NEXT J
  PRINT "  routine FLEG"
  FOR J = 1 TO NPOLY
    PRINT USING "#####.####"; PLGNDR(J - 1, 0, X);
  NEXT J
  PRINT "  routine PLGNDR"
  PRINT
NEXT I
END
```

MRQMIN is used along with MRQCOF to perform nonlinear least-squares fits with the Levenberg-Marquardt method. The artificial data used to try it in sample program D14R8A is computed as the sum of two Gaussians plus noise:

$$Y(I) = A(1) \exp\{-[(X(I) - A(2))/A(3)]^2\}$$
$$+ A(4) \exp\{-[(X(I) - A(5))/A(6)]^2\} + \text{noise.}$$

The A(I) are set up in a DATA statement, as are the initial guesses GUES(I) for these parameters to be used in initiating the fit. Also initialized for the fit are LISTA(I)=I for I=1,..,MFIT to specify that all six of the parameters are to be fit. On the first call to MRQMIN, ALAMDA=-1 to initialize. Then a loop is entered in which MRQMIN is iterated while testing successive values of chi-squared CHISQ. When CHISQ changes by less than 0.1 on two consecutive iterations, the fit is considered complete, and MRQMIN is called one final time with ALAMDA=0.0 so that array COVAR will return the covariance matrix. Uncertainties are derived from the square roots of the diagonal elements of COVAR. Expected results for the parameters are, of course, the values used to generate the "data" in the first place.

Here is the recipe MRQMIN:

```
DECLARE SUB MRQCOF (X!(), Y!(), SIG!(), NDATA!, A!(), MA!, LISTA!(), MFIT!,
ALPHA!(), BETA!(), NALP!, CHISQ!, DUM!)
DECLARE SUB GAUSSJ (A!(), N!, NP!, B!(), M!, MP!)
DECLARE SUB COVSRT (COVAR!(), NCVM!, MA!, LISTA!(), MFIT!)
COMMON SHARED BETA()

SUB MRQMIN (X(), Y(), SIG(), NDATA, A(), MA, LISTA(), MFIT, COVAR(), ALPHA(),
NCA, CHISQ, DUM, ALAMDA) STATIC
    Levenberg-Marquardt method, attempting to reduce the value χ² of a fit between a set of
    NDATA points X(I),Y(I) with individual standard deviations SIG(I), and a nonlinear func-
    tion dependent on MA coefficients A. The array LISTA numbers the parameters A such that the
    first MFIT elements correspond to values actually being adjusted; the remaining MA-MFIT pa-
    rameters are held fixed at their input values. The program returns current best-fit values for the
```

MA fit parameters A, and χ^2 = CHISQ. The arrays COVAR(NCA,NCA), ALPHA(NCA,NCA) with physical dimension NCA ($\geq$ MFIT) are used as working space during most iterations. Supply a subroutine FUNCS(X,A,YFIT,DYDA,MA) that evaluates the fitting function YFIT, and its derivatives DYDA with respect to the fitting parameters A at X. On the first call provide an initial guess for the parameters A, and set ALAMDA<0 for initialization (which then sets ALAMDA=.001). If a step succeeds CHISQ becomes smaller and ALAMDA decreases by a factor of 10. If a step fails ALAMDA grows by a factor of 10. You must call this routine repeatedly until convergence is achieved. Then, make one final call with ALAMDA=0, so that COVAR(I,J) returns the covariance matrix, and ALPHA(I,J) the curvature matrix.

```
DIM ATRY(MA), DA(MA, 1)
IF ALAMDA < 0! THEN                    Initialization.
  KK = MFIT + 1
  FOR J = 1 TO MA                      Does LISTA contain a proper permutation of the coefficients?
    IHIT = 0
    FOR K = 1 TO MFIT
      IF LISTA(K) = J THEN IHIT = IHIT + 1
    NEXT K
    IF IHIT = 0 THEN
      LISTA(KK) = J
      KK = KK + 1
    ELSEIF IHIT > 1 THEN
      PRINT "Improper permutation in LISTA"
      EXIT SUB
    END IF
  NEXT J
  IF KK <> MA + 1 THEN PRINT "Improper permutation in LISTA": EXIT SUB
  ALAMDA = .001
  CALL MRQCOF(X(), Y(), SIG(), NDATA, A(), MA, LISTA(), MFIT, ALPHA(), BETA(),
NCA, CHISQ, DUM)
  OCHISQ = CHISQ
  FOR J = 1 TO MA
    ATRY(J) = A(J)
  NEXT J
END IF
FOR J = 1 TO MFIT                      Alter linearized fitting matrix, by augmenting diagonal ele-
  FOR K = 1 TO MFIT                    ments.
    COVAR(J, K) = ALPHA(J, K)
  NEXT K
  COVAR(J, J) = ALPHA(J, J) * (1! + ALAMDA)
  DA(J, 1) = BETA(J, 1)
NEXT J
CALL GAUSSJ(COVAR(), MFIT, NCA, DA(), 1, 1)        Matrix solution.
IF ALAMDA = 0! THEN                    Once converged evaluate covariance matrix with ALAMDA=0.
  CALL COVSRT(COVAR(), NCA, MA, LISTA(), MFIT)
  ERASE DA, ATRY
  EXIT SUB
END IF
FOR J = 1 TO MFIT                      Did the trial succeed?
  ATRY(LISTA(J)) = A(LISTA(J)) + DA(J, 1)
NEXT J
CALL MRQCOF(X(), Y(), SIG(), NDATA, ATRY(), MA, LISTA(), MFIT, COVAR(), DA(),
NCA, CHISQ, DUM)
IF CHISQ < OCHISQ THEN                 Success, accept the new solution.
  ALAMDA = .1 * ALAMDA
  OCHISQ = CHISQ
```

```
    FOR J = 1 TO MFIT
      FOR K = 1 TO MFIT
        ALPHA(J, K) = COVAR(J, K)
      NEXT K
      BETA(J, 1) = DA(J, 1)
      A(LISTA(J)) = ATRY(LISTA(J))
    NEXT J
  ELSE                            Failure, increase ALAMDA and return.
    ALAMDA = 10! * ALAMDA
    CHISQ = OCHISQ
  END IF
  ERASE DA, ATRY
  END SUB
```

A sample program using MRQMIN is the following:

```
DECLARE SUB MRQMIN (X!(), Y!(), SIG!(), NDATA!, A!(), MA!, LISTA!(), MFIT!, CO-
VAR!(), ALPHA!(), NCA!, CHISQ!, DUM!, ALAMDA!)
DECLARE FUNCTION GASDEV! (IDUM&)
COMMON BETA()

'PROGRAM D14R8A
'Driver for routine MRQMIN
CLS
NPT = 100
MA = 6
SPREAD = .001
DIM BETA(20, 1), X(NPT), Y(NPT), SIG(NPT)
DIM A(MA), LISTA(MA), COVAR(MA, MA), ALPHA(MA, MA), GUES(MA)
FOR I = 1 TO MA
  READ A(I)
NEXT I
DATA 5.0,2.0,3.0,2.0,5.0,3.0
FOR I = 1 TO MA
  READ GUES(I)
NEXT I
DATA 4.5,2.2,2.8,2.5,4.9,2.8
IDUM& = -911
'First try a sum of two Gaussians
FOR I = 1 TO 100
  X(I) = .1 * I
  Y(I) = 0!
  FOR J = 1 TO 4 STEP 3
    Y(I) = Y(I) + A(J) * EXP(-((X(I) - A(J + 1)) / A(J + 2)) ^ 2)
  NEXT J
  Y(I) = Y(I) * (1! + SPREAD * GASDEV(IDUM&))
  SIG(I) = SPREAD * Y(I)
NEXT I
MFIT = MA
FOR I = 1 TO MFIT
  LISTA(I) = I
NEXT I
ALAMDA = -1
FOR I = 1 TO MA
  A(I) = GUES(I)
```

```
NEXT I
CALL MRQMIN(X(), Y(), SIG(), NPT, A(), MA, LISTA(), MFIT, COVAR(), ALPHA(), MA,
CHISQ, DUM, ALAMDA)
K = 1
ITST = 0
DO
  PRINT
  PRINT "Iteration #"; K; " Chi-squared:";
  PRINT USING "#####.####"; CHISQ;
  PRINT " ALAMDA: ";
  PRINT USING ".##^^^^"; ALAMDA
  PRINT " A(1)    A(2)    A(3)    A(4)    A(5)    A(6)"
  FOR I = 1 TO 6
    PRINT USING "###.####"; A(I);
  NEXT I
  PRINT
  K = K + 1
  OCHISQ = CHISQ
  CALL MRQMIN(X(), Y(), SIG(), NPT, A(), MA, LISTA(), MFIT, COVAR(), ALPHA(), MA,
CHISQ, DUM, ALAMDA)
  IF CHISQ > OCHISQ THEN
    ITST = 0
  ELSEIF ABS(OCHISQ - CHISQ) < .1 THEN
    ITST = ITST + 1
  END IF
LOOP WHILE ITST < 2
ALAMDA = 0!
CALL MRQMIN(X(), Y(), SIG(), NPT, A(), MA, LISTA(), MFIT, COVAR(), ALPHA(), MA,
CHISQ, DUM, ALAMDA)
PRINT "Uncertainties:"
FOR I = 1 TO 6
  PRINT USING "###.####"; SQR(COVAR(I, I));
NEXT I
PRINT
PRINT
PRINT "Expected results:"
PRINT USING "####.###"; 5!; 2!; 3!; 2!; 5!; 3!
END
```

The nonlinear least-squares fit makes use of a vector β_k (the gradient of χ^2 in parameter-space) and α_{kl} (the Hessian of χ^2 in the same space). These quantities are produced by MRQCOF, as demonstrated by sample program D14R8B. The function is a sum of two Gaussians with noise added (the same function as in D14R8A) and it is used twice. In the first call, LISTA(I)=I and MFIT=6 so all six parameters are used. In the second call, LISTA(I)=I+3 and MFIT=3 so the first three parameters are fixed and the last three, A(4)...A(6) are fit.

Here is the recipe MRQCOF:

```
DECLARE SUB FUNCS (X!, A!(), Y!, DYDA!(), NA!)
DECLARE SUB FGAUSS (X!, A!(), Y!, DYDA!(), NA!)

SUB FUNCS (X, A(), Y, DYDA(), NA)
CALL FGAUSS(X, A(), Y, DYDA(), NA)
END SUB

SUB MRQCOF (X(), Y(), SIG(), NDATA, A(), MA, LISTA(), MFIT, ALPHA(), BETA(),
NALP, CHISQ, DUM)
```
        Used by MRQMIN to evaluate the linearized fitting matrix ALPHA, and vector BETA as in (14.4.8).
```
DIM DYDA(MA)
FOR J = 1 TO MFIT                     Initialize (symmetric) ALPHA, BETA.
  FOR K = 1 TO J
    ALPHA(J, K) = 0!
  NEXT K
  BETA(J, 1) = 0!
NEXT J
CHISQ = 0!
FOR I = 1 TO NDATA                    Summation loop over all data.
  CALL FUNCS(X(I), A(), YMOD, DYDA(), MA)
  SIG2I = 1! / (SIG(I) * SIG(I))
  DY = Y(I) - YMOD
  FOR J = 1 TO MFIT
    WT = DYDA(LISTA(J)) * SIG2I
    FOR K = 1 TO J
      ALPHA(J, K) = ALPHA(J, K) + WT * DYDA(LISTA(K))
    NEXT K
    BETA(J, 1) = BETA(J, 1) + DY * WT
  NEXT J
  CHISQ = CHISQ + DY * DY * SIG2I     And find $\chi^2$.
NEXT I
FOR J = 2 TO MFIT                     Fill in the symmetric side.
  FOR K = 1 TO J - 1
    ALPHA(K, J) = ALPHA(J, K)
  NEXT K
NEXT J
ERASE DYDA
END SUB
```

A sample program using MRQCOF is the following:

```
DECLARE SUB MRQCOF (X!(), Y!(), SIG!(), NDATA!, A!(), MA!, LISTA!(), MFIT!, AL-
PHA!(), BETA!(), NALP!, CHISQ!, DUM!)
DECLARE FUNCTION GASDEV! (IDUM&)

'PROGRAM D14R8B
'Driver for routine MRQCOF
CLS
NPT = 100
MA = 6
SPREAD = .1
DIM X(NPT), Y(NPT), SIG(NPT), A(MA), LISTA(MA)
DIM COVAR(MA, MA), ALPHA(MA, MA), BETA(MA, 1), GUES(MA)
FOR I = 1 TO MA
```

```
    READ A(I)
NEXT I
DATA 5.0,2.0,3.0,2.0,5.0,3.0
FOR I = 1 TO MA
    READ GUES(I)
NEXT I
DATA 4.9,2.1,2.9,2.1,4.9,3.1
IDUM& = -911
'First try sum of two gaussians
FOR I = 1 TO 100
    X(I) = .1 * I
    Y(I) = 0!
    FOR J = 1 TO 4 STEP 3
      Y(I) = Y(I) + A(J) * EXP(-((X(I) - A(J + 1)) / A(J + 2)) ^ 2)
    NEXT J
    Y(I) = Y(I) * (1! + SPREAD * GASDEV(IDUM&))
    SIG(I) = SPREAD * Y(I)
NEXT I
MFIT = MA
FOR I = 1 TO MFIT
    LISTA(I) = I
NEXT I
FOR I = 1 TO MA
    A(I) = GUES(I)
NEXT I
CALL MRQCOF(X(), Y(), SIG(), NPT, A(), MA, LISTA(), MFIT, ALPHA(), BETA(), MA, CHISQ,
FGAUSS)
PRINT "matrix alpha"
FOR I = 1 TO MA
    FOR J = 1 TO MA
      PRINT USING "#######.####"; ALPHA(I, J);
    NEXT J
    PRINT
NEXT I
PRINT "vector beta"
FOR I = 1 TO MA
    PRINT USING "#######.####"; BETA(I, 1);
NEXT I
PRINT
PRINT "Chi-squared:";
PRINT USING "#######.####"; CHISQ
PRINT
'Next fix one line and improve the other
FOR I = 1 TO 3
    LISTA(I) = I + 3
NEXT I
MFIT = 3
FOR I = 1 TO MA
    A(I) = GUES(I)
NEXT I
CALL MRQCOF(X(), Y(), SIG(), NPT, A(), MA, LISTA(), MFIT, ALPHA(), BETA(), MA, CHISQ,
FGAUSS)
PRINT "matrix alpha"
FOR I = 1 TO MFIT
    FOR J = 1 TO MFIT
```

```
    PRINT USING "#######.####"; ALPHA(I, J);
  NEXT J
  PRINT
NEXT I
PRINT
PRINT "vector beta"
FOR I = 1 TO MFIT
  PRINT USING "#######.####"; BETA(I, 1);
NEXT I
PRINT
PRINT "Chi-squared:";
PRINT USING "#######.####"; CHISQ
END
```

FGAUSS is an example of the type of subroutine that must be supplied to **MRQFIT** (via a call from within **FUNCS()**, above) in order to fit a user-defined function, in this case the sum of Gaussians. **FGAUSS** calculates both the function, and its derivative with respect to each adjustable parameter in a fairly compact fashion. The sample program **D14R9** calculates the same quantities in a more pedantic fashion, just to be sure we got everything right.

Here is the recipe **FGAUSS**:

```
SUB FGAUSS (X, A(), Y, DYDA(), NA)
```
Y(X;A) is the sum of NA/3 Gaussians (14.4.16). The amplitude, center, and width of the Gaussians are stored in consecutive locations of A: $A(I) = B_k$, $A(I+1) = E_k$, $A(I+2) = G_k$, $k = 1, ..., NA/3$.
```
Y = 0!
FOR I = 1 TO NA - 1 STEP 3
  ARG = (X - A(I + 1)) / A(I + 2)
  EX = EXP(-ARG ^ 2)
  FAC = A(I) * EX * 2! * ARG
  Y = Y + A(I) * EX
  DYDA(I) = EX
  DYDA(I + 1) = FAC / A(I + 2)
  DYDA(I + 2) = FAC * ARG / A(I + 2)
NEXT I
END SUB
```

A sample program using **FGAUSS** is the following:

```
DECLARE SUB FGAUSS (X!, A!(), Y!, DYDA!(), NA!)

'PROGRAM D14R9
'Driver for routine FGAUSS
CLS
NPT = 3
NLIN = 2
NA = 3 * NLIN
DIM A(NA), DYDA(NA), DF(NA)
FOR I = 1 TO NA
  READ A(I)
NEXT I
DATA 3.0,0.2,0.5,1.0,0.7,0.3
PRINT "      X        Y      DYDA1    DYDA2    DYDA3    DYDA4    DYDA5    DYDA6"
```

```
FOR I = 1 TO NPT
  X = .3 * I
  CALL FGAUSS(X, A(), Y, DYDA(), NA)
  E1 = EXP(-((X - A(2)) / A(3)) ^ 2)
  E2 = EXP(-((X - A(5)) / A(6)) ^ 2)
  F = A(1) * E1 + A(4) * E2
  DF(1) = E1
  DF(4) = E2
  DF(2) = A(1) * E1 * 2! * (X - A(2)) / A(3) ^ 2
  DF(5) = A(4) * E2 * 2! * (X - A(5)) / A(6) ^ 2
  DF(3) = A(1) * E1 * 2! * (X - A(2)) ^ 2 / A(3) ^ 3
  DF(6) = A(4) * E2 * 2! * (X - A(5)) ^ 2 / A(6) ^ 3
  PRINT "from FGAUSS"
  PRINT USING "###.####"; X; Y;
  FOR J = 1 TO 6
    PRINT USING "###.####"; DYDA(J);
  NEXT J
  PRINT
  PRINT "independent calc."
  PRINT USING "###.####"; X; F;
  FOR J = 1 TO 6
    PRINT USING "###.####"; DF(J);
  NEXT J
  PRINT
  PRINT
NEXT I
END
```

MEDFIT is a subroutine illustrating a more "robust" way of fitting. It performs a fit of data to a straight line, but instead of using the least-squares criterion for figuring the merit of a fit, it uses the least-absolute-deviation. For comparison, sample routine D14R10 fits lines to a noisy linear data set, using first the least-squares routine FIT, and then the least-absolute-deviation routine MEDFIT. You may be interested to see if you can figure out what mean value of absolute deviation you expect for data with Gaussian noise of amplitude SPREAD.

Here is the recipe MEDFIT:

```
DECLARE FUNCTION ROFUNC! (B!)
COMMON SHARED NPT, X(), Y(), ARR(), AA, ABDEV

SUB MEDFIT (XD(), YD(), NDATA, A, B, ABDEVD)
```
> Fits $y = a + bx$ by the criterion of least absolute deviations. The arrays XD and YD, of length NDATA, are the input experimental points. The fitted parameters A and B are output, along with ABDEVD which is the mean absolute deviation (in y) of the experimental points from the fitted line. This routine uses the routine ROFUNC, with communication via a common block.

```
SX = 0!
SY = 0!
SXY = 0!
SXX = 0!
FOR J = 1 TO NDATA           As a first guess for A and B, we will find the least-squares
  X(J) = XD(J)               fitting line.
  Y(J) = YD(J)
  SX = SX + XD(J)
  SY = SY + YD(J)
```

```
  SXY = SXY + XD(J) * YD(J)
  SXX = SXX + XD(J) ^ 2
NEXT J
NPT = NDATA
DEQ = NDATA * SXX - SX ^ 2
AA = (SXX * SY - SX * SXY) / DEQ        Least-squares solutions.
BB = (NDATA * SXY - SX * SY) / DEQ
CHISQ = 0!
FOR J = 1 TO NDATA
  CHISQ = CHISQ + (YD(J) - (AA + BB * XD(J))) ^ 2
NEXT J
SIGB = SQR(CHISQ / DEQ)                  The standard deviation will give some idea of how big an
B1 = BB                                  iteration step to take.
F1 = ROFUNC(B1)
B2 = BB + ABS(3! * SIGB) * SGN(F1)       Guess bracket as 3-σ away, in the downhill direction known
F2 = ROFUNC(B2)                          from F1.
WHILE F1 * F2 > 0!                       Bracketing.
  BB = 2! * B2 - B1
  B1 = B2
  F1 = F2
  B2 = BB
  F2 = ROFUNC(B2)
WEND
SIGB = .01 * SIGB                        Refine until error a negligible number of standard deviations.
DO WHILE ABS(B2 - B1) > SIGB             Bisection.
  BB = .5 * (B1 + B2)
  IF BB = B1 OR BB = B2 THEN EXIT DO
  F = ROFUNC(BB)
  IF F * F1 >= 0! THEN
    F1 = F
    B1 = BB
  ELSE
    F2 = F
    B2 = BB
  END IF
LOOP
A = AA
B = BB
ABDEVD = ABDEV / NDATA
END SUB
```

A sample program using MEDFIT is the following:

```
DECLARE SUB MEDFIT (XD!(), YD!(), NDATA!, A!, B!, ABDEVD!)
DECLARE SUB FIT (X!(), Y!(), NDATA!, SIG!(), MWT!, A!, B!, SIGA!, SIGB!, CHI2!,
Q!)
DECLARE FUNCTION GASDEV! (IDUM&)
COMMON NPT, X(), Y(), ARR(), AA, ABDEV

'PROGRAM D14R10
'Driver for routine MEDFIT
CLS
NPT = 100
SPREAD = .1
DIM XD(NPT), YD(NPT), SIG(NPT)
```

```
DIM X(NPT), Y(NPT), ARR(NPT)
IDUM& = -1984
FOR I = 1 TO NPT
  XD(I) = .1 * I
  YD(I) = -2! * XD(I) + 1! + SPREAD * GASDEV(IDUM&)
  SIG(I) = SPREAD
NEXT I
MWT = 1
CALL FIT(XD(), YD(), NPT, SIG(), MWT, A, B, SIGA, SIGB, CHI2, Q)
PRINT "According to routine FIT the result is:"
PRINT "   A = ";
PRINT USING "###.####"; A;
PRINT "   Uncertainty: ";
PRINT USING "###.####"; SIGA
PRINT "   B = ";
PRINT USING "###.####"; B;
PRINT "   Uncertainty: ";
PRINT USING "###.####"; SIGB
PRINT "   Chi-squared: ";
PRINT USING "###.####"; CHI2;
PRINT " for"; NPT; "points"
PRINT "   Goodness-of-fit: ";
PRINT USING "###.####"; Q
PRINT
PRINT "According to routine MEDFIT the result is:"
CALL MEDFIT(XD(), YD(), NPT, A, B, ABDEVD)
PRINT "   A = ";
PRINT USING "###.####"; A
PRINT "   B = ";
PRINT USING "###.####"; B
PRINT "   Absolute deviation (per data point): ";
PRINT USING "###.####"; ABDEVD
PRINT "   (note: Gaussian spread is";
PRINT USING "###.####"; SPREAD;
PRINT ")"
END
```

ROFUNC is an auxiliary function for MEDFIT. It evaluates the quantity

$$\sum_{i=1}^{N} x_i \, \mathrm{sgn}(y_i - a - bx_i)$$

given arrays x_i and y_i. Data are communicated to and from ROFUNC primarily through the common variables, but the value of the sum above is returned as the value of ROFUNC(B). ABDEV is the summed absolute deviation, and AA (listed below as A) is given the value which minimizes ABDEV. Our results for these quantities are:

B	A	ROFUNC	ABDEV
-2.10	1.51	245.40	25.38
-2.08	1.41	242.20	20.54
-2.06	1.30	242.20	15.69
-2.04	1.20	234.20	10.94
-2.02	1.09	193.20	6.61
-2.00	1.00	22.00	4.07
-1.98	.89	-199.20	5.78

-1.96	.79	-237.40	10.37
-1.94	.68	-246.40	15.25
-1.92	.58	-246.40	20.18
-1.90	.48	-248.40	25.13

Here is the recipe ROFUNC:

```
DECLARE FUNCTION ROFUNC! (B!)
DECLARE SUB SORT (N!, RA!())
COMMON SHARED NPT, X(), Y(), ARR(), AA, ABDEV
```

FUNCTION ROFUNC (B)

> Evaluates the right-hand side of equation (14.6.16) for a given value of B. Communication with the program MEDFIT is through a common block.

```
N1 = NPT + 1
NML = INT(N1 / 2)
NMH = N1 - NML
FOR J = 1 TO NPT
  ARR(J) = Y(J) - B * X(J)
NEXT J
CALL SORT(NPT, ARR())
AA = .5 * (ARR(NML) + ARR(NMH))
SUM = 0!
ABDEV = 0!
FOR J = 1 TO NPT
  D = Y(J) - (B * X(J) + AA)
  ABDEV = ABDEV + ABS(D)
  SUM = SUM + X(J) * SGN(D)
NEXT J
ROFUNC = SUM
END FUNCTION
```

A sample program using ROFUNC is the following:

```
DECLARE FUNCTION GASDEV! (IDUM&)
DECLARE FUNCTION ROFUNC! (B!)
COMMON NPT, X(), Y(), ARR(), AA, ABDEV

'PROGRAM D14R11
'Driver for routine ROFUNC
CLS
NMAX = 1000
DIM X(NMAX), Y(NMAX), ARR(NMAX)
SPREAD = .05
IDUM& = -11
NPT = 100
FOR I = 1 TO NPT
  X(I) = .1 * I
  Y(I) = -2! * X(I) + 1! + SPREAD * GASDEV(IDUM&)
NEXT I
PRINT "        B        A      ROFUNC      ABDEV"
PRINT
FOR I = -5 TO 5
  B = -2! + .02 * I
  RF = ROFUNC(B)
```

```
   PRINT USING "#######.##"; B; AA; ROFUNC(B); ABDEV
NEXT I
END
```

Chapter 15: Ordinary Differential Equations

Chapter 15 of *Numerical Recipes* deals with the integration of ordinary differential equations, restricting its attention specifically to initial-value problems. Three practical methods are introduced: 1) Runge-Kutta methods (RK4, RKDUMB, RKQC, and ODEINT), 2) Richardson extrapolation and the Bulirsch-Stoer method (BSSTEP, MMID, RZEXTR, PZEXTR), 3) predictor-corrector methods. In general, for applications not demanding high precision, and where convenience is paramount, the fourth-order Runge-Kutta with adaptive stepsize control is recommended. For higher precision applications, the Bulirsch-Stoer method dominates. The predictor-corrector methods are covered because of their history of widespread use, but are not regarded (by us) as having an important role today. (For a possible exception to this strong statement, see *Numerical Recipes*.)

$$\star \quad \star \quad \star \quad \star$$

Routine RK4 advances the solution vector Y(N) of a set of ordinary differential equations over a single small interval H in x using the fourth-order Runge-Kutta method. The operation is shown by sample program D15R1 for an array of four variables Y(1),...,Y(4). The first-order differential equations satisfied by these variables are specified by the accompanying routine DERIVS, and are simply the equations describing the first four Bessel functions $J_0(x),...,J_3(x)$. The Y's are initialized to the values of these functions at $x = 1.0$. Note that the values of DYDX are also initialized at $x = 1.0$, because RK4 uses the values of DYDX before its first call to DERIVS. The reason for this is discussed in the text. The sample program calls RK4 with H (the stepsize) set to various values from 0.2 to 1.0, so that you can see how well RK4 can do even with quite sizeable steps.

Here is the recipe RK4:

```
DECLARE SUB DERIVS (X!, Y!(), DYDX!())

SUB RK4 (Y(), DYDX(), N, X, H, YOUT(), DUM)
        Given values for N variables Y and their derivatives DYDX known at X, use the fourth-order
        Runge-Kutta method to advance the solution over an interval H and return the incremented
        variables as YOUT, which need not be a distinct array from Y. The user supplies the subroutine
        DERIVS(X,Y,DYDX) which returns derivatives DYDX at X.
DIM YT(N), DYT(N), DYM(N)
HH = H * .5
H6 = H / 6!
XH = X + HH
FOR I = 1 TO N                          First step
   YT(I) = Y(I) + HH * DYDX(I)
NEXT I
CALL DERIVS(XH, YT(), DYT())            Second step
FOR I = 1 TO N
```

```
   YT(I) = Y(I) + HH * DYT(I)
NEXT I
CALL DERIVS(XH, YT(), DYM())          Third step
FOR I = 1 TO N
   YT(I) = Y(I) + H * DYM(I)
   DYM(I) = DYT(I) + DYM(I)
NEXT I
CALL DERIVS(X + H, YT(), DYT())     Fourth step
FOR I = 1 TO N                      Accumulate increments with proper weights.
   YOUT(I) = Y(I) + H6 * (DYDX(I) + DYT(I) + 2! * DYM(I))
NEXT I
ERASE DYM, DYT, YT
END SUB
```

A sample program using **RK4** is the following:

```
DECLARE SUB RK4 (Y!(), DYDX!(), N!, X!, H!, YOUT!(), DUM!)
DECLARE FUNCTION BESSJ! (A!, X!)
DECLARE FUNCTION BESSJ0! (X!)
DECLARE FUNCTION BESSJ1! (X!)

'PROGRAM D15R1
'Driver for routine RK4
CLS
N = 4
DIM Y(N), DYDX(N), YOUT(N)
X = 1!
Y(1) = BESSJ0(X)
Y(2) = BESSJ1(X)
Y(3) = BESSJ(2, X)
Y(4) = BESSJ(3, X)
DYDX(1) = -Y(2)
DYDX(2) = Y(1) - Y(2)
DYDX(3) = Y(2) - 2! * Y(3)
DYDX(4) = Y(3) - 3! * Y(4)
PRINT "Bessel Function: J0          J1          J3          J4"
PRINT
FOR I = 1 TO 5
   H = .2 * I
   CALL RK4(Y(), DYDX(), N, X, H, YOUT(), DUM)
   PRINT "For a step size of:";
   PRINT USING "###.##"; H
   PRINT "       RK4:";
   FOR J = 1 TO 4
      PRINT USING "#####.######"; YOUT(J);
   NEXT J
   PRINT
   PRINT "    Actual:";
   PRINT USING "#####.######"; BESSJ0(X + H); BESSJ1(X + H); BESSJ(2, X + H);
   PRINT USING "#####.######"; BESSJ(3, X + H)
   PRINT
NEXT I
END

SUB DERIVS (X, Y(), DYDX())
```

```
DYDX(1) = -Y(2)
DYDX(2) = Y(1) - (1! / X) * Y(2)
DYDX(3) = Y(2) - (2! / X) * Y(3)
DYDX(4) = Y(3) - (3! / X) * Y(4)
END SUB
```

RKDUMB is an extension of RK4 which allows you to integrate over larger intervals. It is "dumb" in the sense that it has no adaptive stepsize determination, and no code to estimate errors. Sample program D15R2 works with the same functions and derivatives as the previous program, but integrates from X1=1.0 to X2=20.0, breaking the interval into NSTEP=150 equal steps. The variables VSTART(1),...,VSTART(4) which become the starting values of the Y's, are initialized as before, but their derivatives this time are not initialized; RKDUMB takes care of that. This time only the results for the fourth variable $J_3(x)$ are listed, and only every tenth value is given. The values are passed in the common arrays XX and Y.

Here is the recipe RKDUMB:

```
DECLARE SUB DERIVS (X!, Y!(), DYDX!())
DECLARE SUB RK4 (Y!(), DYDX!(), N!, X!, H!, YOUT!(), DUM!)
COMMON SHARED XX(), Y()

SUB RKDUMB (VSTART(), NVAR, X1, X2, NSTEP, DUM)
    Starting from initial values VSTART for NVAR functions, known at X1 use fourth-order Runge-
    Kutta to advance NSTEP equal increments to X2. The user-supplied subroutine DERIVS(X,V,DVDX)
    evaluates derivatives. Results are stored in the common arrays XX and Y. Be sure to dimension
    the common block appropriately.
DIM V(NVAR), DV(NVAR)
FOR I = 1 TO NVAR                    Load starting values.
  V(I) = VSTART(I)
  Y(I, 1) = V(I)
NEXT I
XX(1) = X1
X = X1
H = (X2 - X1) / CSNG(NSTEP)
FOR K = 1 TO NSTEP                   Take NSTEP steps.
  CALL DERIVS(X, V(), DV())
  CALL RK4(V(), DV(), NVAR, X, H, V(), DUM)
  IF X + H = X THEN PRINT "Stepsize not significant in RKDUMB.": EXIT SUB
  X = X + H
  XX(K + 1) = X                      Store intermediate steps.
  FOR I = 1 TO NVAR
    Y(I, K + 1) = V(I)
  NEXT I
NEXT K
ERASE DV, V
END SUB
```

A sample program using RKDUMB is the following:

```
DECLARE SUB RKDUMB (VSTART!(), NVAR!, X1!, X2!, NSTEP!, DUM!)
DECLARE FUNCTION BESSJ! (A!, X!)
DECLARE FUNCTION BESSJ0! (X!)
DECLARE FUNCTION BESSJ1! (X!)
COMMON XX(), Y()

'PROGRAM D15R2
'Driver for routine RKDUMB
CLS
NVAR = 4
NSTEP = 150
DIM VSTART(NVAR), XX(200), Y(10, 200)
X1 = 1!
VSTART(1) = BESSJ0(X1)
VSTART(2) = BESSJ1(X1)
VSTART(3) = BESSJ(2, X1)
VSTART(4) = BESSJ(3, X1)
X2 = 20!
CALL RKDUMB(VSTART(), NVAR, X1, X2, NSTEP, DUM)
PRINT "        X          Integrated      BESSJ3"
PRINT
FOR I = 1 TO INT(NSTEP / 10)
  J = 10 * I
  PRINT USING "#####.####"; XX(J);
  PRINT "   ";
  PRINT USING "#####.######"; Y(4, J); BESSJ(3, XX(J))
NEXT I
END

SUB DERIVS (X, Y(), DYDX())
DYDX(1) = -Y(2)
DYDX(2) = Y(1) - (1! / X) * Y(2)
DYDX(3) = Y(2) - (2! / X) * Y(3)
DYDX(4) = Y(3) - (3! / X) * Y(4)
END SUB
```

RKQC performs a single step of fifth-order Runge-Kutta integration, this time with monitoring of local truncation error and corresponding stepsize adjustment. Its sample program D15R3 is similar to that for routine RK4, using four Bessel functions as the example, and starting the integration at $x = 1.0$. However, on each pass a value is set for EPS, the desired accuracy, and the trial value HTRY for the interval size is set to 0.1. For the first few passes, EPS is not too demanding and HTRY may be perfectly adequate. As EPS becomes smaller, the routine will be forced to diminish H and return smaller values of HDID and HNEXT. Our results (in single precision) are:

eps	htry	hdid	hnext
.3679E+00	.10	.100000	.400000
.1353E+00	.10	.100000	.354954
.4979E-01	.10	.100000	.287823
.1832E-01	.10	.100000	.233879
.6738E-02	.10	.100000	.190423
.2479E-02	.10	.100000	.155293
.9119E-03	.10	.100000	.126883

.3355E-03	.10	.100000	.103845
.1234E-03	.10	.073460	.066323
.4540E-04	.10	.034162	.031216
.1670E-04	.10	.028686	.025915
.6144E-05	.10	.011732	.010733
.2260E-05	.10	.010758	.009784
.8315E-06	.10	.004284	.003911
.3059E-06	.10	.004460	.004035

Here is the recipe RKQC:

```
DECLARE SUB DERIVS (X!, Y!(), DYDX!())
DECLARE SUB RK4 (Y!(), DYDX!(), N!, X!, H!, YOUT!(), DUM!)

SUB RKQC (Y(), DYDX(), N, X, HTRY, EPS, YSCAL(), HDID, HNEXT, DUM)
```

Fifth-order Runge-Kutta step with monitoring of local truncation error to ensure accuracy and adjust stepsize. Input are the dependent variable vector Y of length N and its derivative DYDX at the starting value of the independent variable X. Also input are the stepsize to be attempted HTRY, the required accuracy EPS, and the vector YSCAL against which the error is scaled. On output, Y and X are replaced by their new values, HDID is the stepsize which was actually accomplished, and HNEXT is the estimated next stepsize. The user must supply a subroutine DERIVS that computes the right-hand side derivatives.

```
FCOR = .0666666667#
ONE = 1!
SAFETY = .9
ERRCON = .0006                            The value ERRCON equals (4/SAFETY)^(1/PGROW), see use
DIM YTEMP(N), YSAV(N), DYSAV(N)           below.
PGROW = -.2
PSHRNK = -.25
XSAV = X                                  Save initial values.
FOR I = 1 TO N
  YSAV(I) = Y(I)
  DYSAV(I) = DYDX(I)
NEXT I
H = HTRY                                  Set stepsize to the initial trial value.
DO
  HH = .5 * H                             Take two half steps.
  CALL RK4(YSAV(), DYSAV(), N, XSAV, HH, YTEMP(), DUM)
  X = XSAV + HH
  CALL DERIVS(X, YTEMP(), DYDX())
  CALL RK4(YTEMP(), DYDX(), N, X, HH, Y(), DUM)
  X = XSAV + H
  IF X = XSAV THEN PRINT "Stepsize not significant in RKQC.": EXIT SUB
  CALL RK4(YSAV(), DYSAV(), N, XSAV, H, YTEMP(), DUM)     Take the large step.
  ERRMAX = 0!                             Evaluate accuracy.
  FOR I = 1 TO N
    YTEMP(I) = Y(I) - YTEMP(I)            YTEMP now contains the error estimate.
    IF ABS(YTEMP(I) / YSCAL(I)) > ERRMAX THEN ERRMAX = ABS(YTEMP(I) / YSCAL(I))
  NEXT I
  ERRMAX = ERRMAX / EPS                   Scale relative to required tolerance.
  IF ERRMAX > ONE THEN                    Truncation error too large, reduce stepsize.
    H = SAFETY * H * (ERRMAX ^ PSHRNK)
    FLAG = 1                              For another try.
  ELSE                                    Step succeeded. Compute size of next step.
    HDID = H
    IF ERRMAX > ERRCON THEN
```

```
      HNEXT = SAFETY * H * (ERRMAX ^ PGROW)
    ELSE
      HNEXT = 4! * H
    END IF
    FLAG = 0
  END IF
LOOP WHILE FLAG = 1
FOR I = 1 TO N                    Mop up fifth-order truncation error.
  Y(I) = Y(I) + YTEMP(I) * FCOR
NEXT I
ERASE DYSAV, YSAV, YTEMP
END SUB
```

A sample program using **RKQC** is the following:

```
DECLARE SUB RKQC (Y!(), DYDX!(), N!, X!, HTRY!, EPS!, YSCAL!(), HDID!, HNEXT!, DUM!)
DECLARE FUNCTION BESSJ! (A!, X!)
DECLARE FUNCTION BESSJ0! (X!)
DECLARE FUNCTION BESSJ1! (X!)

'PROGRAM D15R3
'Driver for routine RKQC
CLS
N = 4
DIM Y(N), DYDX(N), YSCAL(N)
X = 1!
Y(1) = BESSJ0(X)
Y(2) = BESSJ1(X)
Y(3) = BESSJ(2, X)
Y(4) = BESSJ(3, X)
DYDX(1) = -Y(2)
DYDX(2) = Y(1) - Y(2)
DYDX(3) = Y(2) - 2! * Y(3)
DYDX(4) = Y(3) - 3! * Y(4)
FOR I = 1 TO N
  YSCAL(I) = 1!
NEXT I
HTRY = .1
PRINT "      eps        htry       hdid        hnext"
FOR I = 1 TO 15
  EPS = EXP(-CSNG(I))
  CALL RKQC(Y(), DYDX(), N, X, HTRY, EPS, YSCAL(), HDID, HNEXT, DUM)
  PRINT "    ";
  PRINT USING ".####^^^^"; EPS;
  PRINT USING "#####.##"; HTRY;
  PRINT USING "#####.######"; HDID; HNEXT
NEXT I
END

SUB DERIVS (X, Y(), DYDX())
DYDX(1) = -Y(2)
DYDX(2) = Y(1) - (1! / X) * Y(2)
DYDX(3) = Y(2) - (2! / X) * Y(3)
DYDX(4) = Y(3) - (3! / X) * Y(4)
END SUB
```

The full driver routine for RKQC, which provides Runge-Kutta integration over large intervals with adaptive stepsize control, is ODEINT. It plays the same role for RKQC that RKDUMB plays for RK4, and like RKDUMB it stores intermediate results in common variables and arrays. Integration is performed on four Bessel functions from X1=1.0 to X2=10.0, with an accuracy EPS=1.0E-4. Independent of the values of stepsize actually used by ODEINT, intermediate values will be recorded only at intervals greater than DXSAV. The sample program returns values of $J_3(x)$ for checking against actual values produced by BESSJ. It also records how many steps were successful, and how many were "bad." Bad steps are redone, and indicate no extra loss in accuracy. At the same time, they do represent a loss in efficiency, so that an excessive number of bad steps should initiate an investigation.

Here is the recipe ODEINT:

```
DECLARE SUB DERIVS (X!, Y!(), DYDX!())
DECLARE SUB RKQC (Y!(), DYDX!(), N!, X!, HTRY!, EPS!, YSCAL!(), HDID!, HNEXT!,
DUM!)
COMMON SHARED KMAX, KOUNT, DXSAV, XP(), YP()

SUB ODEINT (YSTART(), NVAR, X1, X2, EPS, H1, HMIN, NOK, NBAD, DUM1, DUM2)
    Runge-Kutta driver with adaptive stepsize control. Integrate the NVAR starting values YSTART
    from X1 to X2 with accuracy EPS, storing intermediate results in a common block. H1 should
    be set as a guessed first stepsize, HMIN as the minimum allowed stepsize (can be zero). On
    output NOK and NBAD are the number of good and bad (but retried and fixed) steps taken,
    and YSTART is replaced by values at the end of the integration interval. DERIVS is the user-
    supplied subroutine for calculating the right-hand side derivative, while RKQC is the name of
    the stepper routine to be used. The common variables contain their own information about
    how often an intermediate value is to be stored.
MAXSTP = 10000
TWO = 2!
ZERO = 0!
TINY = 1E-30
DIM YSCAL(NVAR), Y(NVAR), DYDX(NVAR)
X = X1
H = ABS(H1) * SGN(X2 - X1)
NOK = 0
NBAD = 0
KOUNT = 0
FOR I = 1 TO NVAR
  Y(I) = YSTART(I)
NEXT I
IF KMAX > 0 THEN XSAV = X - DXSAV * TWO       Assures storage of first step.
FOR NSTP = 1 TO MAXSTP                        Take at most MAXSTP steps.
  CALL DERIVS(X, Y(), DYDX())
  FOR I = 1 TO NVAR                                          Scaling used to monitor ac-
    YSCAL(I) = ABS(Y(I)) + ABS(H * DYDX(I)) + TINY           curacy. This general-pur-
  NEXT I                                                     pose choice can be modi-
  IF KMAX > 0 THEN                                           fied if need be.
    IF ABS(X - XSAV) > ABS(DXSAV) THEN                       Store intermediate results.
      IF KOUNT < KMAX - 1 THEN
        KOUNT = KOUNT + 1
        XP(KOUNT) = X
        FOR I = 1 TO NVAR
```

```
        YP(I, KOUNT) = Y(I)
      NEXT I
      XSAV = X
    END IF
  END IF
END IF
IF (X + H - X2) * (X + H - X1) > ZERO THEN H = X2 - X
```
If step can overshoot end, cut down stepsize.
```
  CALL RKQC(Y(), DYDX(), NVAR, X, H, EPS, YSCAL(), HDID, HNEXT, DUM)
  IF HDID = H THEN
    NOK = NOK + 1
  ELSE
    NBAD = NBAD + 1
  END IF
  IF (X - X2) * (X2 - X1) >= ZERO THEN         Are we done?
    FOR I = 1 TO NVAR
      YSTART(I) = Y(I)
    NEXT I
    IF KMAX <> 0 THEN
      KOUNT = KOUNT + 1                Save final step.
      XP(KOUNT) = X
      FOR I = 1 TO NVAR
        YP(I, KOUNT) = Y(I)
      NEXT I
    END IF
    ERASE DYDX, Y, YSCAL
    EXIT SUB                    Normal exit.
  END IF
  IF ABS(HNEXT) < HMIN THEN PRINT "Stepsize smaller than minimum.": EXIT SUB
  H = HNEXT
NEXT NSTP
PRINT "Too many steps."
END SUB
```

A sample program using ODEINT is the following:

```
DECLARE SUB ODEINT (YSTART! (), NVAR!, X1!, X2!, EPS!, H1!, HMIN!, NOK!, NBAD!, DUM1!,
DUM2!)
DECLARE FUNCTION BESSJ! (A!, X!)
DECLARE FUNCTION BESSJ0! (X!)
DECLARE FUNCTION BESSJ1! (X!)
COMMON KMAX, KOUNT, DXSAV, XP (), YP ()

'PROGRAM D15R4
'Driver for ODEINT
CLS
NVAR = 4
DIM YSTART(NVAR), XP(200), YP(10, 200)
X1 = 1!
X2 = 10!
YSTART(1) = BESSJ0(X1)
YSTART(2) = BESSJ1(X1)
YSTART(3) = BESSJ(2, X1)
YSTART(4) = BESSJ(3, X1)
EPS = .0001
```

```
H1 = .1
HMIN = 0!
KMAX = 100
DXSAV = (X2 - X1) / 20!
CALL ODEINT(YSTART(), NVAR, X1, X2, EPS, H1, HMIN, NOK, NBAD, DUM, RKQC)
PRINT "Successful steps:        "; NOK
PRINT "Bad steps:               "; NBAD
PRINT "Stored intermediate values:"; KOUNT
PRINT
PRINT "        X            Integral        BESSJ(3,X)"
FOR I = 1 TO KOUNT
  PRINT USING "#####.####"; XP(I);
  PRINT USING "#########.######"; YP(4, I); BESSJ(3, XP(I))
NEXT I
END


SUB DERIVS (X, Y(), DYDX())
DYDX(1) = -Y(2)
DYDX(2) = Y(1) - (1! / X) * Y(2)
DYDX(3) = Y(2) - (2! / X) * Y(3)
DYDX(4) = Y(3) - (3! / X) * Y(4)
END SUB
```

The modified midpoint routine MMID is presented in *Numerical Recipes* primarily as a component of the more powerful Bulirsch-Stoer routine. It integrates variables over an interval HTOT through a sequence of much smaller steps. Sample routine D15R5 takes the number of subintervals I to be $5, 10, 15, \ldots, 50$ so that we can witness any improvements in accuracy that may occur. The values of the four Bessel functions are compared with the results of the integrations.

Here is the recipe MMID:

```
DECLARE SUB DERIVS (X!, Y!(), DYDX!())

SUB MMID (Y(), DYDX(), NVAR, XS, HTOT, NSTEP, YOUT(), DUM)
```
Modified midpoint step. Dependent variable vector Y of length NVAR and its derivative vector DYDX are input at XS. Also input is HTOT, the total step to be made, and NSTEP, the number of substeps to be used. The output is returned as YOUT, which need not be a distinct array from Y; if it is distinct, however, then Y and DYDX are returned undamaged.

```
DIM YM(NVAR), YN(NVAR)
H = HTOT / NSTEP                Stepsize this trip.
FOR I = 1 TO NVAR
  YM(I) = Y(I)
  YN(I) = Y(I) + H * DYDX(I)    First step.
NEXT I
X = XS + H
CALL DERIVS(X, YN(), YOUT())    Will use YOUT for temporary storage of derivatives.
H2 = 2! * H
FOR N = 2 TO NSTEP             General step.
  FOR I = 1 TO NVAR
    SWAQ = YM(I) + H2 * YOUT(I)
    YM(I) = YN(I)
    YN(I) = SWAQ
  NEXT I
  X = X + H
```

```
   CALL DERIVS(X, YN(), YOUT())
NEXT N
FOR I = 1 TO NVAR                    Last step.
   YOUT(I) = .5 * (YM(I) + YN(I) + H * YOUT(I))
NEXT I
ERASE YN, YM
END SUB
```

A sample program using MMID is the following:

```
DECLARE SUB MMID (Y!(), DYDX!(), NVAR!, XS!, HTOT!, NSTEP!, YOUT!(), DUM!)
DECLARE FUNCTION BESSJ! (A!, X!)
DECLARE FUNCTION BESSJ0! (X!)
DECLARE FUNCTION BESSJ1! (X!)

'PROGRAM D15R5
'Driver for routine MMID
CLS
NVAR = 4
X1 = 1!
HTOT = .5
DIM Y(NVAR), YOUT(NVAR), DYDX(NVAR)
Y(1) = BESSJ0(X1)
Y(2) = BESSJ1(X1)
Y(3) = BESSJ(2, X1)
Y(4) = BESSJ(3, X1)
DYDX(1) = -Y(2)
DYDX(2) = Y(1) - Y(2)
DYDX(3) = Y(2) - 2! * Y(3)
DYDX(4) = Y(3) - 3! * Y(4)
XF = X1 + HTOT
B1 = BESSJ0(XF)
B2 = BESSJ1(XF)
B3 = BESSJ(2, XF)
B4 = BESSJ(3, XF)
PRINT "First four Bessel functions"
PRINT
FOR I = 5 TO 50 STEP 5
   CALL MMID(Y(), DYDX(), NVAR, X1, HTOT, I, YOUT(), DUM)
   PRINT "X ="; .0001 * CINT(10000 * X1); "to";
   PRINT .0001 * CINT(10000 * (X1 + HTOT)); "in";
   PRINT I; "steps"
   PRINT "   Integration      BESSJ"
   PRINT USING "#####.######"; YOUT(1); B1
   PRINT USING "#####.######"; YOUT(2); B2
   PRINT USING "#####.######"; YOUT(3); B3
   PRINT USING "#####.######"; YOUT(4); B4
   PRINT
   PRINT "press RETURN to continue..."
   LINE INPUT DUM$
NEXT I
END

SUB DERIVS (X, Y(), DYDX())
DYDX(1) = -Y(2)
```

```
DYDX(2) = Y(1) - (1! / X) * Y(2)
DYDX(3) = Y(2) - (2! / X) * Y(3)
DYDX(4) = Y(3) - (3! / X) * Y(4)
END SUB
```

The Bulirsch-Stoer method, illustrated by routine BSSTEP, is the integrator of choice for higher accuracy calculations on smooth functions. An interval H is broken into finer and finer steps, and the results of integration are extrapolated to zero stepsize. The extrapolation is via a rational function with RZEXTR. BSSTEP monitors local truncation error and adjusts the stepsize appropriately, to keep errors below EPS. From an external point of view, BSSTEP operates exactly as does RKQC: it has the same arguments and in the same order. Consequently it can be used in place of RKQC in routine ODEINT, allowing more efficient integration over large regions of x. For this reason, the sample program D15R6 is of the same form used to demonstrate RKQC.

Here is the recipe BSSTEP:

```
DECLARE SUB MMID (Y!(), DYDX!(), NVAR!, XS!, HTOT!, NSTEP!, YOUT!(), DUM!)
DECLARE SUB RZEXTR (IEST!, XEST!, YEST!(), YZ!(), DY!(), NV!, NUSE!)
COMMON SHARED X(), D()
DATA 2,4,6,8,12,16,24,32,48,64,96

SUB BSSTEP (Y(), DYDX(), NV, X, HTRY, EPS, YSCAL(), HDID, HNEXT, DUM)
        Bulirsch-Stoer step with monitoring of local truncation error to ensure accuracy and adjust
        stepsize. Input are the dependent variable vector Y of length NV and its derivative DYDX at
        the starting value of the independent variable X. Also input are the stepsize to be attempted
        HTRY, the required accuracy EPS, and the vector YSCAL against which the error is scaled. On
        output, Y and X are replaced by their new values, HDID is the stepsize which was actually
        accomplished, and HNEXT is the estimated next stepsize. Subroutine MMID requires a user-
        supplied subroutine DERIVS that computes the right-hand side derivatives.
IMAX = 11
NUSE = 7
ONE = 1!
SHRINK = .95
GROW = 1.2
DIM YERR(NV), YSAV(NV), DYSAV(NV), YSEQ(NV), NSEQ(IMAX)
RESTORE
FOR I = 1 TO IMAX
  READ NSEQ(I)
NEXT I
H = HTRY
XSAV = X
FOR I = 1 TO NV                  Save the starting values.
  YSAV(I) = Y(I)
  DYSAV(I) = DYDX(I)
NEXT I
DO
  FOR I = 1 TO IMAX              Evaluate the sequence of modified midpoint integrations.
    CALL MMID(YSAV(), DYSAV(), NV, XSAV, H, NSEQ(I), YSEQ(), DUM)
    XEST = (H / NSEQ(I)) ^ 2      Squared, since error series is even.
    CALL RZEXTR(I, XEST, YSEQ(), Y(), YERR(), NV, NUSE)   Perform extrapolation.
    IF I > 3 THEN
      ERRMAX = 0!                Check local truncation error.
      FOR J = 1 TO NV
        IF ABS(YERR(J) / YSCAL(J)) > ERRMAX THEN ERRMAX = ABS(YERR(J) / YSCAL(
```

```
      NEXT J
      ERRMAX = ERRMAX / EPS              Scale accuracy relative to tolerance.
      IF ERRMAX < ONE THEN               Step converged.
        X = X + H
        HDID = H
        IF I = NUSE THEN
          HNEXT = H * SHRINK
        ELSEIF I = NUSE - 1 THEN
          HNEXT = H * GROW
        ELSE
          HNEXT = (H * NSEQ(NUSE - 1)) / NSEQ(I)
        END IF
        ERASE NSEQ, YSEQ, DYSAV, YSAV, YERR
        EXIT SUB                         Normal return.
      END IF
    END IF
  NEXT I
```

If here, then step failed, quite unusual for this method. We reduce the stepsize and try again.

```
  H = .25 * H / 2 ^ ((IMAX - NUSE) / 2)
LOOP UNTIL X + H = X
PRINT "Step size underflow."
END SUB
```

A sample program using **BSSTEP** is the following:

```
DECLARE SUB BSSTEP (Y!(), DYDX!(), NV!, X!, HTRY!, EPS!, YSCAL!(), HDID!, HNEXT!,
DERIVS!)
DECLARE FUNCTION BESSJ! (A!, X!)
DECLARE FUNCTION BESSJ0! (X!)
DECLARE FUNCTION BESSJ1! (X!)
COMMON X(), D()

'PROGRAM D15R6
'Driver for routine BSSTEP
CLS
N = 4
DIM Y(N), DYDX(N), YSCAL(N), X(11), D(10, 7)
X = 1!
Y(1) = BESSJ0(X)
Y(2) = BESSJ1(X)
Y(3) = BESSJ(2, X)
Y(4) = BESSJ(3, X)
DYDX(1) = -Y(2)
DYDX(2) = Y(1) - Y(2)
DYDX(3) = Y(2) - 2! * Y(3)
DYDX(4) = Y(3) - 3! * Y(4)
FOR I = 1 TO N
  YSCAL(I) = 1!
NEXT I
HTRY = 1!
PRINT "     eps       htry       hdid       hnext"
PRINT
FOR I = 1 TO 15
  EPS = EXP(-CSNG(I))
```

```
  CALL BSSTEP(Y(), DYDX(), N, X, HTRY, EPS, YSCAL(), HDID, HNEXT, DUM)
  PRINT "    ";
  PRINT USING ".####^^^^"; EPS;
  PRINT USING "#####.##"; HTRY;
  PRINT USING "#####.######"; HDID; HNEXT
NEXT I
END

SUB DERIVS (X, Y(), DYDX())
DYDX(1) = -Y(2)
DYDX(2) = Y(1) - (1! / X) * Y(2)
DYDX(3) = Y(2) - (2! / X) * Y(3)
DYDX(4) = Y(3) - (3! / X) * Y(4)
END SUB
```

RZEXTR performs a diagonal rational function extrapolation for BSSTEP. It takes a sequence of interval lengths and corresponding integrated values, and extrapolates to the value the integral would have if the interval length were zero. Sample routine D15R8 works with a known function

$$F_n = \frac{1 - x + x^3}{(x + 1)^n} \qquad n = 1, .., 4.$$

We extrapolate the vector YEST $= (F_1, F_2, F_3, F_4)$ given a sequence of ten values (only the last NUSE=5 of which are used). The ten values are labeled IEST=1,...,10 and are evaluated at XEST=1.0/IEST. A call to RZEXTR produces extrapolated values YZ, and estimated errors DY, and compares to the true values $(1.0, 1.0, 1.0, 1.0)$ at XEST=0.0.

Here is the recipe RZEXTR:

```
COMMON SHARED X(), D()

SUB RZEXTR (IEST, XEST, YEST(), YZ(), DY(), NV, NUSE)
    Use diagonal rational function extrapolation to evaluate NV functions at x = 0 by fitting a
    diagonal rational function to a sequence of estimates with progressively smaller values x =
    XEST, and corresponding function vectors YEST. This call is number IEST in the sequence of
    calls. The extrapolation uses at most the last NUSE estimates. Extrapolated function values
    are output as YZ, and their estimated error is output as DY.
DIM FX(NUSE)
X(IEST) = XEST                    Save current independent variable.
IF IEST = 1 THEN
  FOR J = 1 TO NV
    YZ(J) = YEST(J)
    D(J, 1) = YEST(J)
    DY(J) = YEST(J)
  NEXT J
ELSE
  M1 = IEST
  IF NUSE < IEST THEN M1 = NUSE   Use at most NUSE previous members.
  FOR K = 1 TO M1 - 1
    FX(K + 1) = X(IEST - K) / XEST
  NEXT K
  FOR J = 1 TO NV                 Evaluate next diagonal in tableau.
    YY = YEST(J)
    V = D(J, 1)
```

```
      C = YY
      D(J, 1) = YY
      FOR K = 2 TO M1
        B1 = FX(K) * V
        B = B1 - C
        IF B <> 0! THEN
          B = (C - V) / B
          DDY = C * B
          C = B1 * B
        ELSE                           Care needed to avoid division by 0.
          DDY = V
        END IF
        IF K <> M1 THEN V = D(J, K)
        D(J, K) = DDY
        YY = YY + DDY
      NEXT K
      DY(J) = DDY
      YZ(J) = YY
    NEXT J
END IF
ERASE FX
END SUB
```

A sample program using **RZEXTR** is the following:

```
DECLARE SUB RZEXTR (IEST!, XEST!, YEST!(), YZ!(), DY!(), NV!, NUSE!)
COMMON X(), D()

'PROGRAM D15R8
'Driver for routine RZEXTR
'Feed values from a rational function
'Fn(x)=(1-x+x^3)/(x+1)^n
CLS
NV = 4
NUSE = 5
DIM YEST(NV), YZ(NV), DY(NV), X(11), D(10, 7)
FOR I = 1 TO 10
  IEST = I
  XEST = 1! / CSNG(I)
  DUM = 1! - XEST + XEST ^ 3
  FOR J = 1 TO NV
    DUM = DUM / (XEST + 1!)
    YEST(J) = DUM
  NEXT J
  CALL RZEXTR(IEST, XEST, YEST(), YZ(), DY(), NV, NUSE)
  PRINT "IEST ="; I; "   XEST =";
  PRINT USING "###.####"; XEST
  PRINT "Extrap. Function: ";
  FOR J = 1 TO NV
    PRINT USING "#####.######"; YZ(J);
  NEXT J
  PRINT
  PRINT "Estimated Error:   ";
  FOR J = 1 TO NV
    PRINT USING "#####.######"; DY(J);
```

```
    NEXT J
    PRINT
    PRINT
NEXT I
PRINT "Actual Values:     ";
PRINT USING "#####.######"; 1!; 1!; 1!; 1!
```

PZEXTR is a less powerful standby for RZEXTR, to be used primarily when some problem crops up with the extrapolation. It performs a polynomial, rather than a rational function, extrapolation. The sample program D15R9 is identical to that for RZEXTR.

Here is the recipe PZEXTR:

```
COMMON SHARED X(), QCOL()

SUB PZEXTR (IEST, XEST, YEST(), YZ(), DY(), NV, NUSE)
DIM D(NV)
X(IEST) = XEST                        Store current dependent value.
FOR J = 1 TO NV
  DY(J) = YEST(J)
  YZ(J) = YEST(J)
NEXT J
IF IEST = 1 THEN                      Store first estimate in first column.
  FOR J = 1 TO NV
    QCOL(J, 1) = YEST(J)
  NEXT J
ELSE
  M1 = IEST
  IF NUSE < IEST THEN M1 = NUSE       Use at most NUSE previous estimates.
  FOR J = 1 TO NV
    D(J) = YEST(J)
  NEXT J
  FOR K1 = 1 TO M1 - 1
    DELTA = 1! / (X(IEST - K1) - XEST)
    F1 = XEST * DELTA
    F2 = X(IEST - K1) * DELTA
    FOR J = 1 TO NV                   Propagate tableau 1 diagonal more.
      Q = QCOL(J, K1)
      QCOL(J, K1) = DY(J)
      DELTA = D(J) - Q
      DY(J) = F1 * DELTA
      D(J) = F2 * DELTA
      YZ(J) = YZ(J) + DY(J)
    NEXT J
  NEXT K1
  FOR J = 1 TO NV
    QCOL(J, M1) = DY(J)
  NEXT J
END IF
ERASE D
END SUB
```

A sample program using **PZEXTR** is the following:

```
DECLARE SUB PZEXTR (IEST!, XEST!, YEST!(), YZ!(), DY!(), NV!, NUSE!)
COMMON X(), QCOL()

'PROGRAM D15R9
'Driver for routine PZEXTR
'Feed values from a rational function
'Fn(x)=(1-x+x^3)/(x+1)^n
CLS
NV = 4
NUSE = 5
DIM YEST(NV), YZ(NV), DY(NV), X(11), QCOL(10, 7)
FOR I = 1 TO 10
  IEST = I
  XEST = 1! / CSNG(I)
  DUM = 1! - XEST + XEST * XEST * XEST
  FOR J = 1 TO NV
    DUM = DUM / (XEST + 1!)
    YEST(J) = DUM
  NEXT J
  CALL PZEXTR(IEST, XEST, YEST(), YZ(), DY(), NV, NUSE)
  PRINT "I = "; I
  PRINT "Extrap. function:";
  FOR J = 1 TO NV
    PRINT USING "#####.######"; YZ(J);
  NEXT J
  PRINT "Estimated error: ";
  FOR J = 1 TO NV
    PRINT USING "#####.######"; DY(J);
  NEXT J
  PRINT
  PRINT
NEXT I
PRINT "Actual values:    ";
PRINT USING "#####.######"; 1!; 1!; 1!; 1!
```

Chapter 16: Two-Point Boundary Value Problems

Two-point boundary value problems, and their iterative solution, is the substance of Chapter 16 of *Numerical Recipes*. The first step is to cast the problem as a set of N coupled first-order ordinary differential equations, satisfying n_1 conditions at one boundary point, and $n_2 = N - n_1$ conditions at the other boundary point. We apply two general methods to the solutions. First are the shooting methods, typified by subroutines SHOOT and SHOOTF, which enforce the n_1 conditions at one boundary and set n_2 conditions freely. Then they integrate across the interval to find discrepancies with the n_2 conditions at the other end. The Newton-Raphson method is used to reduce these discrepancies by adjusting the variable parameters.

The other approach is the relaxation method in which the differential equations are replaced by finite difference equations on a grid that covers the range of interest. Routine SOLVDE demonstrates this method, and is demonstrated "in action" by program SFROID, which uses it to compute eigenvalues of spheroidal harmonics. The program SFROID in *Numerical Recipes* is already self-contained. For the purpose of comparison, we apply these routines to the same problem attacked with SFROID.

$$\star \quad \star \quad \star \quad \star$$

Subroutine SHOOT works as described above. Demonstration program D16R1 uses it to find eigenvalues of both prolate and oblate spheroidal harmonics. The oblate and prolate cases are handled simultaneously, although they actually involve two independent sets of three coupled first-order differential equations, one set with c^2 positive and the other with c^2 negative. The complete set of differential equations is

$$\frac{dy_1}{dx} = y_2$$

$$\frac{dy_2}{dx} = \frac{2x(m+1)y_2 - (y_3 - c^2x^2)y_1}{(1 - x^2)}$$

$$\frac{dy_3}{dx} = 0$$

$$\frac{dy_4}{dx} = y_5$$

$$\frac{dy_5}{dx} = \frac{2x(m+1)y_5 - (y_6 + c^2x^2)y_4}{(1 - x^2)}$$

$$\frac{dy_6}{dx} = 0$$

These are specified in subroutine DERIVS which is called, in turn, by ODEINT in SHOOT. The first three equations correspond to prolate harmonics and the second three to oblate harmonics. Comparing either set of three to equation (16.4.4) in *Numerical Recipes*, you may quickly verify that y_1 and y_4 correspond to the two spheroidal harmonic solutions, y_3 and y_6 correspond to the sought-after eigenvalues (whose derivative with respect to x is of course 0), and y_2 and y_5 are intermediate variables created to change the second-order equations to coupled first-order equations.

Two other subroutines are used by SHOOT. Subroutine LOAD sets the values of all the variables $y_1, \ldots, y_6$ at the first boundary, and SCORE calculates a discrepancy vector F (which will be zero when a successful solution has been reached) at the second boundary. Each of these subroutines has some interesting aspects. In LOAD, y_3 and y_6 are initialized to V(1) and V(2), values calculated in the sample program to give rough estimates of the size of the proper result. We arrived at these estimates just by looking through some tables of values. Also notice that, for example, y_1 is set to FACTR + y_2DX. This is the same as saying that $y_1 = \text{FACTR} + (dy_1/dx)\Delta x$. The quantity FACTR comes from equation (16.4.20) in *Numerical Recipes*, and the term with DX comes from the fact that we placed the lower boundary X1 at $-1.0 + \text{DX}$ (where DX=1.0E-4) rather than at -1.0. This is because dy_2/dx and dy_5/dx cannot be evaluated exactly at $x = -1.0$. The subroutine SCORE follows from equation (16.4.18) in *Numerical Recipes*. For example, if $N - M$ is odd, $y_1 = 0$ at $x = 0$, but if $N - M$ is even, then $y_2 = dy_1/dx = 0$.

That more or less explains things. Now, given M, N, and c^2, sample program D16R1 sets up estimates V(1) and V(2) and iterates the routine SHOOT until changes in the V are less than some preset fraction EPS of their size. Some values of the eigenvalues of the spheroidal harmonics are given in section 16.4 of *Numerical Recipes* if you want to check the results.

Here is the recipe SHOOT:

```
DECLARE SUB LOAD (X1!, V!(), Y!())
DECLARE SUB SCORE (X2!, Y!(), F!())
DECLARE SUB ODEINT (YSTART!(), NVAR!, X1!, X2!, EPS!, H1!, HMIN!, NOK!, NBAD!,
DUM1!, DUM2!)
DECLARE SUB LUDCMP (A!(), N!, NP!, INDX!(), D!)
DECLARE SUB LUBKSB (A!(), N!, NP!, INDX!(), B!())

SUB SHOOT (NVAR, V(), DELV(), N2, X1, X2, EPS, H1, HMIN, F(), DV())
        Improve the trial solution of a two-point boundary value problem for NVAR coupled ODEs
        shooting from X1 to X2. Initial values for the NVAR ODEs at X1 are generated from the N2
        coefficients V, using the user-supplied routine LOAD. The routine integrates the ODEs to X2
        using the Runge-Kutta method with tolerance EPS, initial step size H1, and minimum step size
        HMIN. At X2 it calls the user-supplied subroutine SCORE to evaluate the N2 functions F that
        ought to be zero to satisfy the boundary conditions at X2. Multi-dimensional Newton-Raphson
        is then used to develop a linear matrix equation for the N2 increments DV to the adjustable
        parameters V. These increments are solved for and added before return. The user must supply
        a subroutine DERIVS(X,Y,DYDX) that supplies derivative information to the ODE integrator
        (see Chapter 15).
DIM Y(NVAR), DFDV(NVAR, NVAR), INDX(NVAR)
CALL LOAD(X1, V(), Y())                 Integrate from X1 with best trial values.
CALL ODEINT(Y(), NVAR, X1, X2, EPS, H1, HMIN, NOK, NBAD, DUM, DUM)
CALL SCORE(X2, Y(), F())
FOR IV = 1 TO N2                        Vary boundary conditions at X1.
  SAV = V(IV)
  V(IV) = V(IV) + DELV(IV)              Increment parameter IV.
  CALL LOAD(X1, V(), Y())
  CALL ODEINT(Y(), NVAR, X1, X2, EPS, H1, HMIN, NOK, NBAD, DUM, DUM)
  CALL SCORE(X2, Y(), DV())
  FOR I = 1 TO N2                       Evaluate numerical derivatives of N2 matching conditions.
```

```
      DFDV(I, IV) = (DV(I) - F(I)) / DELV(IV)
    NEXT I
    V(IV) = SAV                      Restore incremented parameter.
  NEXT IV
  FOR IV = 1 TO N2
    DV(IV) = -F(IV)
  NEXT IV
  CALL LUDCMP(DFDV(), N2, NVAR, INDX(), DET) Solve linear equations.
  CALL LUBKSB(DFDV(), N2, NVARP, INDX(), DV())
  FOR IV = 1 TO N2                   Increment boundary parameters.
    V(IV) = V(IV) + DV(IV)
  NEXT IV
  ERASE INDX, DFDV, Y
  END SUB
```

A sample program using **SHOOT** is the following:

```
DECLARE SUB SHOOT (NVAR!, V!(), DELV!(), N2!, X1!, X2!, EPS!, H1!, HMIN!, F!(),
DV!())

'PROGRAM D16R1
'Driver for routine SHOOT
'Solves for eigenvalues of Spheroidal Harmonics. Both
'Prolate and Oblate case are handled simultaneously, leading
'to six first-order equations. Unknown to SHOOT, these are
'actually two independent sets of three coupled equations,
'one set with c^2 positive and the other with c^2 negative.
CLS
NVAR = 6
N2 = 2
DELTA = .001
EPS = .000001
DIM V(2), DELV(2), F(2), DV(2)
DX = .0001
DO
  PRINT "Input M,N,C-Squared (999 to end)"
  INPUT M, N, C2
  IF C2 = 999! THEN END
LOOP WHILE N < M OR M < 0 OR N < 0
FACTR = 1!
IF M <> 0 THEN
  Q1 = N
  FOR I = 1 TO M
    FACTR = -.5 * FACTR * (N + I) * (Q1 / I)
    Q1 = Q1 - 1!
  NEXT I
END IF
V(1) = N * (N + 1) - M * (M + 1) + C2 / 2!
V(2) = N * (N + 1) - M * (M + 1) - C2 / 2!
DELV(1) = DELTA * V(1)
DELV(2) = DELV(1)
H1 = .1
HMIN = 0!
X1 = -1! + DX
X2 = 0!
```

```
PRINT "          Prolate              Oblate"
PRINT "    Mu(M,N)     Error Est.   Mu(M,N)     Error Est."
DO
  CALL SHOOT(NVAR, V(), DELV(), N2, X1, X2, EPS, H1, HMIN, F(), DV())
  PRINT USING "#####.######"; V(1); DV(1); V(2); DV(2)
LOOP WHILE ABS(DV(1)) > ABS(EPS * V(1)) OR ABS(DV(2)) > ABS(EPS * V(2))
END

SUB DERIVS (X, Y(), DYDX())
SHARED C2, M, N, FACTR, DX
DYDX(1) = Y(2)
DYDX(3) = 0!
DYDX(2) = (2! * X * (M + 1!) * Y(2) - (Y(3) - C2 * X * X) * Y(1)) / (1! - X * X)
DYDX(4) = Y(5)
DYDX(6) = 0!
DYDX(5) = (2! * X * (M + 1!) * Y(5) - (Y(6) + C2 * X * X) * Y(4)) / (1! - X * X)
END SUB

SUB LOAD (X1, V(), Y())
SHARED C2, M, N, FACTR, DX
Y(3) = V(1)
Y(2) = -(Y(3) - C2) * FACTR / 2! / (M + 1!)
Y(1) = FACTR + Y(2) * DX
Y(6) = V(2)
Y(5) = -(Y(6) + C2) * FACTR / 2! / (M + 1!)
Y(4) = FACTR + Y(5) * DX
END SUB

SUB SCORE (X2, Y(), F())
SHARED C2, M, N, FACTR, DX
IF (N - M) MOD 2 = 0 THEN
  F(1) = Y(2)
  F(2) = Y(5)
ELSE
  F(1) = Y(1)
  F(2) = Y(4)
END IF
END SUB
```

Another shooting method is shooting to a fitting point. More explicitly, we set values at two boundaries, from both of which we integrate toward an intermediate point. For the spheroidal harmonics, we take the endpoints, in sample program D16R2, to be $-1.0 + $ DX and $1.0 - $ DX, and the intermediate point to be $x = 0.0$. For clarity, we considered only prolate spheroids. The calculation is similar to that in the previous sample program, except for these details:

1. There are only three first-order differential equations in DERIVS because of the restriction to prolate spheroids. (Note: the oblate case requires only that we input c^2 as a negative number.)

2. There are two load routines, LOAD1 and LOAD2, which set values at the two boundaries. At the first boundary Y(3) is initialized to V1(1), which is initially set to our crude guess of the magnitude of the eigenvalue. Y(1), the spheroidal harmonic value itself, is set to FACTR $+ (dy_1/dx)\Delta x$, and Y(2) is also set as before. At boundary two, Y(3) and Y(1) are given guessed values for the eigenvalue and for $y(1 - \Delta x)$ respectively. We treat the

guessed eigenvalue at boundary two as independent of that at boundary one, although they ought certainly to converge to the same value. To verify this point, we make the initial guess that the values differ by 1.0 (i.e. V2(2)=V1(1)+1.0).

Sample program D16R2 otherwise proceeds much as D16R1 did, however with SCORE kept at $x = 0.0$ where the solutions must match up. The subroutine SCORE has been set to a dummy operation equating F_i to y_i so that the condition of success is that the y_i all match at $x = 0$. This is discussed more fully in *Numerical Recipes*. Check the eigenvalue results against the previous routine.

Here is the recipe SHOOTF:

```
DECLARE SUB LOAD1 (X1!, V1!(), Y!())
DECLARE SUB LOAD2 (X2!, V2!(), Y!())
DECLARE SUB SCORE (XF!, Y!(), F!())
DECLARE SUB ODEINT (YSTART!(), NVAR!, X1!, X2!, EPS!, H1!, HMIN!, NOK!, NBAD!,
DUM1!, DUM2!)
DECLARE SUB LUDCMP (A!(), N!, NVAR!, INDX!(), D!)
DECLARE SUB LUBKSB (A!(), N!, NVAR!, INDX!(), B!())

SUB SHOOTF (NVAR, V1(), V2(), DELV1(), DELV2(), N1, N2, X1, X2, XF, EPS, H1,
HMIN, F(), DV1(), DV2())
    Improve the trial solution of a two-point boundary value problem for NVAR coupled ODEs
    shooting from X1 and X2 to a fitting point XF. Initial values for the NVAR ODEs at X1
    (X2) are generated from the N2 (N1) coefficients V1 (V2), using the user-supplied routine
    LOAD1 (LOAD2). The routine integrates the ODEs to XF using the Runge-Kutta method
    with tolerance EPS, initial stepsize H1, and minimum stepsize HMIN. At XF it calls the user-
    supplied subroutine SCORE to evaluate the NVAR functions F that ought to match at XF.
    Multi-dimensional Newton-Raphson is then used to develop a linear matrix equation for the N2
    (N1) increments DV1 (DV2) to the adjustable parameters V1 (V2). These increments are
    solved for and added before return. The user must supply a subroutine DERIVS(X,Y,DYDX)
    that provides derivative information to the ODE integrator (see Chapter 15).
DIM Y(NVAR), F1(NVAR), F2(NVAR), DFDV(NVAR, NVAR), INDX(NVAR)
CALL LOAD1(X1, V1(), Y())            Path from X1 to XF with best trial values V1.
CALL ODEINT(Y(), NVAR, X1, XF, EPS, H1, HMIN, NOK, NBAD, DERIVS, RKQC)
CALL SCORE(XF, Y(), F1())
CALL LOAD2(X2, V2(), Y())            Path from X2 to XF with best trial values V2.
CALL ODEINT(Y(), NVAR, X2, XF, EPS, H1, HMIN, NOK, NBAD, DERIVS, RKQC)
CALL SCORE(XF, Y(), F2())
J = 0
FOR IV = 1 TO N2                     Vary boundary conditions at X1.
  J = J + 1
  SAV = V1(IV)
  V1(IV) = V1(IV) + DELV1(IV)
  CALL LOAD1(X1, V1(), Y())
  CALL ODEINT(Y(), NVAR, X1, XF, EPS, H1, HMIN, NOK, NBAD, DERIVS, RKQC)
  CALL SCORE(XF, Y(), F())
  FOR I = 1 TO NVAR                  Evaluate numerical derivatives of NVAR fitting conditions.
    DFDV(I, J) = (F(I) - F1(I)) / DELV1(IV)
  NEXT I
  V1(IV) = SAV                       Restore boundary parameter.
NEXT IV
FOR IV = 1 TO N1                     Next vary boundary conditions at X2.
  J = J + 1
```

```
    SAV = V2(IV)
    V2(IV) = V2(IV) + DELV2(IV)
    CALL LOAD2(X2, V2(), Y())
    CALL ODEINT(Y(), NVAR, X2, XF, EPS, H1, HMIN, NOK, NBAD, DERIVS, RKQC)
    CALL SCORE(XF, Y(), F())
    FOR I = 1 TO NVAR
      DFDV(I, J) = (F2(I) - F(I)) / DELV2(IV)
    NEXT I
    V2(IV) = SAV
  NEXT IV
  FOR I = 1 TO NVAR
    F(I) = F1(I) - F2(I)
    F1(I) = -F(I)                    F1 used as handy temporary.
  NEXT I
  CALL LUDCMP(DFDV(), NVAR, NVAR, INDX(), DET)      Solve to find increments to free param-
  CALL LUBKSB(DFDV(), NVAR, NVAR, INDX(), F1())          eters.
  J = 0
  FOR IV = 1 TO N2                   Increment adjustable boundary parameters at X1.
    J = J + 1
    V1(IV) = V1(IV) + F1(J)
    DV1(IV) = F1(J)
  NEXT IV
  FOR IV = 1 TO N1                   Increment adjustable boundary parameters at X2.
    J = J + 1
    V2(IV) = V2(IV) + F1(J)
    DV2(IV) = F1(J)
  NEXT IV
  ERASE INDX, DFDV, F2, F1, Y
END SUB
```

A sample program using **SHOOTF** is the following:

```
DECLARE SUB SHOOTF (NVAR!, V1!(), V2!(), DELV1!(), DELV2!(), N1!, N2!, X1!, X2!,
XF!, EPS!, H1!, HMIN!, F!(), DV1!(), DV2!())

'PROGRAM D16R2
'Driver for routine SHOOTF
CLS
NVAR = 3
N1 = 2
N2 = 1
DELTA = .001
EPS = .000001
DXX = .0001
DIM V1(N2), DELV1(N2), V2(N1), DELV2(N1), DV1(N2), DV2(N1), F(NVAR)
DO
  PRINT "INPUT M,N,C-SQUARED (999 TO STOP)"
  INPUT M, N, C2
  PRINT
  IF C2 = 999! THEN END
  IF N < M OR M < 0 THEN PRINT "Improper arguments"
LOOP WHILE N < M OR M < 0
FACTR = 1!
IF M <> 0 THEN
  Q1 = N
```

```
    FOR I = 1 TO M
      FACTR = -.5 * FACTR * (N + I) * (Q1 / I)
      Q1 = Q1 - 1!
    NEXT I
  END IF
  DX = DXX
  V1(1) = N * (N + 1) - M * (M + 1) + C2 / 2!
  IF (N - M) MOD 2 = 0 THEN
    V2(1) = FACTR
  ELSE
    V2(1) = -FACTR
  END IF
  V2(2) = V1(1) + 1!
  DELV1(1) = DELTA * V1(1)
  DELV2(1) = DELTA * FACTR
  DELV2(2) = DELV1(1)
  H1 = .1
  HMIN = 0!
  X1 = -1! + DX
  X2 = 1! - DX
  XF = 0!
  PRINT "                    Mu(-1)              Y(1-dx)              Mu(+1)"
  DO
    CALL SHOOTF(NVAR, V1(), V2(), DELV1(), DELV2(), N1, N2, X1, X2, XF, EPS, H1, HMIN
  F(), DV1(), DV2())
    PRINT
    PRINT "    V ";
    PRINT USING "###########.#####"; V1(1); V2(1); V2(2)
    PRINT "    DV";
    PRINT USING "###########.#####"; DV1(1); DV2(1); DV2(2)
  LOOP WHILE ABS(DV1(1)) > ABS(EPS * V1(1))
  END


  SUB DERIVS (X, Y(), DYDX())
  SHARED C2, M, N
  DYDX(1) = Y(2)
  DYDX(3) = 0!
  DYDX(2) = (2! * X * (M + 1!) * Y(2) - (Y(3) - C2 * X * X) * Y(1)) / (1! - X * X)
  END SUB


  SUB LOAD1 (X1, V1(), Y())
  SHARED C2, M, N, FACTR, DX
  Y(3) = V1(1)
  Y(2) = -(Y(3) - C2) * FACTR / 2! / (M + 1!)
  Y(1) = FACTR + Y(2) * DX
  END SUB


  SUB LOAD2 (X2, V2(), Y())
  SHARED C2, M, N
  Y(3) = V2(2)
  Y(1) = V2(1)
  Y(2) = (Y(3) - C2) * Y(1) / 2! / (M + 1!)
  END SUB


  SUB SCORE (XF, Y(), F())
```

```
SHARED C2, M, N
FOR I = 1 TO 3
  F(I) = Y(I)
NEXT I
END SUB
```

Here is the recipe SOLVDE:

```
DECLARE SUB DIFEQ (K!, K1!, K2!, JSF!, IS1!, ISF!, INDEXV!(), NE!, S!(),
  NSI!, NSJ!, Y!(), NYJ!, NYK!)
DECLARE SUB PINVS (IE1!, IE2!, JE1!, JSF!, JC1!, K!, C!(), NCI!, NCJ!, NCK!,
  S!(), NSI!, NSJ!)
DECLARE SUB RED (IZ1!, IZ2!, JZ1!, JZ2!, JM1!, JM2!, JMF!, IC1!, JC1!, JCF!,
  KC!, C!(), NCI!, NCJ!, NCK!, S!())
DECLARE SUB BKSUB (NE!, NB!, JF!, K1!, K2!, C!(), NCI!, NCJ!, NCK!)
```

```
SUB SOLVDE (ITMAX, CONV, SLOWC, SCALV(), INDEXV(), NE, NB, M, Y(), NYJ, NYK,
  C(), NCI, NCJ, NCK, S())
```
> Driver routine for solution of two-point boundary value problems by relaxation. ITMAX is the maximum number of iterations. CONV is the convergence criterion (see text). SLOWC controls the fraction of corrections actually used after each iteration. SCALV contains typical sizes for each dependent variable, used to weight errors. INDEXV lists the column ordering of variables used to construct the matrix S of derivatives. (The NB boundary conditions at the first mesh point must contain some dependence on the first NB variables listed in INDEXV.) The problem involves NE equations for NE adjustable dependent variables at each point. At the first mesh point there are NB boundary conditions. There are a total of M mesh points. Y is the two-dimensional array of size (NYJ,NYK) that contains the initial guess for all the dependent variables at each mesh point. On each iteration, it is updated by the calculated correction. The arrays C(NCI,NCJ,NCK), S(NSI,NSJ) supply dummy storage used by the relaxation code; the minimum dimensions must satisfy: NCI=NE, NCJ=NE-NB+1, NCK=M+1, NSI=NE, NSJ=2*NE+1. The variables NSI and NSJ have been removed as arguments of the subroutine because most versions of BASIC are limited to 16 arguments and these are not needed.

```
DIM ERMAX(NE), KMAX(NE)
K1 = 1                              Set up row and column markers.
K2 = M
NVARS = NE * M
J1 = 1
J2 = NB
J3 = NB + 1
J4 = NE
J5 = J4 + J1
J6 = J4 + J2
J7 = J4 + J3
J8 = J4 + J4
J9 = J8 + J1
IC1 = 1
IC2 = NE - NB
IC3 = IC2 + 1
IC4 = NE
JC1 = 1
JCF = IC3
FOR IT = 1 TO ITMAX                Primary iteration loop.
  K = K1                           Boundary conditions at first point.
```

```
  CALL DIFEQ(K, K1, K2, J9, IC3, IC4, INDEXV(), NE, S(), NSI, NSJ, Y(), NYJ,
NYK)
  CALL PINVS(IC3, IC4, J5, J9, JC1, K1, C(), NCI, NCJ, NCK, S(), NSI, NSJ)
  FOR K = K1 + 1 TO K2              Finite difference equations at all point pairs.
    KP = K - 1
    CALL DIFEQ(K, K1, K2, J9, IC1, IC4, INDEXV(), NE, S(), NSI, NSJ, Y(), NYJ,
NYK)
    CALL RED(IC1, IC4, J1, J2, J3, J4, J9, IC3, JC1, JCF, KP, C(), NCI, NCJ,
NCK, S())
    CALL PINVS(IC1, IC4, J3, J9, JC1, K, C(), NCI, NCJ, NCK, S(), NSI, NSJ)
  NEXT K
  K = K2 + 1                       Final boundary conditions.
  CALL DIFEQ(K, K1, K2, J9, IC1, IC2, INDEXV(), NE, S(), NSI, NSJ, Y(), NYJ,
NYK)
  CALL RED(IC1, IC2, J5, J6, J7, J8, J9, IC3, JC1, JCF, K2, C(), NCI, NCJ, NCK
S())
  CALL PINVS(IC1, IC2, J7, J9, JCF, K2 + 1, C(), NCI, NCJ, NCK, S(), NSI, NSJ)
  CALL BKSUB(NE, NB, JCF, K1, K2, C(), NCI, NCJ, NCK)       Back substitution
  ERQ = 0!
  FOR J = 1 TO NE                    Convergence check, accumulate average error.
    JV = INDEXV(J)
    ERRJ = 0!
    KM = 0
    VMAX = 0!
    FOR K = K1 TO K2                    Find point with largest error, for each dependent variable.
      VZ = ABS(C(J, 1, K))
      IF VZ > VMAX THEN
        VMAX = VZ
        KM = K
      END IF
      ERRJ = ERRJ + VZ
    NEXT K
    IF SCALV(JV) <> 0! THEN
      ERQ = ERQ + ERRJ / SCALV(JV)    Note weighting for each dependent variable.
      ERMAX(J) = C(J, 1, KM) / SCALV(JV)
    END IF
    KMAX(J) = KM
  NEXT J
  ERQ = ERQ / NVARS
  IF ERQ > SLOWC THEN DUM = ERQ ELSE DUM = SLOWC
  FAC = SLOWC / DUM                  Reduce correction applied when error is large.
  FOR JV = 1 TO NE                   Apply corrections.
    J = INDEXV(JV)
    FOR K = K1 TO K2
      Y(J, K) = Y(J, K) - FAC * C(JV, 1, K)
    NEXT K
  NEXT JV
  PRINT USING "####"; IT;            Summary of corrections for this step.
  PRINT USING "#####.######"; ERQ; FAC
  FOR J = 1 TO NE
    PRINT USING "#########"; KMAX(J);
    PRINT USING "#####.######"; ERMAX(J)
    PRINT
  NEXT J
  IF ERQ < CONV THEN ERASE KMAX, ERMAX: EXIT FOR
```

```
NEXT IT
PRINT "ITMAX exceeded"              Convergence failed.
END SUB
```

Here is the recipe BKSUB:

```
SUB BKSUB (NE, NB, JF, K1, K2, C(), NCI, NCJ, NCK)
        Back substitution, used internally by SOLVDE.
NBF = NE - NB
IM = 1
FOR K = K2 TO K1 STEP -1           Use recurrence relations to eliminate remaining dependences.
  IF K = K1 THEN IM = NBF + 1
  KP = K + 1
  FOR J = 1 TO NBF
    XX = C(J, JF, KP)
    FOR I = IM TO NE
      C(I, JF, K) = C(I, JF, K) - C(I, J, K) * XX
    NEXT I
  NEXT J
NEXT K
FOR K = K1 TO K2                   Reorder corrections to be in column 1.
  KP = K + 1
  FOR I = 1 TO NB
    C(I, 1, K) = C(I + NBF, JF, K)
  NEXT I
  FOR I = 1 TO NBF
    C(I + NB, 1, K) = C(I, JF, KP)
  NEXT I
NEXT K
END SUB
```

Here is the recipe DIFEQ:

```
COMMON SHARED X(), H, MM, N, C2, ANORM

SUB DIFEQ (K, K1, K2, JSF, IS1, ISF, INDEXV(), NE, S(), NSI, NSJ, Y(), NYJ,
NYK)
        Returns matrix S(I,J) for SOLVDE.
M = 41
IF K = K1 THEN                     Boundary condition at first point.
  IF (N + MM) MOD 2 = 1 THEN
    S(3, 3 + INDEXV(1)) = 1!       Equation (16.4.32).
    S(3, 3 + INDEXV(2)) = 0!
    S(3, 3 + INDEXV(3)) = 0!
    S(3, JSF) = Y(1, 1)            Equation (16.4.31).
  ELSE
    S(3, 3 + INDEXV(1)) = 0!       Equation (16.4.32).
    S(3, 3 + INDEXV(2)) = 1!
    S(3, 3 + INDEXV(3)) = 0!
    S(3, JSF) = Y(2, 1)            Equation (16.4.31).
  END IF
ELSEIF K > K2 THEN                 Boundary conditions at last point.
  S(1, 3 + INDEXV(1)) = -(Y(3, M) - C2) / (2! * (MM + 1!))    Equation (16.4.35).
  S(1, 3 + INDEXV(2)) = 1!
```

```
S(1, 3 + INDEXV(3)) = -Y(1, M) / (2! * (MM + 1!))
S(1, JSF) = Y(2, M) - (Y(3, M) - C2) * Y(1, M) / (2! * (MM + 1!))    Equation
S(2, 3 + INDEXV(1)) = 1!           Equation (16.4.36).                (16.4.33).
S(2, 3 + INDEXV(2)) = 0!
S(2, 3 + INDEXV(3)) = 0!
S(2, JSF) = Y(1, M) - ANORM        Equation (16.4.34).
ELSE                               Interior point.
  S(1, INDEXV(1)) = -1!            Equation (16.4.28).
  S(1, INDEXV(2)) = -.5 * H
  S(1, INDEXV(3)) = 0!
  S(1, 3 + INDEXV(1)) = 1!
  S(1, 3 + INDEXV(2)) = -.5 * H
  S(1, 3 + INDEXV(3)) = 0!
  TEMP = H / (1! - (X(K) + X(K - 1)) ^ 2 * .25)
  TEMP2 = .5 * (Y(3, K) + Y(3, K - 1)) - C2 * .25 * (X(K) + X(K - 1)) ^ 2
  S(2, INDEXV(1)) = TEMP * TEMP2 * .5        Equation (16.4.29).
  S(2, INDEXV(2)) = -1! - .5 * TEMP * (MM + 1!) * (X(K) + X(K - 1))
  S(2, INDEXV(3)) = .25 * TEMP * (Y(1, K) + Y(1, K - 1))
  S(2, 3 + INDEXV(1)) = S(2, INDEXV(1))
  S(2, 3 + INDEXV(2)) = 2! + S(2, INDEXV(2))
  S(2, 3 + INDEXV(3)) = S(2, INDEXV(3))
  S(3, INDEXV(1)) = 0!               Equation (16.4.30).
  S(3, INDEXV(2)) = 0!
  S(3, INDEXV(3)) = -1!
  S(3, 3 + INDEXV(1)) = 0!
  S(3, 3 + INDEXV(2)) = 0!
  S(3, 3 + INDEXV(3)) = 1!
  S(1, JSF) = Y(1, K) - Y(1, K - 1) - .5 * H * (Y(2, K) + Y(2, K - 1))
  DUM = (X(K) + X(K - 1)) * .5 * (MM + 1!) * (Y(2, K) + Y(2, K - 1))
  DUM = DUM - TEMP2 * .5 * (Y(1, K) + Y(1, K - 1))
  S(2, JSF) = Y(2, K) - Y(2, K - 1) - TEMP * DUM
  S(3, JSF) = Y(3, K) - Y(3, K - 1)          Equation (16.4.27).
END IF
END SUB
```

Here is the recipe PINVS:

```
SUB PINVS (IE1, IE2, JE1, JSF, JC1, K, C(), NCI, NCJ, NCK, S(), NSI, NSJ)
    Diagonalize the square subsection of the S matrix, and store the recursion coefficients in C;
    used internally by SOLVDE.
ZERO = 0!
ONE = 1!
NMAX = 10
DIM PSCL(NMAX), INDXR(NMAX)
JE2 = JE1 + IE2 - IE1
JS1 = JE2 + 1
FOR I = IE1 TO IE2                      Implicit pivoting, as in §2.1.
  BIG = ZERO
  FOR J = JE1 TO JE2
    IF ABS(S(I, J)) > BIG THEN BIG = ABS(S(I, J))
  NEXT J
  IF BIG = ZERO THEN PRINT "Singular matrix, row all 0": EXIT SUB
  PSCL(I) = ONE / BIG
```

```
    INDXR(I) = 0
NEXT I
FOR ID = IE1 TO IE2
  PIV = ZERO
  FOR I = IE1 TO IE2                    Find pivot element.
    IF INDXR(I) = 0 THEN
      BIG = ZERO
      FOR J = JE1 TO JE2
        IF ABS(S(I, J)) > BIG THEN
          JP = J
          BIG = ABS(S(I, J))
        END IF
      NEXT J
      IF BIG * PSCL(I) > PIV THEN
        IPIV = I
        JPIV = JP
        PIV = BIG * PSCL(I)
      END IF
    END IF
  NEXT I
  IF S(IPIV, JPIV) = ZERO THEN PRINT "Singular matrix": EXIT SUB
  INDXR(IPIV) = JPIV               In place reduction. Save column ordering.
  PIVINV = ONE / S(IPIV, JPIV)
  FOR J = JE1 TO JSF               Normalize pivot row.
    S(IPIV, J) = S(IPIV, J) * PIVINV
  NEXT J
  S(IPIV, JPIV) = ONE
  FOR I = IE1 TO IE2               Reduce nonpivot elements in column.
    IF INDXR(I) <> JPIV THEN
      IF S(I, JPIV) <> ZERO THEN
        DUM = S(I, JPIV)
        FOR J = JE1 TO JSF
          S(I, J) = S(I, J) - DUM * S(IPIV, J)
        NEXT J
        S(I, JPIV) = ZERO
      END IF
    END IF
  NEXT I
NEXT ID
JCOFF = JC1 - JS1                Sort and store unreduced coefficients.
ICOFF = IE1 - JE1
FOR I = IE1 TO IE2
  IROW = INDXR(I) + ICOFF
  FOR J = JS1 TO JSF
    C(IROW, J + JCOFF, K) = S(I, J)
  NEXT J
NEXT I
ERASE INDXR, PSCL
END SUB
```

Here is the recipe RED:

```
SUB RED (IZ1, IZ2, JZ1, JZ2, JM1, JM2, JMF, IC1, JC1, JCF, KC, C(), NCI, NCJ,
NCK, S())
```
> Reduce columns JZ1–JZ2 of the S matrix, using previous results as stored in the C matrix. Only columns JM1–JM2,JMF are affected by the prior results. RED is used internally by SOLVDE. The variables NSI and NSJ have been removed as arguments of the subroutine because most versions of BASIC are limited to 16 arguments and these are not needed.
```
LOFF = JC1 - JM1
IC = IC1
FOR J = JZ1 TO JZ2              Loop over columns to be zeroed.
  FOR L = JM1 TO JM2            Loop over columns altered.
    VX = C(IC, L + LOFF, KC)
    FOR I = IZ1 TO IZ2          Loop over rows.
      S(I, L) = S(I, L) - S(I, J) * VX
    NEXT I
  NEXT L
  VX = C(IC, JCF, KC)
  FOR I = IZ1 TO IZ2            Plus final element.
    S(I, JMF) = S(I, JMF) - S(I, J) * VX
  NEXT I
  IC = IC + 1
NEXT J
END SUB
```

Here is the recipe SFROID:

```
DECLARE SUB SOLVDE (ITMAX!, CONV!, SLOWC!, SCALV!(), INDEXV!(), NE!, NB!, M!,
Y!(), NYJ!, NYK!, C!(), NCI!, NCJ!, NCK!, S!())
DECLARE FUNCTION PLGNDR! (L!, M!, X!)
COMMON X(), H, MM, N, C2, ANORM        Communicates with DIFEQ.

'PROGRAM SFROID
```
> Sample program using SOLVDE. Computes eigenvalues of spheroidal harmonics $S_{mn}(x;c)$ for $m \geq 0$ and $n \geq m$. In the program, m is MM, c^2 is C2, and γ of equation (16.4.20) is ANORM.
```
CLS
NE = 3
M = 41
NB = 1
NCI = NE
NCJ = NE - NB + 1
NCK = M + 1
NSI = NE
NSJ = 2 * NE + 1
NYJ = NE
NYK = M
DIM X(M), SCALV(NE), INDEXV(NE), Y(NE, M), C(NCI, NCJ, NCK), S(NSI, NSJ)
ITMAX = 100
CONV = .000005
SLOWC = 1!
H = 1! / (M - 1)
C2 = 0!
PRINT "ENTER M,N"
INPUT MM, N
```

```
IF (N + MM) MOD 2 = 1 THEN              No interchanges necessary.
  INDEXV(1) = 1
  INDEXV(2) = 2
  INDEXV(3) = 3
ELSE                                    Interchange y₁ and y₂.
  INDEXV(1) = 2
  INDEXV(2) = 1
  INDEXV(3) = 3
END IF
ANORM = 1!                              Compute γ.
IF MM <> 0 THEN
  Q1 = N
  FOR I = 1 TO MM
    ANORM = -.5 * ANORM * (N + I) * (Q1 / I)
    Q1 = Q1 - 1!
  NEXT I
END IF
FOR K = 1 TO M - 1                      Initial guess.
  X(K) = (K - 1) * H
  FAC1 = 1! - X(K) ^ 2
  FAC2 = FAC1 ^ (-MM / 2!)
  Y(1, K) = PLGNDR(N, MM, X(K)) * FAC2           Pₙᵐ from §6.6.
  DUM = (N + 1 - MM) * PLGNDR(N + 1, MM, X(K))
  DERIV = -(DUM - (N + 1) * X(K) * PLGNDR(N, MM, X(K))) / FAC1  Derivative of Pₙᵐ
  Y(2, K) = MM * X(K) * Y(1, K) / FAC1 + DERIV * FAC2          from a recurrence
  Y(3, K) = N * (N + 1) - MM * (MM + 1)                        relation.
NEXT K
X(M) = 1!                              Initial guess at x = 1 done separately.
Y(1, M) = ANORM
Y(3, M) = N * (N + 1) - MM * (MM + 1)
Y(2, M) = (Y(3, M) - C2) * Y(1, M) / (2! * (MM + 1!))
SCALV(1) = ABS(ANORM)
IF Y(2, M) > ABS(ANORM) THEN SCALV(2) = Y(2, M) ELSE SCALV(2) = ABS(ANORM)
IF Y(3, M) > 1! THEN SCALV(3) = Y(3, M) ELSE SCALV(3) = 1!
DO
  PRINT "ENTER C^2 OR 999 TO END"
  INPUT C2
  IF C2 = 999! THEN EXIT DO
  CALL SOLVDE(ITMAX, CONV, SLOWC, SCALV(), INDEXV(), NE, NB, M, Y(), NYJ, NYK,
C(), NCI, NCJ, NCK, S())
  PRINT " M = "; MM; "      N = "; N; "      C^2 = ";
  PRINT USING "#.######^^^^"; C2;
  PRINT "      LAMBDA = ";
  PRINT USING "#####.######"; Y(3, 1) + MM * (MM + 1)
LOOP                                    for another value of c².
END
```

Chapter 17: Partial Differential Equations

Several methods for solving partial differential equations by numerical means are treated in Chapter 17 of *Numerical Recipes*. All are finite differencing methods, including forward time centered space differencing, the Lax method, staggered leapfrog differencing, the two-step Lax-Wendroff scheme, the Crank-Nicholson method, Fourier analysis and cyclic reduction (FACR), Jacobi's method, the Gauss-Seidel method, simultaneous over-relaxation (SOR) with and without Chebyshev acceleration, and operator splitting methods as exemplified by the alternating direction implicit (ADI) method. There are so many methods, in fact, that we have not provided each topic with a subroutine of its own. In many cases the nature of such subroutines follows naturally from the description. In other cases, you will have to consult other references. The subroutines that do appear in the chapter, SOR and ADI, show two of the more useful and efficient methods for elliptic equations in application.

$$\star \quad \star \quad \star \quad \star$$

Subroutine SOR incorporates simultaneous over-relaxation with Chebyshev acceleration to solve an elliptic partial differential equation. As input it accepts six arrays of coefficients, an estimate of the spectral radius of Jacobi iteration, and a trial solution which is often just set to zero over the solution grid. In program D17R1 the method is applied to the model problem

$$\frac{\partial^2 u}{\partial x^2} + \frac{\partial^2 u}{\partial y^2} = \rho$$

which is treated as the relaxation problem

$$\frac{\partial u}{\partial t} = \frac{\partial^2 u}{\partial x^2} + \frac{\partial^2 u}{\partial y^2} - \rho$$

Using FTCS differencing, this becomes

$$u^n_{j+1,l} + u^n_{j-1,l} + u^n_{j,l+1} + u^n_{j,l-1} - 4u^{n+1}_{j,l} = \rho_{jl}\Delta^2$$

(The notation is explained in Chapter 17 of *Numerical Recipes*.) This is a simple form of the general difference equation to which SOR may be applied, with

$$A_{jl} = B_{jl} = C_{jl} = D_{jl} = 1.0 \text{ and } E_{jl} = -4.0$$

for all j and l. The starting guess for u is $u_{jl} = 0.0$ for all j, l. For a source function $F_{j,l}$ we took $F_{j,l} = 0.0$ except directly in the center of the grid where $F(\text{MIDL}, \text{MIDL}) = 2.0$. The value of ρ_{Jacobi}, which is called RJAC, is taken from equation (17.5.24) of *Numerical Recipes*,

$$\rho_{Jacobi} = \frac{\cos\dfrac{\pi}{J} + \left(\dfrac{\Delta x}{\Delta y}\right)^2 \cos\dfrac{\pi}{L}}{1 + \left(\dfrac{\Delta x}{\Delta y}\right)^2}$$

In this case, J=L=JMAX and $\Delta x = \Delta y$ so RJAC $= \cos(\pi/\text{JMAX})$. A call to SOR leads to the solution shown below. As a test that this is indeed a solution to the finite difference equation, the program plugs the result back into that equation, calculating

$$F_{j,l} = u^n_{j+1,l} + u^n_{j-1,l} + u^n_{j,l+1} + u^n_{j,l-1} - 4u^{n+1}_{j,l}$$

The test is whether $F_{j,l}$ is almost everywhere zero, but equal to 2.0 at the very centerpoint of the grid.

```
SOR solution grid:
 .00   .00   .00   .00   .00   .00   .00   .00   .00   .00   .00
 .00  -.02  -.04  -.06  -.08  -.09  -.08  -.06  -.04  -.02   .00
 .00  -.04  -.09  -.13  -.17  -.19  -.17  -.13  -.09  -.04   .00
 .00  -.06  -.13  -.20  -.28  -.32  -.28  -.20  -.13  -.06   .00
 .00  -.08  -.17  -.28  -.41  -.55  -.41  -.28  -.17  -.08   .00
 .00  -.09  -.19  -.32  -.55 -1.05  -.55  -.32  -.19  -.09   .00
 .00  -.08  -.17  -.28  -.41  -.55  -.41  -.28  -.17  -.08   .00
 .00  -.06  -.13  -.20  -.28  -.32  -.28  -.20  -.13  -.06   .00
 .00  -.04  -.09  -.13  -.17  -.19  -.17  -.13  -.09  -.04   .00
 .00  -.02  -.04  -.06  -.08  -.09  -.08  -.06  -.04  -.02   .00
 .00   .00   .00   .00   .00   .00   .00   .00   .00   .00   .00
```

Here is the recipe SOR:

```
SUB SOR (A#(), B#(), C#(), D#(), E#(), F#(), U#(), JMAX!, RJAC#)
```
Simultaneous over-relaxation solution of equation (17.5.25) with Chebyshev acceleration. A#, B#, C#, D#, E# and F# are input as the coefficients of the equation, each dimensioned to the grid size JMAX × JMAX. U# is input as the initial guess to the solution, usually zero, and returns with the final value. RJAC# is input as the spectral radius of the Jacobi iteration, or an estimate of it.

```
DEFDBL A-H, O-Z                    Double precision is a good idea for JMAX bigger than about
MAXITS = 1000                      25.
EPS = .00001#
ZERO = 0#
HALF = .5#
QTR = .25#
ONE = 1#
ANORMF = ZERO                      Compute initial norm of residual and terminate iteration when
FOR J = 2 TO JMAX - 1              norm has been reduced by a factor EPS.
  FOR L = 2 TO JMAX - 1
    ANORMF = ANORMF + ABS(F(J, L))              Assumes initial U# is zero.
  NEXT L
NEXT J
OMEGA = ONE
FOR N = 1 TO MAXITS
  ANORM = ZERO
  FOR J = 2 TO JMAX - 1
    FOR L = 2 TO JMAX - 1
      IF (J + L) MOD 2 = N MOD 2 THEN        Odd-even ordering.
        RESID = A(J, L) * U(J + 1, L) + B(J, L) * U(J - 1, L)
        RESID = RESID + C(J, L) * U(J, L + 1) + D(J, L) * U(J, L - 1)
        RESID = RESID + E(J, L) * U(J, L) - F(J, L)
        ANORM = ANORM + ABS(RESID)
        U(J, L) = U(J, L) - OMEGA * RESID / E(J, L)
```

```
      END IF
    NEXT L
  NEXT J
  IF N = 1 THEN
    OMEGA = ONE / (ONE - HALF * RJAC ^ 2)
  ELSE
    OMEGA = ONE / (ONE - QTR * RJAC ^ 2 * OMEGA)
  END IF
  IF N > 1 AND ANORM < EPS * ANORMF THEN EXIT SUB
NEXT N
PRINT "MAXITS exceeded"
END SUB
```

A sample program using SOR is the following:

```
DECLARE SUB SOR (A#(), B#(), C#(), D#(), E#(), F#(), U#(), JMAX!, RJAC#)

'PROGRAM D17R1
'Driver for routine SOR
DEFDBL A-H, O-Z
CLS
JMAX = 11
PI = 3.1415926#
DIM A(JMAX, JMAX), B(JMAX, JMAX), C(JMAX, JMAX), D(JMAX, JMAX)
DIM E(JMAX, JMAX), F(JMAX, JMAX), U(JMAX, JMAX)
FOR I = 1 TO JMAX
  FOR J = 1 TO JMAX
    A(I, J) = 1!
    B(I, J) = 1!
    C(I, J) = 1!
    D(I, J) = 1!
    E(I, J) = -4!
    F(I, J) = 0!
    U(I, J) = 0!
  NEXT J
NEXT I
MIDL = INT(JMAX / 2) + 1
F(MIDL, MIDL) = 2!
RJAC = COS(PI / JMAX)
CALL SOR(A(), B(), C(), D(), E(), F(), U(), JMAX, RJAC)
PRINT "SOR Solution:"
FOR I = 1 TO JMAX
  FOR J = 1 TO JMAX
    PRINT USING "###.##"; U(I, J);
  NEXT J
  PRINT
NEXT I
PRINT
PRINT "Test that solution satisfies Difference Eqns:"
FOR I = 2 TO JMAX - 1
  FOR J = 2 TO JMAX - 1
    DUM = 4! * U(I, J)
    F(I, J) = U(I + 1, J) + U(I - 1, J) + U(I, J + 1) + U(I, J - 1) - DUM
  NEXT J
  PRINT "      ";
```

```
FOR J = 2 TO JMAX - 1
  PRINT USING "###.##"; F(I, J);
NEXT J
PRINT
NEXT I
END
```

Routine **ADI** uses the alternating direction implicit method for solving partial differential equations. This method can be considerably more efficient than the **SOR** calculation, and is preferred among relaxation methods when the shape of the grid and the boundary conditions allow its use. It is admittedly slightly more difficult to program, and sometimes does not converge, but it is the recommended "first-try" algorithm. Sample program **D17R2** uses the same model problem outlined above. When it is subjected to operator splitting and put in the form of equations (17.6.22) of *Numerical Recipes*, the coefficient arrays become

$$A_{jl} = C_{jl} = D_{jl} = F_{jl} = -1.0$$

$$B_{jl} = E_{jl} = 2.0$$

Again the trial solution is set to zero everywhere, and the source term is zeroed except at the centerpoint of the grid. As given in the text (equation 17.6.20) bounds on the eigenvalues are

$$\texttt{ALPHA} = 2\left[1 - \cos\left(\frac{\pi}{\texttt{JMAX}}\right)\right]$$
$$\texttt{BETA} = 2\left[1 - \cos\frac{(\texttt{JMAX} - 1)\pi}{\texttt{JMAX}}\right]$$

where JMAX × JMAX is the dimension of the grid. The number of iterations 2^k is minimized by choosing it to be about $\ln(4\texttt{JMAX}/\pi)$. As in routine SOR, the solution for u is printed out and may be compared with the copy listed before program **D17R1**. Also, this solution is substituted into the difference equation and should give a zero result everywhere except at the centerpoint of the grid, where its value is 2.0. Notice that **ADI** makes calls to **TRIDAG** and requires a double precision version of that routine. For this reason we have included such a version with the sample program.

Here is the recipe **ADI**:

```
DECLARE SUB TRIDAG (A#(), B#(), C#(), R#(), U#(), N!)

SUB ADI (A#(), B#(), C#(), D#(), E#(), F#(), G#(), U#(), JMAX!, K!, ALPHA#,
BETA#, EPS#)
```
   ADI solution of equations (17.6.12) and (17.6.13), with the operators as defined in equation (17.6.22). On input, A#, B#, C#, D#, E# and F# contain the coefficients of the equation. G# contains the right-hand side, while U# is input as the initial guess, usually zero. All these arrays are dimensioned to the grid size, JMAX × JMAX. The routine carries out 2∧K iterations with different values of r, and then repeats. ALPHA# and BETA# are user-supplied bounds for the eigenvalues of $\mathcal{L}_x$ and $\mathcal{L}_y$, while EPS# is the desired reduction in the norm of the residual. Note that the routine as given requires a double precision version of TRIDAG from §2.6.

```
DEFDBL A-H, O-Z
JJ = 50
KK = 6
NRR = 32
MAXITS = 100
```
Double precision is a good idea for JMAX bigger than about 25.

```
ZERO = 0#
TWO = 2#
HALF = .5#
DIM AA(JJ), BB(JJ), CC(JJ), RR(JJ), UU(JJ), PSI(JJ, JJ)
DIM ALPH(KK), BET(KK), R(NRR), S(NRR, KK)
IF JMAX > JJ THEN PRINT "Increase JJ": EXIT SUB
IF K > KK - 1 THEN PRINT "Increase KK": EXIT SUB
K1 = K + 1
NR = 2 ^ K
ALPH(1) = ALPHA                    Determine r's from (17.6.15)-(17.6.19).
BET(1) = BETA
FOR J = 1 TO K
  ALPH(J + 1) = SQR(ALPH(J) * BET(J))
  BET(J + 1) = HALF * (ALPH(J) + BET(J))
NEXT J
S(1, 1) = SQR(ALPH(K1) * BET(K1))
FOR J = 1 TO K
  AB = ALPH(K1 - J) * BET(K1 - J)
  FOR N = 1 TO 2 ^ (J - 1)
    DISC = SQR(S(N, J) ^ 2 - AB)
    S(2 * N, J + 1) = S(N, J) + DISC
    S(2 * N - 1, J + 1) = AB / S(2 * N, J + 1)
  NEXT N
NEXT J
FOR N = 1 TO NR
  R(N) = S(N, K1)
NEXT N
ANORMG = ZERO                      Compute initial residual, assuming U# is zero.
FOR J = 2 TO JMAX - 1
  FOR L = 2 TO JMAX - 1
    ANORMG = ANORMG + ABS(G(J, L))
    DUM = -D(J, L) * U(J, L - 1)   Equation (17.6.23).
    PSI(J, L) = DUM + (R(1) - E(J, L)) * U(J, L) - F(J, L) * U(J, L + 1)
  NEXT L
NEXT J
NITS = MAXITS / NR
FOR KITS = 1 TO NITS
  FOR N = 1 TO NR                  Start cycle of 2^K iterations.
    IF N = NR THEN
      NEXQ = 1
    ELSE
      NEXQ = N + 1
    END IF
    RFACT = R(N) + R(NEXQ)         This is "2r" in (17.6.27).
    FOR L = 2 TO JMAX - 1
      FOR J = 2 TO JMAX - 1        Solve (17.6.24).
        AA(J - 1) = A(J, L)
        BB(J - 1) = B(J, L) + R(N)
        CC(J - 1) = C(J, L)
        RR(J - 1) = PSI(J, L) - G(J, L)
      NEXT J
      CALL TRIDAG(AA(), BB(), CC(), RR(), UU(), JMAX - 2)
      FOR J = 2 TO JMAX - 1
        PSI(J, L) = -PSI(J, L) + TWO * R(N) * UU(J - 1)   Equation (17.6.25).
      NEXT J
```

```
      NEXT L
      FOR J = 2 TO JMAX - 1
        FOR L = 2 TO JMAX - 1           Solve (17.6.26).
          AA(L - 1) = D(J, L)
          BB(L - 1) = E(J, L) + R(N)
          CC(L - 1) = F(J, L)
          RR(L - 1) = PSI(J, L)
        NEXT L
        CALL TRIDAG(AA(), BB(), CC(), RR(), UU(), JMAX - 2)
        FOR L = 2 TO JMAX - 1
          U(J, L) = UU(L - 1)           Store current value of solution.
          PSI(J, L) = -PSI(J, L) + RFACT * UU(L - 1)          Equation (17.6.27).
        NEXT L
      NEXT J
    NEXT N
    ANORM = ZERO                        Check residual for convergence every 2^K iterations.
    FOR J = 2 TO JMAX - 1
      FOR L = 2 TO JMAX - 1
        RESID = A(J, L) * U(J - 1, L) + (B(J, L) + E(J, L)) * U(J, L)
        RESID = RESID + C(J, L) * U(J + 1, L) + D(J, L) * U(J, L - 1)
        RESID = RESID + F(J, L) * U(J, L + 1) + G(J, L)
        ANORM = ANORM + ABS(RESID)
      NEXT L
    NEXT J
    IF ANORM < EPS * ANORMG THEN
      ERASE S, R, BET, ALPH, PSI, UU, RR, CC, BB, AA
      EXIT SUB
    END IF
  NEXT KITS
  PRINT "MAXITS exceeded"
END SUB
```

A sample program using **ADI** is the following:

```
DECLARE SUB ADI (A#(), B#(), C#(), D#(), E#(), F#(), G#(), U#(), JMAX!, K!,
    ALPHA#, BETA#, EPS#)

'PROGRAM D17R2
'Driver for routine ADI
CLS
DEFDBL A-H, O-Z
JMAX = 11
PI = 3.1415926#
DIM A(JMAX, JMAX), B(JMAX, JMAX), C(JMAX, JMAX), D(JMAX, JMAX)
DIM E(JMAX, JMAX), F(JMAX, JMAX), G(JMAX, JMAX), U(JMAX, JMAX)
FOR I = 1 TO JMAX
  FOR J = 1 TO JMAX
    A(I, J) = -1!
    B(I, J) = 2!
    C(I, J) = -1!
    D(I, J) = -1!
    E(I, J) = 2!
    F(I, J) = -1!
    G(I, J) = 0!
    U(I, J) = 0!
```

```
  NEXT J
NEXT I
MID = JMAX / 2 + 1
G(MID, MID) = 2!
ALPHA = 2! * (1! - COS(PI / JMAX))
BETA = 2! * (1! - COS((JMAX - 1) * PI / JMAX))
ALIM = LOG(4! * JMAX / PI)
K = 0
DO
  K = K + 1
LOOP WHILE 2 ^ K < ALIM
EPS = .0001
CALL ADI(A(), B(), C(), D(), E(), F(), G(), U(), JMAX, K, ALPHA, BETA, EPS)
PRINT "ADI Solution:"
FOR I = 1 TO JMAX
  FOR J = 1 TO JMAX
    PRINT USING "####.##"; U(I, J);
  NEXT J
  PRINT
NEXT I
PRINT
PRINT "Test that solution satisfies Difference Eqns:"
FOR I = 2 TO JMAX - 1
  FOR J = 2 TO JMAX - 1
    DUM = -4! * U(I, J)
    G(I, J) = DUM + U(I + 1, J) + U(I - 1, J) + U(I, J - 1) + U(I, J + 1)
  NEXT J
  PRINT "        ";
  FOR J = 2 TO JMAX - 1
    PRINT USING "####.##"; G(I, J);
  NEXT J
  PRINT
NEXT I
END


SUB TRIDAG (A#(), B#(), C#(), R#(), U#(), N!)
DEFDBL A-H, O-Z
'This is a double precision version for use with ADI
DIM GAM(N)
IF B(1) = 0! THEN EXIT SUB
BET = B(1)
U(1) = R(1) / BET
FOR J = 2 TO N
  GAM(J) = C(J - 1) / BET
  BET = B(J) - A(J) * GAM(J)
  IF BET = 0! THEN EXIT SUB
  U(J) = (R(J) - A(J) * U(J - 1)) / BET
NEXT J
FOR J = N - 1 TO 1 STEP -1
  U(J) = U(J) - GAM(J + 1) * U(J + 1)
NEXT J
ERASE GAM
END SUB
```

Index of Programs

Recipes and demonstration programs are here identified by their names on the *Numerical Recipes BASIC Diskette*. Page numbers refer to where they are listed in this book.